Other titles in the series

Series A

Published

Volume 1

M.F. Greaves (London), Cell Surface Receptors: a Biological Perspective

F. Macfarlane Burnet (Melbourne), The Evolution of Receptors and Recognition in the Immune System

K. Resch (Heidelberg), Membrane Associated Events in Lymphocyte Activation

K.N. Brown (London), Specificity in Host-Parasite Interaction

Volume 2

D. Givol (Jerusalem), A Structural Basis for Molecular Recognition: the Antibody Case

M. de Sousa (Glasgow), Cell Traffic

D. Lewis (London), Incompatibility in Flowering Plants

B. Gomperts (London), Calcium and Cell Activation

A. Levitski (Jerusalem), Catecholamine Receptors

In preparation

Volume 3 (1977)

J. Lindstrom (Salk, California), Antibodies to Receptors for Acetylcholine and other Hormones

H. Furthmayr (New Haven), Erythrocyte Proteins

M. Silverman (Toronto), Specificity of Transport Processes across Membranes

M. Crandall (Kentucky), Mating-Type Interactions in Microorganisms

D. Chapman (London), Lipid-Protein Interactions

Volume 4 (1977)

M. Sonenberg and A. Schneider (New York), Membrane Changes of State Associated with Specific Ligand-Membrane Receptor Interactions

G. Palade (New Haven), Biosynthesis of Cellular Membranes

T. Stossel (Boston), Phagocytosis

C. Hughes (London), and A. Meager (Warwick), Virus Receptors

H. Metzger (NIH, Bethesda), IgE Receptors

K. Weber (Gottingen), The Actin System

M.E. Eldefrawi and A.T. Eldefrawi (Baltimore), Acetylcholine Receptors

Series B (Volumes in preparation)

Specificity in Embryological Interactions
edited by D. Garrod (University of Southampton)

Microbial Interactions
edited by J. Reissig (Long Island University, New York)

Intercellular Junctions and Synapses in Development
edited by J. Feldman, (University of London) N.B. Gilula (Rockefeller University) and J.D. Pitts (University of Glasgow)

Taxis Responses
edited by G.L. Hazelbauer (Wallenberg Laboratory, Uppsala)

Receptors and Recognition

General Editors: P. Cuatrecasas and M.F. Greaves

About the series

Cellular recognition – the process by which cells interact with, and respond to, molecular signals in their environment – plays a crucial role in virtually all important biological functions. These encompass fertilization, infectious interactions, embryonic development, the activity of the nervous system, the regulation of growth and metabolism by hormones and the immune response to foreign antigens. Although our knowledge of these systems has grown rapidly in recent years, it is clear that a full understanding of cellular recognition phenomena will require an integrated and multidisciplinary approach.

This series aims to expedite such an understanding by bringing together accounts by leading researchers of all biochemical, cellular and evolutionary aspects of recognition systems. The series will contain volumes of two types. Firstly, there will be volumes containing about five reviews from different areas of the general subject written at a level suitable for all biologically oriented scientists (Receptors and Recognition, series A). Secondly, there will be more specialized volumes, (Receptors and Recognition, series B), each of which will be devoted to just one particularly important area.

Advisory Editorial Board

Receptors and Recognition

Series B Volume 1

The Specificity and Action of Animal, Bacterial and Plant Toxins

Edited by
P. Cuatrecasas

Wellcome Research Laboratories,
Research Triangle Park, North Carolina

LONDON
CHAPMAN AND HALL

A Halsted Press Book
John Wiley & Sons, Inc., New York

First published 1977
by Chapman and Hall Ltd.,
11 New Fetter Lane, London EC4P 4EE

Typeset by Josée Utteridge of Red Lion Setters
and printed in Great Britain
at the University Printing House, Cambridge

ISBN 0 412 09730 3

Distributed in the U.S.A. by Halsted Press,
a Division of John Wiley & Sons, Inc., New York

Contents

Contributors

E.X. Albuquerque, Department of Pharmacology and Experimental Therapeutics, University of Maryland, Baltimore, Maryland 21201, U.S.A.

J.E. Alouf, Institut Pasteur, 25 Rue du Dr. Roux, 75015 Paris, France.

V. Bennett, The Biological Laboratories, Harvard University, Cambridge. Massachusetts 02138, U.S.A.

B. Bizzini, Institut Pasteur, 25 Rue du Dr. Roux, 75015 Paris, France.

R.J. Collier, Department of Bacteriology, University of California at Los Angeles, California 90024, U.S.A.

P. Cuatrecasas, Wellcome Research Laboratories, 3030 Cornwallis Rd., Research Triangle Park, North Carolina 27709, U.S.A.

J. Daly, National Institutes of Health, Bethesda, Maryland, 20014, U.S.A.

I.B. Holland, Department of Genetics, University of Leicester, Leicester, LEI 7RH, U.K.

S. Olsnes, Norsk Hydro's Institute for Cancer Research, Montebello, Oslo, 3, Norway.

A. Pihl, Norsk Hydro's Institute for Cancer Research, Montebello, Oslo 3, Norway.

L.L. Simpson, Department of Pharmacology, College of Physicians and Surgeons, University of Columbia, New York, N.Y. 10032, U.S.A.

Preface

Although throughout the history of medicine toxic substances have been of great concern to scientists and society, it has only been in recent years that the primary scientific emphasis toward naturally occurring toxins has shifted from the concern for health, pathology and clinical management to their use as powerful tools for the study of complex biological phenomena. This volume deals with recent advances in our understanding of the mechanism of action of various classes of extraordinarily potent 'toxic' substances which exist in nature. The primary orientation of this volume and the studies it describes is not toward understanding the traditionally 'toxic' nature of these substances, in the sense of causing disease, but rather toward the use of these substances to probe and elucidate fundamental processes in biology. Thus, although the substances described are indeed powerful 'toxins', it is not their 'toxic' properties *per se* which are of primary concern here.

A powerful 'poison' is by definition a substance which has potent biological activity in miniscule quantities. It follows therefore that its action must be highly *selective* and of *high affinity*, and that it must involve crucial and essential functions of cells and tissues. These features predict the potential value of toxins in probing receptors and important biological interactions and regulatory functions of cells. The topics described in this volume illustrate vividly and with certainty the extraordinary investigational utility of several such toxins. We are now seeing the realization of Claude Bernard's vision, expressed in 1875 in his classic work, *Experimental Science*, when he said that to the physiologist ' . . . the poison becomes an instrument which dissociates and analyzes the most delicate phenomenon of living structures, and by attending carefully to their mechanism in causing death, he can learn indirectly much about the physiological processes of life.'

Rather than trying to cover exhaustively (but, necessarily, superficially) the wide and enormous spectrum of naturally occurring toxic substances, this volume presents highly detailed and critical analyses of a relatively small number of representative classes of the better studied and understood toxins. In all cases presented, substantial information exists concerning the chemical structure of the toxin as well as the pharmacologic, physiologic and biochemical bases of action, and understanding of the latter has already shed new insights into normal control mechanisms. The protein toxin elaborated by *Vibrio cholerae* (choleragen) stimulates ubiquitously the adenylate cyclase activity of eukaryotic plasma cell membranes by mechanisms relevant to the normal processes by which natural hormones and regulatory substances modulate the activity of this important enzyme system. The extraordinary selectivity of its action is initiated by interaction with a highly specific, chemically defined receptor known to be a cell membrane glycolipid (ganglioside). Other protein toxins described (diptheria, *Pseudomonas,* abrin and ricin) present some interesting and provocative analogies with choleragen in that the molecular region of the toxin molecule involved in recognition and binding to cell surface receptors is structurally distinct (i.e., different subunits) from that portion which endows the molecule with the ability to subsequently exert specific biological effects. Despite these general similarities, important and illuminating differences exist. The specific receptors for all these toxins are different and unique, and their precise loci of action are different.

Although the primary action of several of the toxins (diptheria, *Pseudomonas,* ricin, abrin and colicin E_3) described appears to be inhibition of protein synthesis, different mechanisms of achieving this exist. In all cases, however, special mechanisms exist for translocating the toxin molecule, or a portion thereof, to the interior of the cell where protein synthesis is interrupted by catalytic, toxin-directed mechanisms. Not only have these studies shed light on important features of the regulation of protein synthesis, but they also have important implications for the structure and permeability of cell membranes and the possible mechanisms by which certain macromolecules may normally traverse these seemingly impervious cell barriers.

The target cell specificity of some potent toxins is also illustrated by the exquisite selectivity of tetanus toxin for the central nervous system, where it affects certain types of inhibitory synapses. The mechanism by which this toxin ascends from the nerve terminals up the spinal cord, and by which it subsequently interacts with receptors and interrupts function, will no doubt provide important new insights into fundamental functions of the central nervous system. Botulinum toxin and β-bungarotoxin (from snake venom) are also proteins with selectivity for nervous tissues, but in their case the principal action is at peripheral cholinergic nerve endings where neuromuscular transmission is affected. Both of these toxins act presynapticly (but by different mechanisms) to interfere with the normal release of acetylcholine, whereas α-bungarotoxin blocks neuromuscular transmission by binding to and thus inactivating the acetylcholine receptor of the post-synaptic membrane. The action of α-bungarotoxin will be considered separately and in depth in a chapter dealing with acetylcholine receptors in a forthcoming volume of this series, Receptors and Recognition,(M.E. Eldefrawi and A.T. Eldefrawi in Series A, Vol. 4)

The only non-protein toxic substances considered in this volume are the animal steroidal alkaloids, which demonstrate highly selective modulation of ion transport mechanisms and have thus been of invaluable assistance in the study of the structure-function relationships of electrogenic membranes.

An important class of protein toxins, the cytolytic bacterial toxins, exert their primary effect by physical or chemical dissolution of eukaryote and prokaryote cell membranes. This fascinating but heterogeneous group of toxin molecules disrupts membrane function by acting as surfactants, enzymes (i.e., hydrolysis of lipids or proteins) or by binding to specialized membrane molecules. These toxins have served as excellent probes for elucidating biomembrane organization and function, and for studying factors that control membrane integrity, permeability and function.

June, 1976 *Pedro Cuatrecasas*

1 Cholera Toxin: Membrane Gangliosides and Activation of Adenylate Cyclase

V. BENNETT *and* P. CUATRECASAS

Acknowledgements
We are grateful to Drs Susan Craig, Edward O'Keefe, Naji Sahyoun and Indu Parikh for helpful discussions and collaboration in much of the work presented here.

1. Abbreviations used: cyclic AMP: 3′, 5′ adenosine monophosphate; ACTH: adrenocorticotropic hormone; gangliosides are referred to by the nomenclature of Svennerholm (1963).
2. Estimates of molecular weights of the toxin and its component subunits vary depending on the laboratory. The values used here represent average of those quoted in the literature.

The Specificity and Action of Animal, Bacterial and Plant Toxins
(Receptors and Recognition, Series B, Volume 1)
Edited by P. Cuatrecasas
Published in 1976 by Chapman and Hall, 11 New Fetter Lane, London EC4P 4EE

1.1 INTRODUCTION

Asiatic cholera is a disease caused by the local action of a specific bacterial exotoxin on the small intestine and is characterized by a profuse outpouring of fluid at rates up to 20 liters per day (Pierce *et al.* 1971). The victims rapidly exhibit symptoms of hypovolemic shock, and if unassisted, may succomb within 8 hours of the onset of diarrhea. One of the earliest descriptions of cholera by a European records epidemics near Goa in 1503 and 1545:

> . . . so grievous was the throe and of so bad an aspect that the very worst poison seemed there to take effect as proved by vomiting and cramps that fixed the sinew to the joints . . . the eyes dimmed to sense . . . and the nailes of the hands and feet black and arched. (Pollitzer, 1959).

Cholera was confined primarily to Asia until the nineteenth century when the disease erupted in a series of pandemics which involved most of the countries in Europe as well as the United States and several South American nations (Pollitzer, 1959; Rosenberg, 1969). The disease lessened or disappeared by 1923, and did not spread beyond Asia again until 1947 when a localized epidemic broke out in Egypt. In 1959, however, an outbreak of cholera began in Thailand and expanded into a major pandemic involving 42 countries by 1971 (Finkelstein, 1973). The present international spread of cholera is continuing, and it is anticipated that South America may soon become infected (Finkelstein, 1973; Goodgame and Greenough, 1975).

The recent epidemics have been credited with renewing the scientific interest in cholera (Finkelstein, 1973), and may be indirectly responsible for the significant advances in the understanding of the molecular basis of this disease in the last few years. The purification of a protein exotoxin which completely reproduced the cholera syndrome (Finkelstein and LoSpalluto, 1969; Richardson and Noftle, 1970), and the exciting discovery of the role of cyclic AMP[1] and activation of adenylate cyclase in the pathogenesis of cholera (Field, 1971; Sharp and Hynie, 1971; Kimberg *et al.*, 1971), provided a sound basis for the work reviewed in this chapter. Direct measurements of the interaction of cholera toxin with cell surfaces have since led to the elucidation of the chemical nature of the membrane receptor for the toxin, GM_1 monosialoganglioside, a molecule which is present in the plasma membrane of all eukaryotic cells and which may be involved in very distinct processes such as contact inhibition and viral transformation in tissue culture systems (see below).

The biological effects of cholera toxin are not restricted to stimulation of secretion in the intestine, but have been observed in diverse tissues, and in all cases

Table 1.1 Biological effects of choleragen

Effects	Reference
↑ intestinal secretion in many species	Sharp (1973)
↑ lipolysis in rat fat cells	Vaughan *et al.* (1970)
↓ insulin-stimulated glucose oxidation in rat fat cells	Hollenberg and Cuatrecasa (1975)
↑ pancreatic electrolyte secretion in cats and rats	Case and Smith (1975); Kempen *et al.* (1975)
↑ glycogenolysis in liver and platelets	Zieve *et al.* (1970)
↑ hormone release from cultured rat pituitary cells	Rappaport and Grant (1974)
↑ steroid secretion in cultured mouse adrenal cells, rat testis, and rat adrenal cells	Donta *et al.* (1973); Wolff *et al.* (1973); Sato *et al.* (1975); Haksar *et al.* (1975)
↓ fluid absorption in rabbit gall bladder	Mertens *et al.* (1974)
↑ alkaline phosphatase activity in rat liver	Baker *et al.* (1973)
mimics dopamine in CNS of rats	Miller and Kelly (1975)
↓ IgE-mediated release of histamine from human leukocytes and ↓ cytolytic activation of mouse lymphocytes	Lichtenstein *et al.* 1973)
↓ DNA and RNA biosynthesis in human fibroblasts, mouse spleen lymphocytes, human peripheral lymphocytes, mouse thymocytes and transformed mouse epithelial cells	Hollenberg and Cuatrecasas (1973); Sultzer and Craig (1973); DeRobertis *et al.* (1974); Holmgren *et al.* (1974); Hollenberg *et al.* (1974)
↓ expression of F_c receptors in mouse lymphocytes	Zuckerman and Douglas (1975)
↑ tyrosinase activity in cultured mouse melanocytes	O'Keefe and Cuatrecasas (1974)
modulation of *in vivo* immune response in mice	Henney *et al.* (1973): Northrup and Fauci (1972); Chisari *et al.* (1974); Kately *et al.* (1975)
modulation of *in vitro* early anamnestic antibody response in rabbits	Cook *et al.* (1975)
alters T-cell helper function in mice	Kately and Friedman (1975)
↓ lymphocytes, ↑ corticosterone levels, ↑ serum calcium, ↑ serum glucose, ↓ liver glycogen following intravenous injection	Morse *et al.* (1975); Chisari and Northrup (1974); Hynie *et al.* (1974)
↑ serotonin N-acetyltransferase and cyclic AMP phosphodiesterase activities in cultured pineal glands of rats	Minneman and Iversen (1976)

they mimic the action of cyclic AMP (Table 1.1). Cholera toxin also activates adenylate cyclase ubiquitously, and recent studies of the mechanism of stimulation indicate important similarities to the regulation of cyclase by naturally occurring hormones. Thus, cholera toxin, at one time of interest to a selected group of clinicians and microbiologists, now appears to be a potent new tool for investigation of membrane structure and function.

1.1.1 Pathogenesis of cholera and purification of cholera toxin

Vibrio cholerae, the bacterium responsible for the symptoms of clinical cholera, was isolated by Pacini in 1854 (Hugh, 1964) and by Koch in 1884, and was found to be a motile, comma-shaped, gram negative rod. Koch proposed (1884) that the vibrios cause disease by secretion of a soluble exotoxin in analogy to the formation of tetanus and botulinum toxins. Koch speculated further that the cholera toxin necrotized the intestinal epithelium and induced a systemic intoxication.

Unfortunately, the erroneous concepts were accepted that the pathogenesis of cholera requires damage to the intestinal epithelium and that extraintestinal tissues play an important role in the disease, while Koch's suggestion of an exotoxin was abandoned until 1960. Various attempts were made between 1885 and 1950 to isolate a toxin, which was felt to be derived from the bacterial cell wall or cytoplasmic membrane (Pollitzer, 1959). These investigations were severely hampered, however, by the use of animal models which were designed on the assumption that cholera symptoms would be elicited by the systemic application of bacterial products. De and Chatterje (1953) reintroduced the rabbit ileal loop model (Viole and Crendiropoulo, 1915), in which the ileum was ligated at intervals and cholera vibrios were injected, with a resulting outpouring of fluid similar to that observed in the disease. Dutta and Habu (1955) revived the suckling rabbit system of Issaeff and Kolle (1894) which involved administration of bacteria to young rabbits which subsequently developed diarrhea. These were simple and valid bioassay systems which permitted a rational assessment of the factors involved in the disease. De (1959) and Dutta *et al.* (1959) reported that bacteria-free culture filtrates were enterotoxic, thus implicating some soluble factor. The leading candidates for a toxin material included a mucinase activity detected in culture filtrates (Burnet and Stone, 1947), a hemolytic activity (Koch, 1887; Shottmuller, 1904), and a poorly defined endotoxin (Pollitzer, 1959). De *et al.* reported in a seminal paper (1960) that the symptoms of cholera could be elicited with a heat-labile factor, distinct from mucinase or hemolytic activities, which was clearly not an endotoxin. Also, at this time, the misconception that the disease involved gross anatomical damage to the intestinal epithelium, which had been refuted in an overlooked study in 1923 (Goodpasture), was re-evaluated by Gangarosa and co-workers. No detectable intestinal epithelial mucosal cell lesions were observed in ileal biopsies from cholera victims (Gangarosa *et al.,* 1960), a finding subsequently confirmed (Formal *et al.,* 1961; Fresh *et al.,* 1964). Furthermore, Gordon (1962) found that intravenously administered

^{131}I-polyvinylpyrrolidone was retained during cholera, and thus extended the concept of an intact eipthelium to include the vasculature as well. These findings also strongly suggest that the hemolytic and mucinase activities were not significant in production of diarrhea, and provided a rational basis for the purification of the exotoxin postulated by De *et al.* (1960).

The impetus for the isolation of the toxin was provided by Finkelstein, who developed a simple growth medium for production of a 'choleragen' (Finkelstein *et al.*, 1964) which produced clinical cholera in humans (Benyajati, 1966). Enterotoxic activity was subsequently separated from endotoxin by gel filtration (Finkelstein *et al.*, 1966a, b) and the toxin was eventually isolated in a highly purified form (Finkelstein and LoSpalluto, 1969). An immunologically identical protein was also purified which was biologically inactive, and termed choleragenoid. Subsequent studies demonstrated that choleragenoid was derived from choleragen (Finkelstein *et al.*, 1971). Pure choleragen was described as a protein of about 84 000 MW which contained no detectable lipid or carbohydrate (LoSpalluto and Finkelstein, 1972) and did not exhibit the following enzymatic activities: neuraminidase, protease, elastase, lipase, lecithinase, hyaluronidase, chondroitin sulphatase, DNase, RNase, mucinase, or staphylolytic enzymes (unpublished data cited by Finkelstein, 1973). Technical aspects of the purification were improved (Finkelstein and LoSpalluto, 1970) and choleragen and choleragenoid have since been crystallized (Finkelstein and LoSpalluto, 1972). Significant contributions to the techniques of large-scale purification have been made by other workers (Richardson and Noftle, 1970; Spyrides and Feeley, 1970; Rappaport *et al.*, 1974). Affinity adsorbents have been developed which quantitatively adsorb choleragen and choleragenoid, and may provide a single-step method of purification (Cuatrecasas *et al.*, 1973).

1.1.2 Discovery of activation of adenylate cyclase by choleragen

The availability of pure preparations of choleragen made possible unambiguous studies on its mechanism of action at the molecular level. Field and co-workers (1968a, b) reported that addition of cyclic AMP to the basal surface of isolated ileal mucosa rapidly induced a large increase in the short circuit current which then gradually declined. After 30 to 60 min the net sodium flux in the absence of glucose fell to zero, and the direction of chloride transport was reversed such that Cl^- was actively secreted from the mucosal side of the preparation. These effects could quite conceivably cause a passive eflux of water, and thus diarrhea. The potential significance of these findings was quickly realized, and it was soon reported that dialysed filtrates of *Vibrio cholerae* cultures could mimic exactly the effects of cyclic AMP (Field *et al.*, 1969). Further evidence for a possible role of cyclic AMP in the pathogenesis of cholera was provided by the demonstration that injection of prostaglandins or theophylline into the superior mesenteric artery of dogs caused diarrhea (Pierce *et al.*, 1970).

It was concluded that active secretion of chloride ion into the intestinal lumen could explain the loss of fluid during cholera, and that choleragen acted via some mechanism related to cyclic AMP (Field, 1971). It was subsequently discovered that choleragen was also a potent activator of lipolysis in isolated rat fat cells (Vaughan *et al.*, 1970; Greenough *et al.*, 1970), and also that it stimulated glycogenolysis in liver and platelets (Zieve *et al.*, 1971). Choleragen has since been found to mimic the biological effects of cyclic AMP in every tissue so far examined.

Choleragen was subsequently demonstrated to increase the tissue content of cyclic AMP (Shafer *et al.*, 1970), and to stimulate adenylate cyclase activity in intestinal mucosa with no demonstrable effect on cyclic AMP phosphodiesterase activity (Sharp and Hynie, 1971; Kimberg *et al.*, 1971). Choleragen was unusual in that it was effective only if incubated with intact cells, and activation occurred after a latency phase of 15 to 60 min. Stimulation of adenylate cyclase by choleragen, once obtained, persisted in washed membrane preparations. Choleragen was interpreted as acting on the existing adenylate cyclase molecules rather than stimulating *de novo* synthesis of the enzyme since fluoride ion activated the toxin-strimulated cyclase activity to a greatly reduced extent compared to that of the control preparation (Sharp and Hynie, 1971; Kimberg *et al.*, 1971). Measurements of adenylate cyclase activity in biopsy tissue from humans during the diarrheal phase of cholera revealed a similar activation as obtained in animal models, which decreased in the convalescent period (Chen *et al.*, 1971, 1972). Guerrant *et al.*, (1972) analyzed carefully the relationships between the net water and sodium fluxes and the stimulation of adenylate cyclase by choleragen in the small bowel, and obtained a good correlation between the time course and extent of cyclase activation and the changes in secretion.

Considerable evidence has thus accumulated that the clinical symptoms of cholera could be explained solely by stimulation of the adenylate cyclase activity of the intestinal mucosa by choleragen (reviewed by Sharp, 1973). Activation of adenylate cyclase by choleragen has also been demonstrated in such diverse tissues as rat (Beckman *et al.*, 1974; Bennett *et al.*, 1975a) and mouse (Gorman and Bitensky, 1971), liver, human neutrophils (Bourne *et al.*, 1973), cultured mouse adrenal cells (Wolfe *et al.*, 1973; Donta *et al.*, 1973), and melanoma cells (O'Keefe and Cuatrecasas, 1974), erythrocytes of turkeys (Field, 1974; Bennett and Cuatrecasas, 1975b), pigeons (Gill and King, 1975), rats (Bennett and Cuatrecasas, 1975a), and toads (Bennett and Cuatrecasas, 1975b), as well as many others (Table 1.2). As a universal activator of adenylate cyclase, choleragen assumes general importance as a means of elucidating the diverse biological effects of cyclic AMP, and as a potential tool in deciphering the mechanism of regulation of this enzyme.

1.1.3 Subunit structure of choleragen

The first study on the subunit structure of choleragen (LoSpalluto and Finkelstein, 1972) reported that both toxin and its biologically inactive derivative, choleragenoid, dissociated reversibly to 15 000 MW subunits in the presence of mild acid, as

Table 1.2 Tissues which demonstrate activation of adenylate cyclase by choleragen

Tissue	Reference
intestinal epithelial cells of many species	Sharp (1973)
liver from mice and rats	Gorman and Bitensky (1972); Beckman *et al.* (1974); Bennett *et al.* (1975a)
adipocytes from rats, mice, and guinea pigs	Evans *et al.* (1972); Hewlett *et al.* (1974); Bennett *et al.* (1975b); unpublished data
cultured mouse adrenal tumor cells	Donta *et al.* (1973); Wolff *et al.* (1973)
mouse melanocytes	O'Keefe and Cuatrecasas (1974)
thyroid tissue	Mashiter *et al.* (1973)
spleen lymphocytes, thymocytes, and mesenteric lymphocytes from rats and mice	Chisari *et al.* (1974); DeRobertis *et al.* (1974); Boyle and Gardner (1974); Craig and Cuatrecasas (1975); Zenser and Metzger (1974)
human neutrophils	Bourne *et al.* (1973)
erythrocytes from rats, rabbits, turkeys, pigeons, and toads	Field (1974); Bennett and Cuatrecasas (1975a, b); Gilland King (1975); van Heyningen and King (1975)
transformed mouse epithelial cells	Hollenberg *et al.* (1974)
rat renal tubules	Kurokawa *et al.* (1975)
fetal intestinal cells from humans and rats	Grand *et al.* (1973)
rat brain	Miller and Kelly (1975)
liver, thyroid, kidney, spleen of rats following intravenous injection	Hynie *et al.* (1974)
mouse ascites tumor cells	Bitensky *et al.* (1975)

determined by ultracentrifugation. No difference in subunit molecular weight was detected between toxin and choleragenoid, although a partial amino acid analysis did indicate some differences between these molecules. The molecular weight of intact toxin was estimated at about 84 000 Daltons on the basis of sedimentation velocity-diffusion and sedimentation equilibrium measurements, while the value for choleragenoid was 58 000. Toxin and choleragenoid had previously been found to be immunologically identical (Finkelstein and LoSpalluto, 1969), and consequently no strong statements were made concerning a marked difference in subunit composition between these molecules (LoSpalluto and Finkelstein, 1972).

These proteins were subsequently examined by sodium dodecylsulfate disc gel electrophoresis (Finkelstein *et al.*, 1972). If toxin was electrophoresed without prior heating, two bands were detected with molecular weights corresponding to about 56 000 and 28 000, while choleragenoid migrated under these conditions as a single band of 56 000 molecular weight. However, if toxin or choleragenoid were heated in detergent or exposed to acid, the 56 000 MW component was replaced by a rapidly migrating band of 8 000–15 000, which corresponded well with the previous ultracentrifugation studies. The 28 000 MW component, which apparently was unique to toxin, remained intact after simple heating or acidification, but decomposed to a 23 000 MW fragment when heated with a sulfydryl reducing agent. In view of the immunologic evidence for the similarity of toxin and toxoid, the 28 000 MW subunit was interpreted as an aggregate, possibly a dimer, of the smaller peptides, which, for unexplained reasons, was stable in detergent or acid (Finkelstein *et al.*, 1972). This feeling was reinforced by the finding that the 28 000 MW fragment, isolated after gel filtration in detergent, generated antibodies which reacted identically with choleragen and choleragenoid (Finkelstein *et al.*, 1972).

Subsequent reports confirmed the basic features of the behaviour of toxin and toxoid described by Finkelstein *et al.*, (1972) and noted the presence of a subunit of roughly 30 000 MW[2] following electrophoresis on polyacrylamide disc gels in sodium dodecylsulfate or acid urea which was unique to toxin (Lonnroth and Holmgren, 1973; Cuatrecasas *et al.*, 1973; van Heyningen, 1974). It was proposed that choleragen was composed of two noncovalently-linked oligomeric subunits, one of about 60 000 MW[2] which was an aggregate of 8 000–10 000 MW[2] peptides, and another major subunit of approximately 30 000 MW which itself contained components of about 25 000 MW and 5 000 MW possibly linked by disulfide bonds (Cuatrecasas *et al.*, 1973) (Fig. 1.1). Choleragenoid contained only a 60 000 MW component which also dissociated into peptides of about 10 000 MW, and it was suggested that choleragenoid was formed from toxin by loss of the 30 000 MW subunit (Lonnroth and Holmgren, 1973; Cuatrecasas *et al.*, 1973; van Heyningen, 1974). Strong evidence for this view was provided by the demonstration that the 60 000 MW subunit of either toxin or choleragenoid could be dissociated into 10 000 MW peptides and recombined with the 30 000 MW component to restore the original biological activity and molecular weight of the parent molecule (Finkelstein *et al.*, 1974).

The apparent immunologic similarity between toxin and choleragenoid may be explained by the low degree of antigenicity of the 30 000 MW subunit (unpublished data; Finkelstein *et al.*, 1974). Only recently has it been possible to prepare precipitating antibodies against this molecule (Finkelstein *et al.*, 1974). The fact that the 30 000 MW component isolated by Finkelstein *et al.*, (1972) elicited antibodies reactive with choleragenoid could easily be explained by a small contamination with other portions of the toxin molecule which dominated the immune response. Antisera directed against the 30 000 MW subunit have been purified by immunoadsorption with choleragenoid-agarose, and have been demonstrated to form precipitin bands with choleragen and its unique subunit, but not with choleragenoid (Finkelstein

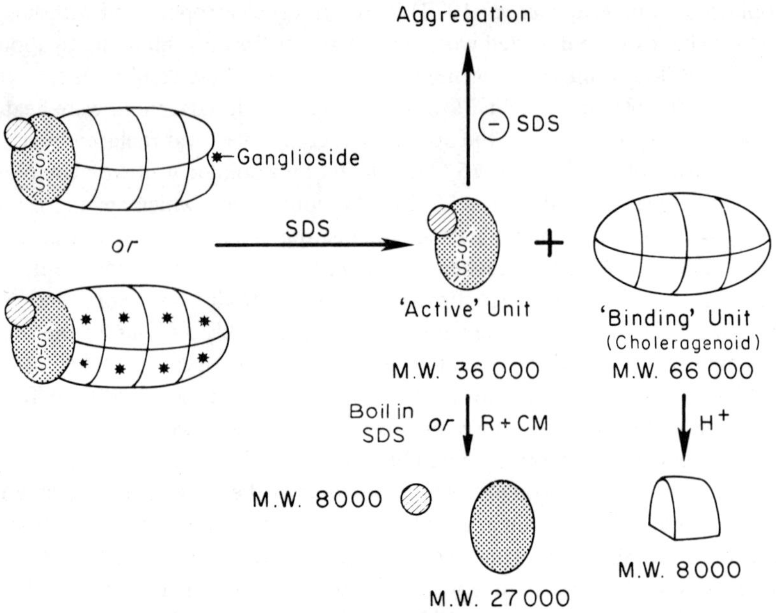

Fig. 1.1 Schematic representation of the structure of cholera toxin. The 'binding subunit (choleragenoid) can be dissociated into 8000 MW components by heating (60°C, 30 min) in SDS or by exposure to 0.1 N NCl. Reduction and carboxymethylation (R and CM) of the 'active' subunit (36 000 MW) results in the appearance of 8000 MW and 27 000 MW components, but these two polypeptides are not joined by disulfide bonds since heating for 50 min at 100°C in 1% SDS in the absence of reducing agents also results in similar dissociation. Of the two proposed structures of the native toxin the bottom one is much more likely since it is now known (Craig and Cuatrecasas, 1975) that the toxin (and choleragenoid) is multivalent with respect to ganglioside binding. Data from Cuatrecasas *et al,* 1973 and unpublished.

et al., 1974).

The absence of the 30 000 MW subunit in the intitial study could be explained by the marked tendency for this molecule to aggregate and precipitate when isolated from the 60 000 MW component (unpublished data; van Heyningen, 1974; Holmgren and Lonnroth, 1975). In re-evaluating the ultracentrifugation experiments, Finkelstein *et al.* (1974) reported a rapidly sedimently material which formed following exposure of toxin to acid, and which presumably represents the aggregated 30 000 MW subunit.

In our laboratory (Cuatrecasas *et al.*, 1973, unpublished; Fig. 1) the MW of the toxin components (after heating the sample in 1 to 2% SDS) have been consistently observed to be 36 000 and about 8000. Upon reduction and carboxymethylation the 36 000 MW subunit is reduced to 27 000 MW. The release of the smaller peptide by this procedure, however, does not mean that it is coupled covalently by disulfide linkage to the larger component since boiling (50 min) the sample in the absence of reducing agents also induces this dissociation. The two components are therefore associated by extrarodinarily tight non-covalent forces. Both the 36 000 MW as well as the 27 000 MW components are unusually insoluble in aqueous media. Antisera against the former (36 000 MW) produce antibodies which cross-react with native choleragen, choleragenoid, as well as (but weakly) the 27 000 MW subunit. On the other hand, antisera against the 27 000 MW subunit do not cross react with the native toxin or with choleragenoid but does react with the 36 000 MW component. These studies suggest that the 36 000 MW subunit is composed of a unique peptide, the 27 000 MW subunit, associated with a smaller (about 8000 mW) subunit which is similar or identical to the subunit of choleragenoid. For the purposes of the discussion in this chapter, the '30 000' MW subunit is used to refer to the 36 000 MW subunit, called 'active' subunit, with the emphasis on the 27 000 MW component which is believed to represent the unique, biologically active portion or subunit of the molecule.

Although choleragenoid is biologically inert, this molecule is a potent antagonist of the action of choleragen on intestinal loops (Pierce, 1973; Holmgren, 1973), isolated fat cells (Cuatrecasas, 1973e), and human neutrophils (Bourne *et al.*, 1973). Furthermore, choleragenoid binds with equal avidity to the identical membrane receptor as choleragen and is in all respects a true *competitive* antagonist of binding and action (Cuatrecasas, 1973e) (see below). These properties led to the proposal (Cuatrecasas *et al.*, 1973) that the 60 000 MW component of choleragen and choleragenoid was responsible for the initial interaction with the cell surface, while the 30 000 MW subunit represented the biologically active portion of the molecule. Thus, the 60 000 MW component or 'binding' subunit would function as a carrier which delivers the active subunit to the plasma membrane in a highly selective manner. Essentially identical models have since been suggested by others (van Heyningen, 1974; Finkelstein *et al.*, 1974; Holmgren and Lonnroth, 1975). This view is supported by the fact that the isolated 60 000 MW component of toxin is biologically inert but binds to cell membranes and to the membrane receptor for toxin (GM1 ganglioside; see below) in an identical manner as the intact molecule (O'Keefe and Cuatrecasas, unpublished data; Holmgren and Lonnroth, 1975). The active subunit alone, however, can activate adenylate cyclase to a similar extent as choleragen although at much higher concentrations than required for the intact molecule, and this activity is not blocked by gangliosides or choleragenoid (van Heyningen and King, 1975; Sahyoun and Cuatrecasas, 1975; Gill and King, 1975). Direct measurement of the interaction of ^{125}I-labeled 'active' subunit with fat cell membranes provides further evidence that this molecule does not interact with the membrane receptors for choleragen or choleragenoid (Sahyoun and Cuatrecasas, 1975). Choleragen thus exhibits features in common with diptheria toxin (Gill *et al.*, 1973) and the toxic plant lectins abrin and ricin (Olsnes *et al.*, 1973a, b) which also have separate subunits responsible for binding and for biological activity.

While the basic features of the subunit structure of choleragen are well established, certain aspects are still unclear. Estimates of the molecular weight of the peptide components composing the binding subunit vary from 15 000 (LoSpalluto and Finkelstein, 1972; van Heyningen, 1974) to 8000–10 000 (Cuatrecasas *et al.*, 1973; Holmgren and Lonnroth, 1975), while the value for the binding subunit has been reported to be from 56 000 (LoSpalluto and Finkelstein, 1972) to 66 000 Daltons (Cuatrecasas *et al.*, 1973). The binding subunit thus may contain from four to eight constituent peptides. Quantitation of certain amino acids in the amino-terminal sequence of the isolated subunits indicates a ratio of four small peptides to one 30 000 MW subunit (Jacobs *et al.*, 1974), although this study remains to be confirmed.

The close similarity in molecular weight suggests that each small peptide may be closely related or identical, although this has not been demonstrated. If such were the case, then each peptide may be capable of binding one molecule of GM1 ganglioside (see below), and the 60 000 MW subunit is thus multivalent. Direct measurements of the stoichiometry of ganglioside binding to choleragen are not presently available. However, the observations that fluorescent derivatives of choleragen form 'caps' on lymphocytes, and that choleragen-treated cells bind specifically to GM1 covalently attached to agarose beads (Craig and Cuatrecasas, 1975) demonstrate that the binding is at least bivalent.

The 30 000 MW active subunit, but not the small peptides, has consistently been observed to be cleaved into two fragments by reduction and alkylation (Finkelstein *et al.*, 1972; Cuatrecasas *et al.*, 1973; Lonnroth and Holmgren, 1973; van Heyningen, 1974; Holmgren and Lonnroth, 1975). These fragments have been reported to have a molecular weight of approximately 8000 and 25 000 by most investigators. Holmgren and Lonnroth (1975), however, present evidence that the subunit is cleaved by reduction into two peptides with an identical molecular weight of roughly 18 000 (on sodium dodecylsulfate disc gels). The small fragment observed in other reports is felt to be a contaminating peptide from the binding subunit (Holmgren and Lonnroth, 1975). The basis for these descrepancies is unclear at this time.

1.2 INTERACTION OF CHOLERAGEN AND CHOLERAGENOID WITH CELL MEMBRANES

The evidence available in 1972 strongly suggested that choleragen should interact with high affinity with the plasma membrane in target tissues. Only a brief exposure to low concentrations of toxin (in the range of 10^{-10}–10^{-8} M) are required for full biological effects, and these cannot be reversed by removal of toxin from the medium (reviewed by Pierce *et al.*, 1971). Moreover, adenylate cyclase activity in mammalian tissues is almost exclusively associated with particulate fractions (Sutherland *et al.*, 1962), and a major proportion of the enzyme is localized in the plasma membrane (Davoren and Sutherland, 1963; Oye and Sutherland, 1966). The methodology for

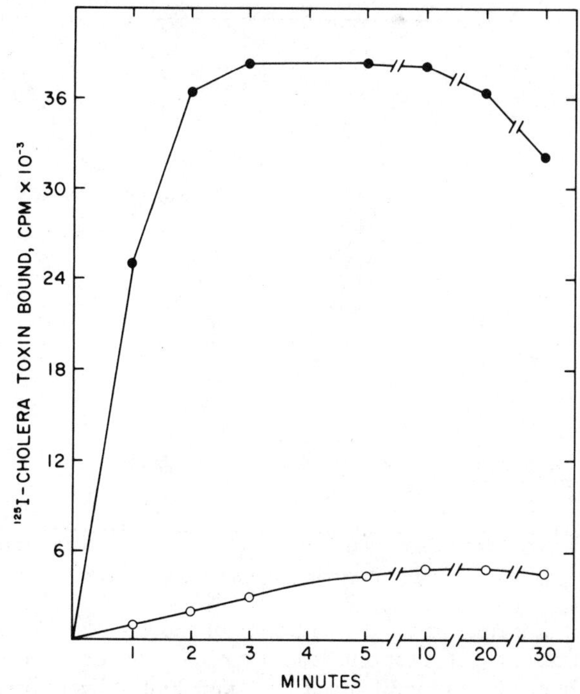

Fig. 1.2 Time course of binding of ^{125}I-labeled cholera toxin to liver membranes at 24°C. Liver membranes (40 μg of protein) were incubated at 24°C in 0.2 ml of Krebs-Ringer bicarbonate buffer containing 0.1% (w/v) albumin and 1.3×10^4 cpm (○) or 8.2×10^4 cpm (●) of ^{125}I-labeled cholera toxin (sp. act. 10 μCi μg^{-1}). Specific binding was determined by filtration as described (Cuatrecasas, 1973b).

quantitatively examining membrane-ligand interactions of high affinity had fortuitously been developed for polypeptide hormones (Lefkowitz *et al.*, 1970; Tomasi *et al.*, 1970; Rodbell *et al.*, 1971; Cuatrecasas, 1971a–d; 1972) and had been extended to the study of the membrane-binding properties of plant lectins (Cuatrecasas, 1973a). The essential feature of such studies was preparation of radioiodinated ligands with carrier-free Na ^{125}I with specific activities in the range of 10^6 cpm per picomole, and which retained full biological activity.

^{125}I-labeled choleragen and choleragenoid were first prepared (Cuatrecasas, 1973b–e) using chloramine-t as an oxidant (Hunter and Greenwood, 1962), with specific activities of 10^6 to 10^7 cpm per picomole. The biological activity of iodocholeragen is apparently not altered since derivatives containing an average of one atom of iodine per molecule (2.2×10^6 cpm picomole^{-1}) stimulate lipolysis (Cuatrecasas, 1973b) and adenylate cyclase (Bennett and Cuatrecasas, 1976a) in rat fat cells to an equal extent and with an identical apparent affinity as native toxin. Subsequent studies have demonstrated that about 80 per cent of the label of

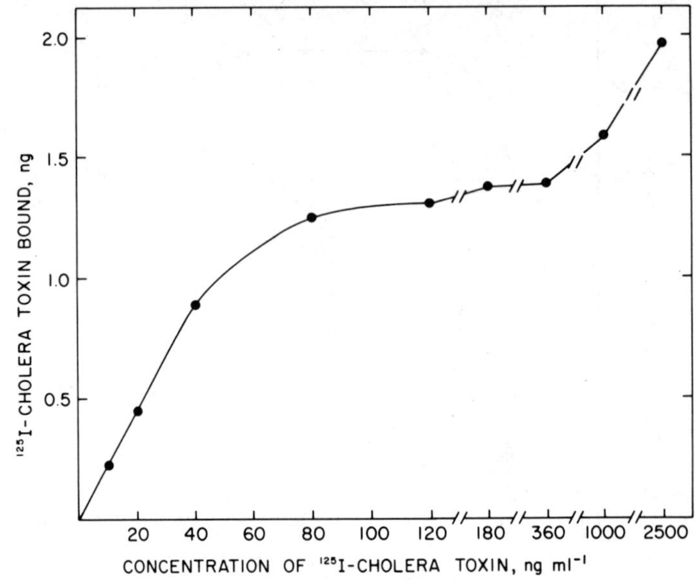

Fig. 1.3 Effect of increasing the concentration of cholera toxin on the binding to isolated fat cells. Isolated fat cells (about 4 x 10^5 cells) obtained from 110-g rats were incubated for 30 min at 24°C in 0.2 ml of Krebs-Ringer-bicarbonate buffer containing 0.1% (w/v) albumin and the indicated concentration of ^{125}I-labeled cholera toxin (1.2 μCi μg^{-1}). Specific binding was determined by filtration over EHWP Millipore filters as described in the text. The samples used to correct for nonspecific binding contained 100 μg of native toxin ml^{-1} (Cuatrecasas, 1973b).

iodocholeragen is confined to the 25 000 MW portion of the active subunit (Cuatrecasas *et al.,* 1973; Holmgren and Lonnroth, 1975). The availability of these radiolabeled derivatives permitted a detailed and quantitative evaluation of the biologically relevant interaction of choleragen with cell membranes, which occurs at extremely low concentrations and frequently with only 1000 to 10 000 molecules of toxin bound per cell.

1.2.1 Characteristics of membrane-^{125}I-choleragen interaction

General features

Detailed studies of the binding of ^{125}I-labeled cholera toxin to rat liver microsomal membranes and to isolated rat epidydimal fat cells (Cuatrecasas, 1973–e) demonstrated that this molecule is nearly quantitatively adsorbed when initially present at concentrations of 10^{-10} M or less. Binding is complete within 15 min at 4°C or 5 min at 24°C, with levels of toxin less than 10^{-10} M (Fig. 1.2). The binding sites

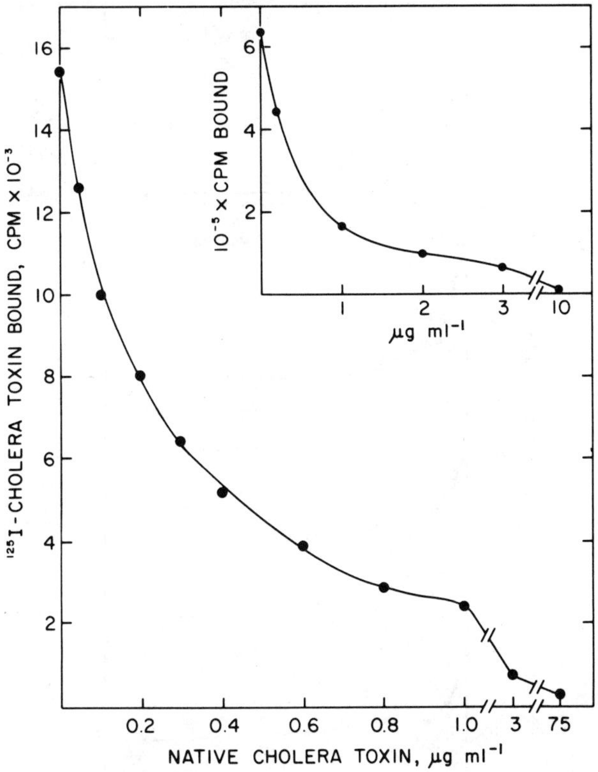

Fig. 1.4 Effects of native cholera toxin on the binding of ^{125}I-labeled cholera toxin to liver membranes. The indicated concentration of native toxin was added to 0.2 ml of Krebs-Ringer-bicarbonate buffer containing 0.1% (w/v) albumin and 50 μg of membrane protein. After incubating for 5 min at 24° ^{125}I-labeled cholera toxin (15 ng/ml, 9 μCi μg^{-1}) was added and the samples were incubated for 15 min at 24°C. Binding to the membranes was determined by filtration over cellulose acetate Millipore filters. The inset describes a similar experiment in which the concentration of ^{125}I-labeled cholera toxin in the incubation medium was increased by more than 20-fold (0.4 μg ml^{-1}); the concentration of protein in this experiment was 0.75 mg ml^{-1} (Cuatrecasas, 1973b).

for ^{125}I-labeled choleragen are saturable (Fig. 1.3), and binding is almost completely prevented by previous addition of the unlabeled protein at a concentration of 10^{-8} M (Fig. 1.4). The apparent dissociation constant for the initial toxin-membrane interaction was estimated at about 5×10^{-10}–10^{-9} M, which is similar to values for membrane interactions with polypeptide hormones (reviewed by Cuatrecasas, 1974). The true affinity of the binding of choleragen is difficult to measure, since following the initial binding event, a portion of the membrane-bound choleragen becomes

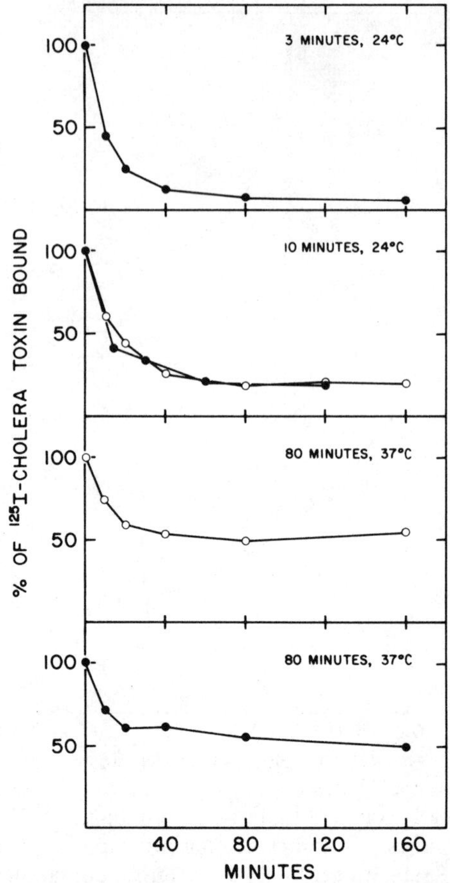

Fig. 1.5 Effect of time and temperature of incubation of isolated fat cells (○) and liver membranes (●) with ^{125}I-labeled cholera toxin on the subsequent rate and extent of dissociation of the complex at 37°C. Fat cells (about 5×10^5 cells) or liver membranes (210 μg of proteins ml^{-1}) were incubated for various time periods at 24 or 37° in 0.2 ml of Krebs-Ringer-bicarbonate buffer containing 0.1% albumin and ^{125}I-labeled cholera toxin (10 ng, 12 μCi μg^{-1}). The dissociation reaction was begun by adding native cholera toxin (20 μg ml^{-1}) to all the samples and immediately incubating at 37°C. At the indicated time intervals the specific binding of ^{125}I-labeled cholera toxin was determined by filtration procedures (Cuatrecasas, 1973d).

permanently adsorbed to the membrane (Fig. 1.5). This property of choleragen is discussed in more detail below in terms of its mechanism of action.

Essentially identical binding behavior of ^{125}I-choleragen has been reported for cultured mouse melanoma cells (O'Keefe and Cuatrecasas, 1974), and epithelial cells (Hollenberg *et al.*, 1974), human lymphocytes (Boyle and Bardener, 1974;

Holmgren *et al.*, 1974d) intestinal mucosal membranes (Walker *et al.*, 1974; Peterson and Verwy, 1974; Holmgren *et al.*, 1975) and toad erythrocytes (Bennett and Cuatrecasas, 1975a). The number of binding sites per cell for ^{125}I-choleragen varies widely depending on the tissue and species studied. Rat epidydimal fat cells (Cuatrecasas, 1973b) and human intestinal epithelial cells (Holmgren *et al.*, 1975) bind maximally about 20 000 molecules of choleragen per cell, while cultured mouse melanoma cells (O'Keefe and Cuatrecasas, 1974), human fibroblasts (Bennett *et al.*, 1975a) and bovine intestinal epithelial cells (Holmgren *et al.*, 1975) each contain 10^6 or more receptor sites. The concentration-response relationship for activation of adenylate cyclase, however, occurs in the range of 100 to 10 000 molecules of choleragen bound per cell, irrespective of the cell type (O'Keefe and Cuatrecasas, 1974; Hollenberg *et al.*, 1974; Bennett and Cuatrecasas, 1975b) (see below).

Identity of binding properties of choleragen and choleragenoid

Choleragenoid, the biologically inactive derivative of choleragen, is a potent competitive inhibitor of the action of choleragen, with a K_i of about 7×10^{-10} M in rat cells (Cuatrecasas, 1973d), and it has been demonstrated in morphological studies (Peterson *et al.*, 1972) to bind to intestinal mucosa in a qualitatively similar manner as choleragen. These results suggested that choleragenoid may bind to the same membrane receptor as choleragen, although in an inactive fashion.

The binding properties of choleragenoid and choleragen have been compared directly with radioiodinated derivatives (Cuatrecasas, 1973e). Iodocholeragenoid binds to fat cell and liver membranes with the same affinity and time course as ^{125}I-choleragen, and the binding is displaced completely and at equivalent concentrations by native choleragen and choleragenoid. Conversely, choleragenoid is at least as effective as native choleragen in blocking the binding of ^{125}I-choleragen. Similar studies have been reported for lymphocytes (Holmgren *et al.*, 1974a). ^{125}I-choleragenoid binding is enhanced to an identical extent as that of choleragen by incorporation of the toxin receptor, GM1 monosialoganglioside (see below) into cell membranes, and binding is abolished by incubation of labeled choleragenoid with low concentrations of this glycolipid (Cuatrecasas, 1973e).

These findings demonstrate conclusively that choleragenoid and choleragen interact in an identical fashion with the same membrane receptor. The choleragenoid-membrane complex, however, is biologically inert, indicating that simple occupation of receptor sites is insufficient to activate adenylate cyclase. The availability of a competitive antagonist of the binding of choleragen has been useful in evaluating the possibility of excess or 'spare' membrane receptor sites (Bennett *et al.*, 1975a) (see below).

Restriction of bound toxin to the plasma membrane

It is important for understanding the mechanism of action of choleragen to know the location of toxin in the cell after binding, and in particular to determine whether toxin must enter the cytoplasm of cells for expression of biological activity.

Table 1.3 Possible entry of [^{125}I] cholera toxin into the cytoplasm of toad erythrocytes[a]

Incubation time (min)	% of cell-bound [^{125}I] cholera toxin		Molecules of [^{125}I] toxin per cell[b]	
	Lysate[c]	Medium[d]	Lysate[c]	Medium[d]
10	0.1		6	
20	0.3		17	
30	1.8		103	
45	0.8	0.6	46	34
60	0.6		34	
90	1.5	1.3	85	74

[a]Cells were washed and suspended in 5 ml of amphibian-Ringer's, 0.5% albumin (w/v), pH 7.5, at a concentration of 4.6 x 10^7 cells ml^{-1}. [^{125}I] Cholera toxin (0.22 μg, 11.8 μCi μg^{-1}) was added, and the suspension was incubated for 30 min at 0°C. The cells were washed twice, suspended in the same buffer (10 ml), and placed in a 30°C bath. The entry of bound [^{125}I] toxin into the cells was assayed at various time by filtering an aliquot of cells (0.2 ml, 0.4 μCi of bound [^{125}I] toxin) under negative pressure across an EG cellulose acetate (Millipore) membrane (0.25 μm pore size), which was then washed with ice-cold amphibian-Ringer's, 0.1% BSA (w/v), pH 7.5 (2.0 ml). The filtrate, which contained over 90% of the hemoglobin, and the filter membrane, were assayed for ^{125}I in a well-type gamma counter operating at 45% efficiency. Dissociation of bound [^{125}I] toxin into the medium was estimated independently by centrifugation of samples of the erythrocytes through an oil phase (dibutyl phthalate, dinonyl phthalate (2:1 v/v) with a Beckman microfuge (Chang and Cuatrecasas, 1974); the radioactivity in the resulting cell pellet and supernatant was thus determined separately. The values for entry (lysate) and dissociation (medium) were determined in triplicate, and are expressed as the percent of the total radio-activity, corrected for the value at zero time.

[b]5700 Molecules of ^{125}I-labeled cholera toxin were bound per cell initially.

[c]Uncorrected for [^{125}I] toxin which has spontaneously dissociated into the medium during incubation or cell lysis; these data thus express a maximum and clearly exaggerated figure for the possible entrance of toxin into the cell.

[d]Estimate of amount of [^{125}I] toxin which has dissociated from the cells during the incubation period. From Bennett and Cuatrecasas, 1975b.

Preparations of ^{125}I-labeled choleragen with the majority of the iodine localized in the 'active' portion of the molecule provide an excellent tool for such localization studies. Endocytosis of the toxin-membrane complex seems unlikely as an obligatory step, since erythrocytes of rats, which are incapable of such a process (Hirsch *et al.*, 1973), respond well to choleragen (Bennett and Cuatrecasas, 1975a). Direct measurements demonstrate that ^{125}I-labeled choleragen enters toad erythrocytes at a nearly negligible rate under conditions where the toxin readily activates adenylate cyclase in these cells (Table 1.3). The amounts of toxin which may exist in the cytoplasm (10–20 molecules per cell out of a total of 6000 bound) could be of significance if choleragen had intrinsic catalytic activity or triggered some catalytic process in the cell. The kinetics of activation by choleragen in these erythrocytes clearly indicate that the toxin does not act by a catalytic mechanism (see below).

The findings in erythrocytes are in good agreement with a morphological study of the localization of choleragen and choleragenoid in the small intestine (Peterson *et al.*, 1972) and with studies of the distribution of ^{125}I-choleragen in sub-cellular fractions of fat cells (Chang *et al.*, 1975). Cellular uptake of choleragen has been reported (Grady *et al.*, 1975) although the choleragen in these studies was impure and was used at concentrations as high as 200 μg ml^{-1} (about 2000-fold higher than the maximal dose of pure toxin).

1.2.2 Nature of the membrane receptor for choleragen

Presumptive identification of GM1 monosialoganglioside

The observation that choleragen was inactivated by incubation with a mixture of brain gangliosides (van Heyningen *et al.*, 1971; Pierce, 1973) suggested that these glycolipids might be related to the tissue receptor for choleragen, but it was impossible to exclude the possibility that inactivation occurred by nonspecific lipid-protein adsorption or binding to glycolipid micelles. It was subsequently demonstrated that the inhibition of biological activity occurred specifically with low concentrations of a monosialoganglioside which was resistant to digestion by neuraminidase (GM1) (see Fig. 6) (Cuatrecasas, 1973a–c; Holmgren *et al.*, 1973a, b; King and van Heyningen, 1973). It was also observed that precipitin bands were formed between choleragen and gangliosides in double diffusion gels analogous to the Ouchterlony technique for antigen-antibody interactions, which provided a means of detecting binding of toxin and glycolipids independent of biological assays (Holmgren *et al.*, 1973a, b; Staerk *et al.*, 1974). Another approach to detecting a choleragen-glycolipid interaction has been to prepare insoluble aggregates of ganglioside and cerebroside, which have been shown to adsorb choleragen (van Heyningen, 1974).

The loss of biological activity following formation of a choleragen-GM1 complex could be explained equally well by two possibilities: (a) GM1 competes with membrane sites for binding of choleragen, and thus prevents the interaction of choleragen with the cell surface or (b) the choleragen-GM1 complex binds to membranes, but in a biologically ineffective fashion. Measurements with ^{125}I-labeled choleragen (Cuatrecasas, 1973b, c) clearly demonstrated that gangliosides block the binding of choleragen to cell membranes. Such competition experiments provided a rapid and quantitative assay for evaluation of the binding specificity of choleragen. Although high concentrations of certain complex sugars and glycoproteins block binding (Table 1.4), ^{125}I-choleragen is remarkably selective for glycosphingolipids

Table 1.4 Effect of special glycopeptides and oligosaccharides in the specific binding of cholera toxin to liver membranes. From Cuatrecasas, 1973b.

Oligosaccharide	% ^{125}I-labeled cholera toxin bound
Fetuin glycopeptide I, 85 μM	85
Fetuin glycopeptide II, 85 μM	98
Thyroglobulin glycopeptide, I, 85 μM	86
Thyroglobulin glycopeptide II, 85 μM	84
γG-Glycopeptide, 0.9 μM	100
γM-Glycopeptide, 0.1 μM	96
Lacto-N-tetraose (Gal $\xrightarrow{\beta 1,3}$ GlcNAc $\xrightarrow{\beta 1,3}$ Gal $\xrightarrow{\beta 1,4}$ GlcNAc),[b] 1 mM	80
Lacto-N-neotetraose (Gal $\xrightarrow{\beta 1,4}$ GlcNAc $\xrightarrow{\beta 1,3}$ Gal $\xrightarrow{\beta 1,4}$ GlcNAc),[b] 1 mM	92

[a] ^{125}I-labeled cholera toxin (50 ng/ml) was preincubated at 24°C for 60 min in Krebs-Ringer-bicarbonate buffer containing 0.1% (w/v) albumin and the compound indicated in the table. Samples of these solutions were then incubated at 24°C for 15 min with Krebs-Ringer-bicarbonate containing 0.1% (w/v) albumin and 40 μg/ml of liver membrane protein. Specific toxin binding was determined by filtration procedures as described in the text.

[b] Gal, galactose; GlyNAc, N-acetyl-glucosamine.

containing sialic acid (gangliosides) (see Fig. 1.6), particularly G_{M1} (Tables 1.5, 1.6). No binding competition is observed with sulfatides, sphingomyelin, psychosine, or crude brain cerebrosides (Table 1.4). The simple glycosphingolipids, such as glucose-ceramide, galactose-ceramide, and galactose-galactose-ceramide, are also relatively inert, although more complex derivatives formed by sequential addition of terminal glactose and N-acetylgalactosamine residues result in an increased potency in blocking binding of ^{125}I-choleragen (Table 1.5). The addition of sialic acid in 3,2 linkage with the galactose of galactose-glucose-ceramide to form G_{M3} monosialoganglioside results in at least a 20-fold increase in inhibitory potency (Table 1.6). Sequential additions of N-acetylgalactosamine to G_{M3} to form G_{M2}, and of galactose to G_{M2} to form G_{M1} result in an increase of 100-fold and 50-fold, respectively, in the apparent affinity for choleragen (Table 1.6). Further addition of N-acetylneuraminic acid to G_{M1} reduces the inhibitory ability by 40–to 100-fold. The approximate concentrations of gangliosides required for half-maximal inhibition of [^{125}I] choleragen binding are: G_{M1} (0.02 μg/ml), G_{D1a} (0.7 μg ml^{-1}), G_{M2} (1 μg/ml), G_{T1} (1.8 μg ml^{-1}) and G_{M3} (50 μg ml^{-1}) (Cuatrecasas, 1973b).

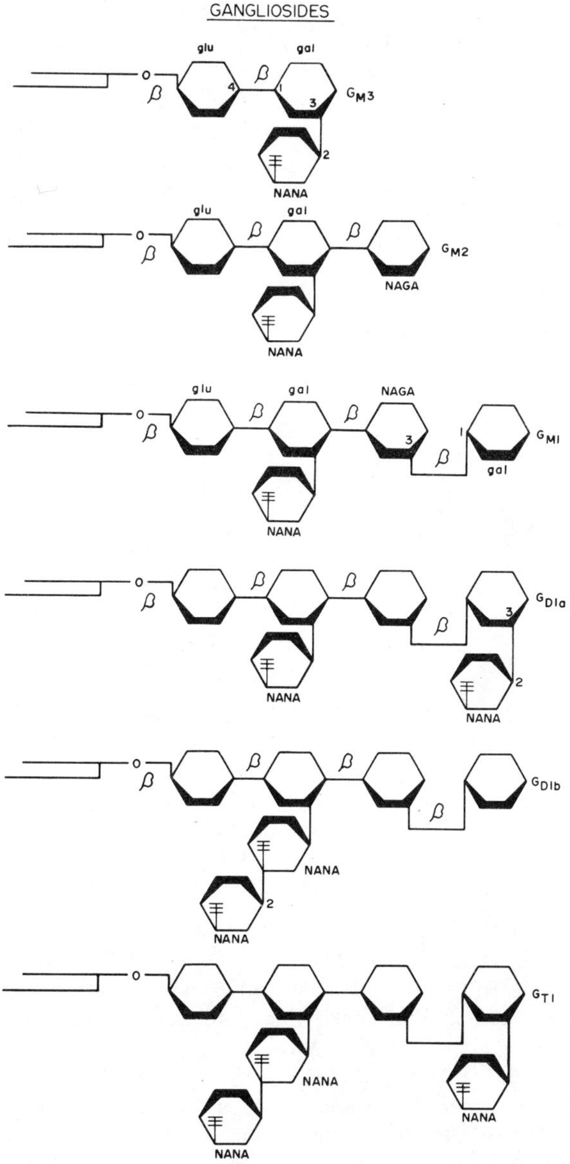

Fig. 1.6 Structures of some commonly occurring gangliosides. The nonmenclature is that of Svennerholm (1963).

These findings stress the specificity of choleragen for the carbohydrate portion of GM1 and are essentially in agreement with conclusions based on inhibition of biological activity (King and van Heyningen, 1973; Holmgren *et al.*, 1973b). Derivatives

Table 1.5 Effect of various glycophingolipids and other glycolipids on the specific binding of ^{125}I-labeled cholera toxin to liver membranes[a]. From Cuatrecasas, 1973b.

Compund[b]	Concentration in pre-incubation mixture (μg ml^{-1})	% Binding of ^{125}I-labeled cholera toxin
None		100
Gall → Cer	100	100
	5	100
Glcl → Cer	100	100
	5	100
Gall $\xrightarrow{\beta}$ 4Glcl → Cer	100	100
	5	100
Gall $\xrightarrow{\beta}$ 4Gall $\xrightarrow{\beta}$ 4Glcl → Cer	100	80
GalNAcl $\xrightarrow{\beta}$ 3Gall $\xrightarrow{\alpha}$ 4Gall $\xrightarrow{\beta}$ 4Glcl $\xrightarrow{\beta}$ Cer	50	35
	20	44
	5	68
HSO_3 → 3Gal → Cer (sulfatide)	150	100
Gangliosides (crude, bovine brain)	50	0
	10	0
	1	2
	0.25	5
	0.05	73
Psychosine (Gall → sphingosine)	100	100
Cerebrosides	500	100
Sphingomyelin	100	100

[a] ^{125}I-labeled cholera toxin (0.1 μg ml^{-1}) was preincubated at 24°C for 60 min in 0.2 ml of Krebs-Ringer-bicarbonate buffer containing 0.1% (w/v) albumin and the compound indicated in the table. Samples (50 μl) were then added to incubation mixtures consisting of 0.2 ml of Krebs-Ringer-bicarbonate buffer, 0.1% (w/v) albumin, and 50 μg of liver membrane protein. After 20 min at 24°C the specific binding of the toxin to the membranes was determined by filtration on cellulose acetate filters. In the absence of liver membranes none of these compounds caused adsorption of the toxin to the filters. The final concentration of the various glycolipids in the membrane incubation mixture is five times lower than that indicated in the table.

[b] Cer, ceramide; Gal, galactose; Glc, glucose; GalNAc, N-acetylgalactosamine.

of GM1 have been prepared in which the fatty acid portion is deleted, or which have lost the entire ceramide portion of the molecule (Staerk *et al.*, 1974; Holmgren *et.al.*, 1974b). The molecules which contain only a single hydrocarbon chain apparently retain the capacity to bind to choleragen, while the oligosaccharide fragment of GM1 binds with an apparently reduced affinity.

Table 1.6 Effect of specific gangliosides on the binding of ^{125}I-labeled cholera toxin to liver membranes[a]. From Cuatrecasas, 1973b.

Ganglioside[b]	Concentration in pre-incubation mixture (μg ml^{-1})	% Binding of ^{125}I-labeled cholera toxin
None		100
G_{M3} (Cer $\rightarrow$ 1Glc4 $\xrightarrow{\beta}$ 1Gal3 $\xrightarrow{\beta}$ 2NANA)	50	71
	25	95
	6	100
G_{M3} (Cer $\rightarrow$ 1Glc4 $\xrightarrow{\beta}$ 1Gal3 $\xrightarrow{\beta}$ 2N-glycolyl-NA)	50	50
	20	60
	5	82
G_{M2} (Cer $\rightarrow$ 1Glc4 $\xrightarrow{\beta}$ 1Gal4 $\xrightarrow{\beta}$ 1GalNAc)	5	16
3	1.6	28
$\uparrow \alpha$	0.5	65
2NANA	0.2	84
G_{M1} (Cer $\rightarrow$ 1Glc4 $\xrightarrow{\beta}$ 1Gal4 $\xrightarrow{\beta}$ 1GalNAc3 $\xrightarrow{\beta}$ 1Gal)	0.35	0
3	79 ng/ml	18
$\uparrow \alpha$	20 ng/ml	45
2NANA	6 ng/ml	74
	1 ng/ml	90
G_{D1a}(Cer $\rightarrow$ 1Glc4 $\xrightarrow{\beta}$ 1Gal4 $\xrightarrow{\beta}$ GalNAc3 $\xrightarrow{\beta}$ 1Gal3	5	20
3 $\qquad\qquad$ $\uparrow \alpha$		
$\uparrow \alpha$ $\qquad\qquad$ 2NANA	1.6	34
2NANA	0.5	43
	0.2	70
G_{T1} (Cer $\rightarrow$ 1Glc4 $\xrightarrow{\beta}$ 1Gal4 $\xrightarrow{\beta}$ 1GalNAc3 $\xrightarrow{\beta}$ 1Gal3	10	43
3 $\qquad\qquad$ $\uparrow \alpha$	2.5	31
$\uparrow \alpha$ $\qquad\qquad$ 2NANA	0.8	85
2NANA8 $\xrightarrow{\alpha}$ 2NANA	0.2	90

[a]These experiments were performed as described in Table 1.4.

[b]Nomenclature according to Svennerholm (1964); Cer, ceramide; Glc, glucose; Gal, galactose; GalNAc, N-acetylgalactosamine; NANA, N-acetylneuraminic acid.

[c]After digesting G_{M1} (0.1 μg ml^{-1}) with neuraminidase (20 μg ml^{-1}) for 3 h at 37° in 0.1 M sodium acetate (pH 6.5) 2mM $CaCl_2$, the binding of toxin in the presence of 1 ng ml^{-1} of the ganglioside was 80%. Under similar conditions digestion of G_{D1a} and G_{T1} resulted in a very large enhance of inhibitor activity.

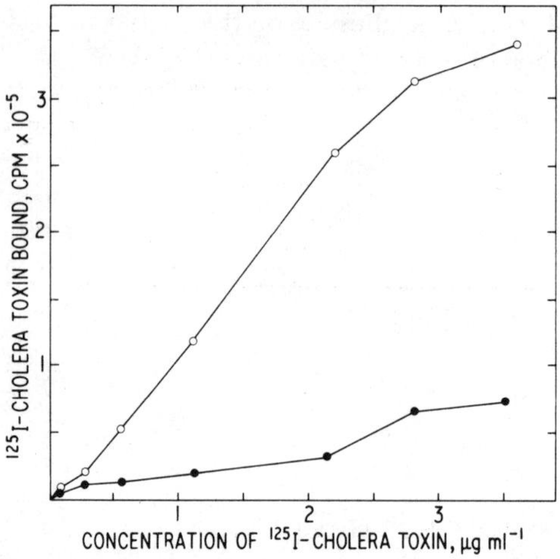

Fig. 1.7 Effect of increasing the concentration of cholera toxin on the binding of the toxin to ganglioside-treated liver membranes. Liver membrane suspensions (3 mg of protein ml^{-1}) were incubated for 50 min at 24°C in Krebs-Ringer bicarbonate buffer 0.1% (w/v) albumin, in the absence (●) and presence (○) of 0.1 mg ml^{-1} of bovine brain gangliosides. The suspensions were then dilated ten-fold with the same buffer (without gangliosides) and centrifuged for 20 min at 30 000 g. The pellets were resuspended in the same buffer and 0.2 ml fractions containing 14 μg of membrane protein were incubated for 15 min at 24°C with varying concentrations of ^{125}I-labeled cholera toxin (2 μCi μg^{-1}) (Cuatrecasas, 1973c).

Reconstitution of GM1 into cell membranes

The demonstration of high affinity interaction between choleragen and GM1 provides circumstantial evidence that GM1 may represent the biologically relevant receptor for choleragen. A receptor role for this glycolipid is strongly supported by the demonstration that GM1 can be reconstituted into cell membranes with a concomitant increase in binding of iodocholeragen, and an enhanced sensitivity to biological action of choleragen (Cuatrecasas, 1973c). Pre-incubation of cells or membranes with ganglioside followed by thorough washing results in a substantial elevation in the capacity to bind choleragen (Fig. 1.7). The binding properties of these sites appear to be very similar to those which occur normally (Cuatrecasas, 1973c). The membrane-ganglioside complex is quite stable, since erythrocytes treated with the glycolipid and then washed, retain fully the enhanced capacity to bind choleragen for at least 24 h at 24°C (Table 1.7).

These studies suggest that the ganglioside is spontaneously incorporated into the membrane and assumes an orientation such that the molecule may still interact with choleragen. It is pertinent that gangliosides are amphipathic molecules (Fig. 1.6) with a hydrophilic carbohydrate portion and a hydrophobic domain which is presumably involved in the micellar and lamellar complexes formed between

Table 1.7 Enhancement of binding of cholera toxin by treating erythrocytes with gangliosides[a]. From Cuatrecasas, 1973c.

Cells	Concentration of ^{125}I-labeled cholera toxin (sp cpm bound)[b]		
	1.5×10^4 cpm (5×10^{-11} M)	6.0×10^4 cpm (2×10^{-10} M)	2.4×10^5 cpm (8×10^{-10} M)
Fresh Cells			
Untreated	4500	11 500	24 000
Treated with gangliosides	5400	19 000	66 500
Cells after 24 h			
Untreated		16 000	24 400
Treated with gangliosides		51 000	77 000

[a]Human erythrocytes (1.5×10^9 cells ml^{-1}) were incubated at 24°C for 60 min in Krebs-Ringer-bicarbonate buffer, 0.1% (w/v) albumin, in the absence or presence of 0.125 mg ml^{-1} of crude bovine brain gangliosides. The cells were washed five times by centrifugation using five times the original volume of buffer. The cells (7.5×10^8 cells ml^{-1}) were incubated at 24°C for 20 min in 0.2 ml of Krebs-Ringer-bicarbonate buffer, 0.1% (w/v) albumin, containing the indicated concentration of ^{125}I-labeled cholera toxin (sp act., 20 μCi μg^{-1}). To examine the stability of the ganglioside-cell complex the cells were left standing at 24°C for 24 h, at which time they were washed, resuspended in the same buffer, and reexamined for binding of ^{125}I-labeled cholera toxin. In all cases the ability of native toxin (20 μg ml^{-1}) to block more than 90% of the binding of iodotoxin to these cells was confirmed.

[b]Counts per minute per 0.2 ml of incubation mixture.

gangliosides and phospholipids (Hill and Lester, 1972; Ohki and Hono, 1970; Haywood, 1974). It is thus reasonable to speculate that the sphingosine portion of the ganglioside partitions into the membrane bilayer, as predicted from thermodynamic considerations (Tanford, 1972), and that the oligosaccharide moiety is exposed to the medium and available to complex with choleragen.

The incoporation of gangliosides into membranes has been measured directly with ^{3}H-labeled GM1 by Holmgren *et al.* (1975) who reported that the increment in exogenously incorporated GM1 is matched by a nearly stoichiometric increase in toxin binding sites, without any detectable change in the affinity of the binding process.

Prior incubation of cells with gangliosides enhances the biological effects of choleragen in addition to increasing the number of receptor sites (Fig. 1.8) (Cuatrecasas, 1973c; Holmgren *et al.*, 1975; Gill and King, 1975; King and van Heyningen, 1975; Moss *et al.*, 1976. It is important to note that these effects can be demonstrated only in cells which have limited numbers of receptors, or a special means must be devised to decrease the existing binding sites to

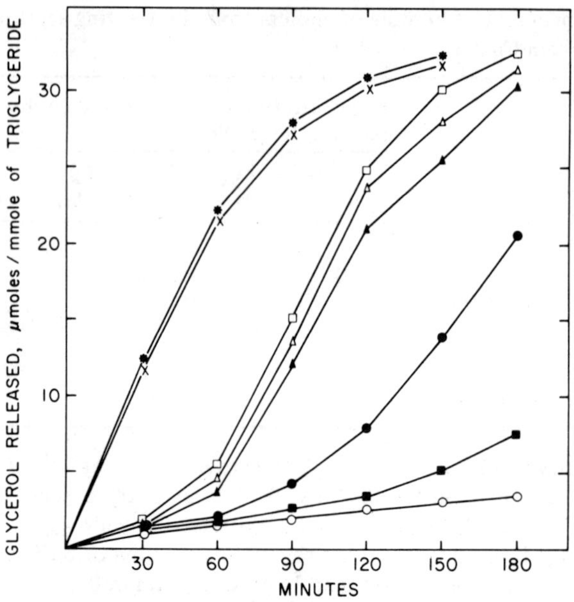

Fig. 1.8 Effect of pre-incubating fat cells with bovine brain gangliosides and with ganglioside G_{M1} followed by washing of the cells on cholera toxin and epinephrine induced lipolysis. Isolated fat cells from rats weighing 190 g were incubated for 30 min at 24°C in Krebs-Ringer-bicarbonate buffer, 0.1% (w/v) albumin, in the absence of other compounds (○, ■, ●, X), in the presence of 0.1 mg ml^{-1} of brain gangliosides (△, ▲, +), or in the presence of 3 μg ml^{-1} of ganglioside G_{M1} (□). The cell suspensions were washed four times with the same buffer. The cells were then incubated at 37°C for varying time periods in Krebs-Ringer-bicarbonate buffer, 3% (w/v) albumin containing no other additions (○) or containing 20 ng ml^{-1} (■, ▲) or 1 μg ml^{-1} (●, △, □) of cholera toxin, or 2 μg ml^{-1} of L-epinephrine (*, X). In the absence of lipolytic agents the cells pretreated with gangliosides or G_{M1} produced the same amount of glycerol as the control cells (○). The response of the untreated cells to very high concentrations (50 μg ml^{-1}) of toxin is equal to the response described by □. The effect of pre-incubating cells with 0.5 mg ml^{-1} of gangliosides was not significantly greater than that of 0.1 mg ml^{-1} (△). 40 μg ml^{-1} of gangliosides produces nearly maximal effects. The sensitivity of fat cells to cholera toxin is dependent on the size of the rats from which the cells are obtained; with cells from smaller (100 g) animals nearly maximal lipolytic responses can be achieved with about 50–100 ng of toxin ml^{-1} (Cuatrecasas, 1973d).

incorporating new receptors. These findings demonstrate that exogenous GM1 is functional, and strongly implicate this glycolipid as the natural receptor.

It is conceivable that certain glycoproteins may contain an oligosaccharide

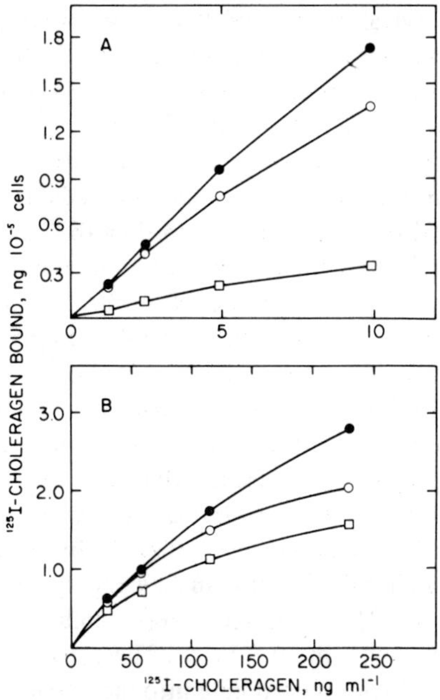

Fig. 1.9 Binding of cholera toxin to transformed AL/N cells and 0, TAL/N, P = 46; O TAL/N, P= 269, SVS ALN. (A) Binding to 8 x 10^4 cells in 0.4 ml of buffer. (B) Binding to 3.2 x 10^5 cells in 0.2 ml of buffer. From Hollenberg, Fishman, Bennett and Cuatrecasas (1974).

sequence similar to GM1 and may also bind choleragen. Evidence that proteins are not directly responsible for toxin binding is provided by the finding that chloroform-methanol extraction of liver membranes removes almost all binding sites for choleragen, and binding activity can be partially recovered in the ganglioside fraction of these extract (Cuatrecasas, 1973b). Also, extensive digestion of membranes (90 min at 37°C) with 1 mg ml^{-1} of trypsin or pronase has no influence on their capacity to bind choleragen (Cuatrecasas, 1973b). Moreover, Hart (1975a) has reported that choleragen does not interact with radioiodinated surface proteins of lymphoid cells which were labeled by lactoperoxidase-catalysed iodination, a procedure known to label almost all of the cell surface proteins (Marchalonis and Cone, 1973). Furthermore, the carbohydrate sequence of GM1 has not been described in glycoproteins.

Further evidence in support of GM1 as the toxin receptor is provided by a study of transformed mouse cell lines from the same parent strain which exhibit loss of specific enzymes involved in the *de novo* biosynthesis of membrane gangliosides (Hollenberg *et al.*, 1974). The binding of ^{125}I-choleragen and the sensitivity to stimulation of adenylate cyclase and inhibition of DNA synthesis vary in parellel with the ganglioside composition of these cells. The SV40-transformed cell line which lacks chemically detectable GM2, GM1 and GD1A as well as UDP-N-acetylgalactosamine-GM3 N-acetylgalactosaminyltransferase activity, is the least sensitive to the metabolic effects and binds the least amount of ^{125}I-choleragen (Fig. 1.9).

A direct and unequivocal means of proving that GM1 is indeed the membrane receptor for choleragen would be to block the binding of ^{125}I-choleragen by exposure of membranes to antibody directed against GM1. Such efforts have been only partially successful (Craig and Cuatrecasas, unpublished data), although negative results must be interpreted with caution. It is possible that gangliosides are not readily accessible to the relatively bulky and charged immunoglobulin molecules, or that the configuration of gangliosides on the cell surface is somehow different from that of the same molecules in solution. Unsuccessful attempts have also been made to destroy GM1 with periodate oxidation and enzymatic digestion with galactose oxidase, B-galactosidase, and by lysosyme (Cuatrecasas, unpublished data).

A practical application of the affinity of choleragen for GM1 has been preparation of agarose derivatives of gangliosides which are capable of adsorbing choleragen nearly quantitatively at concentrations as low as 10^{-11} M (Cuatrecasas *et al.*, 1973). Soluble ganglioside polymers synthesized by coupling the terminal sialic acid residue to the amino groups of branched copolymers of lysine and alanine also prevent the binding of ^{125}I-choleragen to liver membranes and abolish the lipolytic activity of choleragen with rat fat cells. These macromolecular derivatives of gangliosides apparently do not incorporate into membranes as do the free glycolipids, and thus may provide useful therapeutic tools for the management of clinical cholera (Cuatrecasas *et al.*, 1973).

Nonidentity of receptors for choleragen and hormones

Although choleragen activates an enzyme which is normally regulated by hormones, abundant evidence has accumulated that the membrane receptors for choleragen and hormones are clearly different. Direct measurements reveal no competition by choleragen for the binding of ^{125}I-labeled chorionic gonadotrophin to rat testis (Sato *et al.*, 1975) or for the binding of ^{125}I-glucagon to rat liver membranes (Bennett *et al.*, 1975b), although these hormones and choleragen stimulated adenylate cyclase in the respective tissues. Conversely, ^{125}I-labeled choleragen binds to fat cell membranes with an unaltered affinity or number of sites in the presence of high concentrations of epinephrine and glucagon (Bennett *et al.*, 1975b). Gangliosides, which effectively block the activity of choleragen, have no influence on the action of epinephrine (Cuatrecasas, 1973c) or ACTH (Haksar *et al.*, 1975). Similarly, choleragenoid blocks activation of adenylate cyclase by choleragen (see above) but has no influence on the extent or affinity of stimulation by catecholamines (Fig. 16).

Another line of evidence is provided by studies with a mutant line of cultured adrenal cells which have lost the ability to respond to ACTH, but retain fully their sensitivity to choleragen (Donta, 1974; Wishnow and Fiest, 1974; Wolff and Cook, 1975). It has also been demonstrated with isolated rat adrenal cells that digestion with neuraminidase completely abolishes the response to ACTH, but actually enhances the effect of choleragen (Hasksar, *et al.*, 1974).

These findings indicate clearly that choleragen does not interact initially with some membrane molecule which is a ubiquitous component of hormone receptors, but do not exclude the possibility that choleragen binds to the receptor of a heretofore unrecognized hormone. Perhaps the best evidence against such a mechanism is to be found in the striking differences in the nature of activation of adenylate cyclase by choleragen and hormones, which is discussed in detail below. Briefly, the stimulation obtained by hormones is nearly instantaneous in onset, and readily reversible by removing the agent or by addition of an antagonist. Choleragen, on the other hand, acts only after an absolute delay of 15–30 min, and then the extent of stimulation increases for up to six hours. The activated state is not reversed by washing, or even by solubilization of the membranes in detergent (see below). Furthermore, stimulation, once it has occured, is not affected by gangliosides, choleragenoid, or specific antibody against choleragen.

1.2.3 Applications of ^{125}I-labeled choleragen and choleragenoid

Choleragen and choleragenoid are potentially useful as probes for detection and quantitation of cell surface glycolipids which most probably are GM1 monosialogangliosides. These molecules are of particular interest since the complement of cell surface gangliosides can be altered significantly following virus or carcinogen-induced transformation (Cumar *et al.*, 1970; Robbins and MacPherson, 1971; Den *et al.*, 1974; Fishman *et al.*, 1974; Brady *et al.*, 1970), after extensive culturing of spontaneously transformed cell lines (Fishman *et al.*, 1973; Hakamori, 1970) and may vary with cell density (Hakomori, 1975). Choleragen and choleragenoid can be radioiodinated to specific activities as high as 10^7 cpm/picomole, and may be used to detect as few as 100 sites per cell in simple assays requiring less than a milligram of tissue. These markers thus provide a sensitivity exceeding by 10^3–10^4 that of chemical methods now available.

The complex between choleragen or choleragenoid and membrane glycolipids is stable following dissolution of the membrane with nonionic detergents (Bennett *et al.*, 1975a; Hart, 1975b) (Fig. 1.25). The solubilized ligand-ganglioside complexes exhibit much larger stokes radii on gel filtration than native choleragen, which suggests that the glycolipids may be associated with the membrane components. Pre-incubation of intact cells with ^{125}I-choleragen prior to solubilization provides a sensitive means of labeling the membrane-ganglioside complex. It should, in principle, be possible with this tool to isolate and characterize the membrane macromolecule(s) which are associated with these glycolipids in detergent extracts, and presumably in the intact membrane.

The rapidity of the binding of choleragen and choleragenoid to cell membranes, the stability of the resulting membrane-ligand complex and the fact that binding sites are ubiquitous and relatively aundant in most cell surfaces permit the use of these molecules as general membrane markers (Chang *et al.*, 1975). The basic

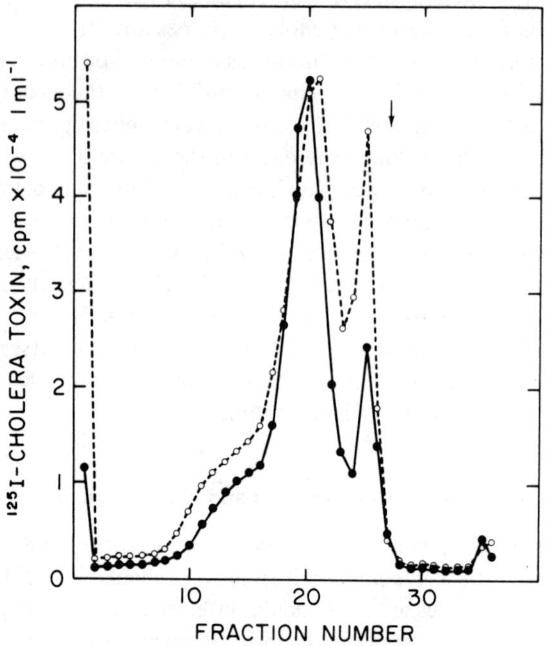

Fig. 1.10 Comparison of the distribution of ^{125}I-cholera toxin on sucrose density centrifugation when membrane labeling is performed in intact fat cells and in whole particulate prepparations from homogenates. Isolated fat cells from 8 rats (150–175 g) were divided into e equal parts. One part (A) was incubated with ^{125}I-cholera toxin (5.1 x 10^6 cpm) for 5 min at 24°C in 5 ml of KRB-1% albumin. The other (B) was treated the same way but without ^{125}I-cholera toxin. Both sets were washed and homogenized as described in Fig. 1. A total particulate pellet (40 000 x g, 30 min) was resuspended in 5 ml of KRB-1% albumin with a Polytron (set at 2.6, 10 s). ^{125}I-cholera toxin was added to Part B, incubated for 5 min at 24°C and washed twice by centrifugation. The distribution of ^{125}I-cholera toxin from the samples labeled in the intact cells (●) and in the particulate fractions (○) are determined on linear sucrose gradients. The same amount of protein material was applied to each gradient. Arrow indicates the interphase of linear gradient and isotonic sucrose-Tris buffer. From Chang, Bennett and Cuatrecasas (1975).

approach of this procedure is to label cells with low concentrations of radiolabeled protein (10–50 ng ml^{-1}, 10 000 to 20 000 cpm ng^{-1}) prior to homogenization by a five to ten min incubation with intact cells at 4°C, or by *in situ* perfusion of an organ such as liver. These conditions allow nearly complete binding of choleragen, but are not sufficient for activation of adenylate cyclase. The unbound ^{125}I-labeled ligands are removed by washing at 4°C, and the distribution and recovery of plasma membrane-bound label are followed quantitatively during subsequent homogenization and fractionation procedures (Fig. 1.10). The spontaneous dissociation of membrane-bound iodotoxin is negligible at 4°C, which permits prolonged manipulations without a significant loss or redistribution of label. The specificity of this procedure for the plasma membrane is indicated by the fact that the distribution of membrane-bound ^{125}I-choleragen from labeled fat cells is nearly identical on sucrose gradients to that of ^{125}I-insulin, which has been shown to bind to receptors localized exclusively at the cell surface (Cuatrecasas, 1972). Furthermore, negligible intracellular toxin can be detected even after prolonged incubation with intact cells (Table 1.3). Other radiolabeled ligands such as insulin and wheat germ agglutinin interact with high affinity with cell surface receptors, and are also useful as plasma membrane markers (Chang *et al.*, 1975; Bennett and Cuatrecasas, 1973).

The rapidity and accuracy of the binding assays with ^{125}I-labeled choleragen and choleragenoid make a 'radioreceptor' binding competition assay a practical means of quantitating levels of native choleragen or *E. coli* enterotoxin (Agre, Parikh and Cuatrecasas, 1976) and for following the titers of choleragen antisera in immunized animals. Liver microsomal membranes (Cuatrecasas, 1973b) provide an easily prepared, standardized source of membranes which retain their binding properties for at least two years when stored at –20°C.

1.2.4 Detection of interaction of choleragen and choleragenoid with membranes by morphological methods

Although studies with radioactive ligands provide quantitative information, the values reflect only averages of a population of binding sites. The localization of choleragen on cell surfaces and the dynamics of receptor-ligand distribution in the plane of the membrane can be examined best by direct visualization of the bound molecules. The first reliable morphological study reported the distribution of choleragen and choleragenoid in the small intestine using as markers fluorescein- and peroxidase-labeled antibodies and ^{3}H-labeled choleragen (Peterson *et al.*, 1972). A similar approach was utilized by Holmgren *et al.*, (1974a) to visualize choleragen on mouse thymocytes.

The pattern of choleragen distribution on lymphocyte membranes has recently been reported to change from diffuse at 4°C to micropatches or 'caps' following incubation at 37°C (Revesz and Greaves, 1975; Craig and Cuatrecasas, 1975). The redistribution of the initial complex into aggregates is observed with derivatives of choleragen or choleragenoid labeled directly with fluorescein (Craig and Cuatrecasas, 1975) as well as with the triple-layer immunofluorescence procedure (Revesz and Greaves, 1975). These observations demonstrate the possibility of rapid lateral mobility of the choleragen-ganglioside complex in the plane of the plasma membrane, and strongly suggest that choleragen is capable of binding more than one ganglioside and of cross-linking these molecules.

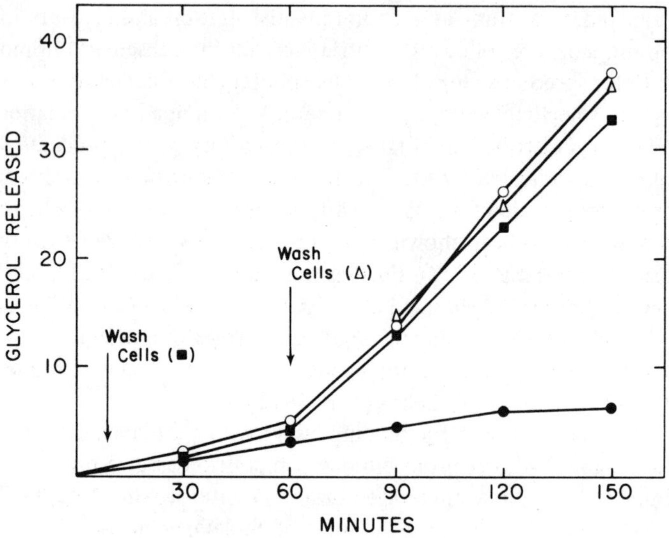

Fig. 1.11 Effect of washing fat cells after exposure to cholera toxin on the lipolytic response of the cells. The cells were incubated at 37°C in Krebs-Ringer-bicarbonate buffer, 3% albumin (w/v), in the absence (●) and presence of 2 μg ml^{-1} of cholera toxin. After 9 (■) and 60 (△) min some of the toxin-treated cells were washed extensively with the same buffer and the incubation was continued at 37°C in the absence of added toxin to the medium. In some cell suspensions (○) the toxin in the medium was not removed. Lipolysis is described as the micromoles of glycerol released per millimole of triglyceride. (From Cuatrecasas, 1973d).

1.3 BIOLOGICAL EFFECTS OF CHOLERAGEN

Although in clinical cholera the action of choleragen is restricted to stimulation of secretory processes in the small intestine, the toxin is capable of stimulating ubiquitously metabolic responses which are mediated by cyclic AMP (Table 1.1). In no instance has choleragen been observed to kill cells directly, and the term 'toxin' applies only to its effects on the whole organism. Several unusual features have been observed consistently in the action of choleragen. A delay of 30–60 min is always obtained between exposure of intact cells to choleragen and any biological effects. Only a five minute incubation is necessary, however, for a complete response, and the effects cannot be reversed by extensive washing (Fig. 1.11), or by addition of known inhibitors of choleragen binding and the metabolic alterations may persist for days in the absence of additional choleragen. This section will cover the work in selected areas, with emphasis on the mechanism of action of choleragen and the application of choleragen in elucidating the role of cyclic AMP in cell growth and differentiation.

1.3.1 Activation of lipolysis in isolated adipocytes

The remarkable ability of an enterotoxin to activate lipolysis in isolated rat fat cells was one of the first demonstrations that the action of choleragen was not confined to the intestine (Vaughan *et al.*, 1970; Greenough *et al.*, 1970; van Heyningen *et al.*, 1971). Isolated rat fat cells also provided a practical system in which to investigate the peculiar features of the action of choleragen and to exclude a number of possible explanations for the lag phase (Cuatrecasas, 1973d).

The delay is most likely not due to the slow conversion of toxin to a soluble, active form or to accumulation of active metabolies in the medium, since cells washed after 60 min exhibit an unaltered rate of lipolysis (Fig. 1.11). Also, choleragen does not act initially as an inhibitor of lipolysis, since the lipolytic response to epinephrine and glucagon is unaltered during the latency phase. The possibility that choleragen acts indirectly by stimulating the biosynthesis of prostaglandins also is unlikely since indomethacin, sodium salicylate, and diphloretin phosphate have no effect on the lipolytic response. Furthermore, no detectable levels of prostaglandins (Kimberg *et al.*, 1974; Hudson *et al.*, 1974) or of 'prostaglandin-like substances' (Bedwani and Okpaho, 1974) in intestinal mucosal preparations exposed to choleragen have been observed by radioimmunoassay and bioassay methods.

It is also unlikely that the metabolic effects of choleragen are mediated by processes requiring biosynthesis of RNA or protein, since high concentrations of actinomycin D, cycloheximide, and puromycin had no effect on the extent or rate of stimulation of lipolysis. Cycloheximide has been demonstrated to block the toxin-induced hypersecretion in rabbit intestinal mucosa (Serebro *et al.*, 1969; Grayer *et al.*, 1970), although the stimulation of adenylate cyclase and increase in cyclic AMP levels are not affected by this compound (Kimberg *et al.*, 1973).

The lag period is not markedly affected by the concentration of choleragen, even at levels as high as 100 $\mu g\ ml^{-1}$. Furthermore, incorporation of exogenous GM1, which increases the binding capacity of fat cells by as much as 10-fold (Cuatrecasas, 1973c) and enhances the sensitivity to low concentrations of choleragen (Fig. 1.8) does not significantly decrease the length of the delay (Fig. 1.8). This is in marked contrast to the lag phase oberved with diptheria toxin, which varies from 30 min to more than 6 h depending on the toxin concentration (Uchida *et al.*, 1973).

It was concluded from these results with fat cells (Cuatrecasas, 1973d) that the initial toxin-ganglioside complex was inactive, and that some structural change occurred during the lag period possibly involving membrane relocation of the complex, or some reorganization of the membrane itself. These possibilities are discussed in more detail below. The measurements of lipolysis, as well as other biological assays, suffer from the limitation that the event measured is separated by multiple steps from the presumably initial process of activation of adenylate cyclase. Since it is not known which is the rate-limiting event, kinetics and concentration-response relationships must be interpreted cautiously in such bioassays.

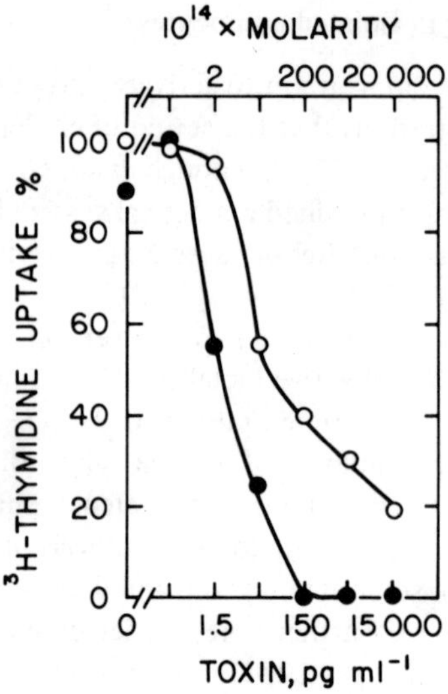

Fig. 1.12 Inhibition by cholera toxin of DNA synthesis in human (HF) fibroblasts stimulated by serum or epidermal growth factor. HF cells at confluency (100 μg of protein) were refed with fresh growth medium containing 0.1% bovine albumin and either 10% fetal-calf serum or 10 ng ml^{-1} of EGF in the presence of increasing amounts of cholera toxin. [^{3}H] Thymidine incorporation, measured at 23 h, is expressed as a percent of the maximum incorporation. Maximum thymidine incorporation was 12 900 ± 155 cpm for serum and 20 900 ± 1862 cpm for EGF. Serum, ○–○; EGF, ●–●. (Data from Hollenberg and Cuatrecasas, 1973).

1.3.2 Inhibition of DNA biosynthesis and cell proliferation

Elevation of cyclic AMP levels has been demonstrated to inhibit cell division (Gilman and Nurenberg, 1971; Masui and Garren, 1971; Johnson and Pastan, 1971), and it is therefore not surprising that choleragen should also block DNA synthesis and cell proliferation. In the first reports, choleragen was effective at concentrations as low as 10^{-14} M in blocking incorporation of ^{3}H-thymidine by human fibroblasts stimulated by epidermal growth factor (Fig. 1.12) (Hollenberg and Cuatrecasas, 1973) and by mouse spleen lymphocytes exposed to Con A or lipopolysaccharide (Sultzer and Craig, 1973). Similar findings have been reported for human peripheral lymphocytes (DeRubertis *et al.*, 1974; Hart and Finkelstein, 1975), mouse thymocytes (Holmgren *et al.*, 1974a), mouse spleen cells (Vischer and LoSpalluto, 1975; DeRubertis *et al.*, 1974) and transformed mouse epithelial cells (Hollenberg *et al.*, 1974). In no instance did choleragen alter the viability of these cells. Furthermore, choleragen was not acting by competing for mitogen or growth factor receptor sites (Holmgren *et al.*, 1974a; Hollenberg and Cuatrecasas, 1973; Hart and Finkelstein, 1975).

The effects of choleragen on thymidine incorporation have been usually determined after a 24–36 h pre-incubation period, which greatly enhances the sensitivity of these assays. Thus, as few as 10–20 molecules of choleragen per cell are sufficient to significantly alter DNA synthesis, making such measurements extraordinarily sensitive. Inhibition of thymidine incorporation by human fibroblasts has recently been utilized as a means of assaying *E. coli* enterotoxin (Agre and Cuatrecasas, 1976), which also activates adenylate cyclase (Evans *et al.*, 1973).

The potential therapeutic application of choleragen in disease states characterized by excessive cell proliferation such as malignancy, psoriasis, and keloid scar formation has not yet been explored. A possible limitation in such a use of choleragen is that its action on cell growth is expressed primarily as an inhibition of a stimulus for mitosis, and may be relatively ineffective in controlling cells which divide in the absence of growth factors. Cells genetically or phenotypically defective in adenylate cyclase are also not likely to respond. Choleragen may be useful as a tool for selecting mutant cell lines which have lost the inhibitory response to cyclic AMP, such as those studied by Tompkins and co-workers (Hochman *et al.*, 1975).

1.3.3 Induction of differentiated functions

Choleragen can induce differentiation of cells as well as inhibit cell division. Two lines of cultured tumor cells, mouse melanoma cells (O'Keefe and Cuatrecasas, 1974) and mouse adrenal cells (Wolff *et al.*, 1973; Donta *et al.*, 1973; Wishnow and Feist, 1974), have been reported to regain some specialized functions and to undergo morphological changes following incubation with choleragen. Melanoma cells demonstrate an increase in tyrosinase activity and in deposition of melanin pigment about three days after exposure ot 10^{-10} M choleragen or to melanocyte stimulating hormone (MSH). These cells also exhibit changes in shape, initially becoming spherical, and after a week develop increased size and assume a fibroblastic growth pattern. Adrenal tumor cells undergo similar intitial morphological changes in response to choleragen or ACTH, and the stimulated cells also demonstrate an increase in the excretion of Δ^4, 3-ketosteroids into the medium.

These findings strongly suggest that the action of MSH and ACTH are mediated by adenylate cyclase and cyclic AMP. The mechanism of induction of enzymatic activities by cyclic AMP is, of course, unknown, but would appear to be essential in understanding normal development and the dedifferentiation which occurs in malignant transformation. Choleragen provides a useful tool in this regard since it activates adenylate cyclase promiscuously and will affect embryonic tissues (Grand *et al.*, 1973) and cells which have lost the capacity to respond to hormones (Wishonow and Feist, 1974; Donta, 1974; Wolff and Cook, 1975).

1.3.4 Applications

Choleragen has been used in *in vitro* systems to provide confirmatory evidence for the role of cyclic AMP in inhibiting antigen-induced, IgE-mediated release of histamine from human basophils and cytotoxic activities of mouse spleen lymphocytes (Lichtenstein *et al.*, 1973). Administration of choleragen *in vivo* may enhance (Northrup and Fauci, 1972) or inhibit the antibody response depending on the time of treatment relative to exposure to antigen and on the dose of choleragen (Kately *et al.*, 1975; Chisari *et al.*, 1974). Injection of choleragen also affects the function of cytolytic thymus-derived lymphocytes (Henney *et al.*, 1973).

These systems are obviously quite complicated, although a recent report suggests that the action of choleragen may in part be explained by modulation of the helper function of thymic lymphocytes (Kately and Friedman, 1975). Choleragen has also been demonstrated to affect *in vitro* the early anamnestic antibody response (Cook *et al.*, 1975) and the function of Fc receptors as measured by rosette formation (Zuckerman and Douglas, 1975).

Choleragen has been exploited in a careful study to discriminate the role of adenylate cyclase in regulating the secretion of water and electrolytes and that of proteins in rat pancreas (Kempen *et al.*, 1975). A novel application of choleragen in neurobiology has been the demonstration that injection of this protein into the nucleus accumbens of rats induces subsequent behavioral changes which are very similar to the effects of dopamine (Miller and Kelly, 1975).

1.4 ACTIVATION OF ADENYLATE CYCLASE BY CHOLERAGEN

The activation of adenylate cyclase by choleragen exhibits some unusual properties that make this protein unique among other known modulators of cyclase activity. Although the initial toxin-membrane interaction is complete within 8 min at 37°C, a period of 20–30 min must pass before any response is observed in intact cells, including stimulation of adenylate cyclase activity (Kimberg *et al.*, 1971; O'Keefe and Cuatrecasas, 1974; Bennett and Cuatrecasas, 1975a). Stimulation of cyclase by choleragen has previously been detectable only in intact cells, although it has recently been possible to observe activation in cell homogenates and washed membrane preparations (van Heyningen and King, 1975; Bitensky *et al.*, 1975; Gill and King, 1975; Sahyoun and Cuatrecasas, 1975). Once activation of adenylate cyclase has occured, the stimulated state of the enzyme persists during cell lysis and membrane isolation procedures (reviewed by Sharp, 1973) and is retained even after solubilization of membranes with nonionic detergents (Beckman *et al.*, 1974; Bennett *et al.*, 1975a). The metabolic effects and stimulation of adenylate cyclase by choleragen exhibit a remarkable duration in living cells and are observed for days following a single brief exposure (Kimberg *et al.*, 1971; Donta *et al.*, 1973; Guerrant *et al.*, 1972; O'Keefe and Cuatrecasas, 1974).

In this section several issues are examined: the nature of the kinetic and regulatory properties of the choleragen-activated enzyme, the relationship between the stimulation obtained with choleragen and that with hormones and fluoride ion, the

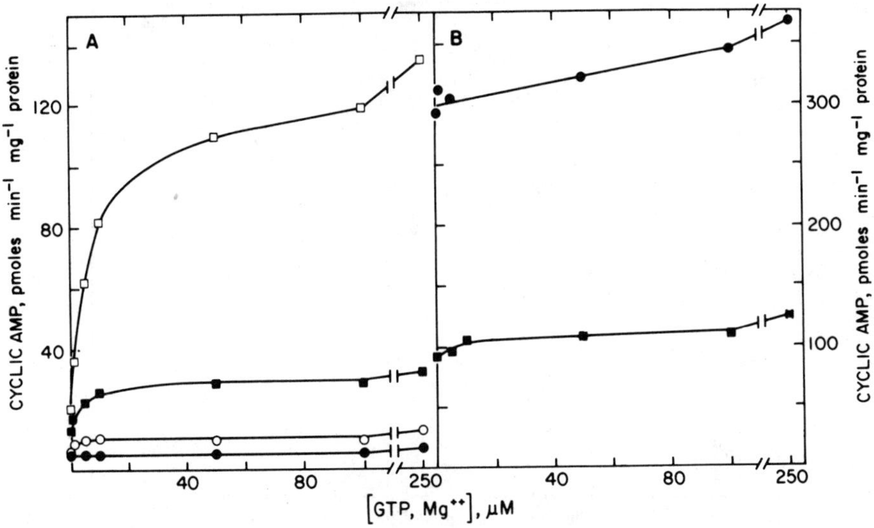

Fig. 1.13 Effect of increasing the concentration of GTP, Mg^{2+} on the adenylate cyclase activity of control (●, ○) and cholera toxin-treated toad erythrocyte (■, □) plasma membranes assayed in the presence (○, □) and absence (●, ■) of 20 μM (−)-epinephrine (A), and in the presence of 10 mM sodium fluoride (B). Erythrocytes were incubated in the presence and absence of cholera toxin (0.25 μg ml^{-1}) for 4 h at 30° in amphibian Ringer's, pH 7.5. The plasma membrane adenylate cyclase activity was determined (15 min, 30°C) in the presence of $MgCl_2$ (6.2 mM), [α-^{32}P]-ATP (0.5 mM, 96 cpm/pmole), aminophylline (5 mM), phosphoenolpyruvate (5 mM), pyruvate kinase (50 μg ml^{-1}, Tris-HC1 (50 mM, pH 8.0), 200–210 μg of membrane protein and increasing concentrations of GTP, Mg^{2+}. From Bennett and Cuatrecasas, 1975a.

properties of the process of activation and, in particular, whether the action of choleragen involves a catalytic step. These inquiries lead to specific predictions which are tested with detergent-solubilized membranes. Finally, a model for the mechanism of action of choleragen is proposed, and certain features are extended to the mode of regulation of adenylate cyclase by hormones.

1.4.1 Properties of choleragen-stimulated adenylate cyclase

Initial studies reported no differences in the K_m for ATP or dependence on divalent cations between control and choleragen-activated adenylate cyclase (Sharp *et al.*, 1973; Kimberg *et al.*, 1971; Sharp and Hynie, 1971; Chen *et al.*, 1972). The characteristics of the toxin-stimulated enzyme have been examined more recently in erythrocytes of turkeys (Field, 1974), toads and rats (Bennett and Cuatrecasas, 1975a), rat fat cells (Hewlett *et al.*, 1974; Bennett *et al.*, 1975b), and rat liver (Beckman *et al.*, 1974; Bennett *et al.*, 1975a) and found to exhibit marked changes in regulatory properties,

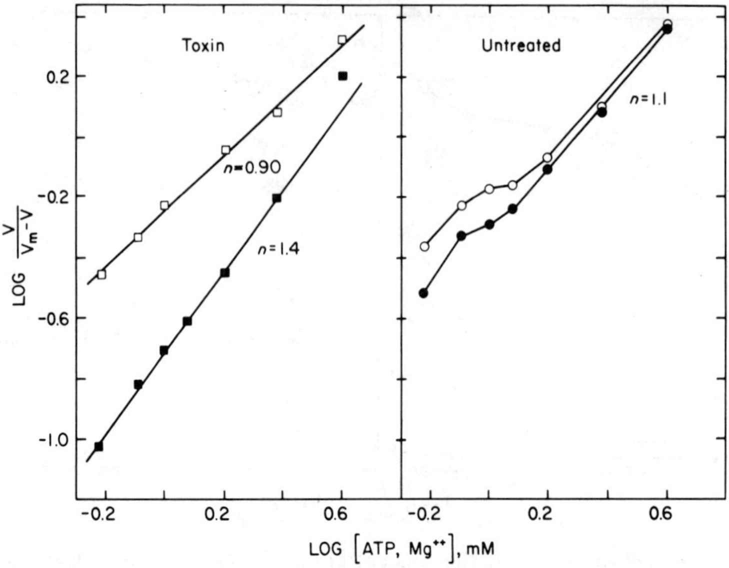

Fig. 1.14 Hill plot of data of Bennett and Cuatrecasas, 1975a). Adenylate cyclase activity of control (●, ○) and choleragen treated (■, □) toad erythrocyte plasma membranes was measured in the presence (○, □) or absence (●, ■) of GTP, Mg^{++} (0.2 mM) with increasing concentrations of substrate, ATP, Mg^{++}.

some of which resemble specifically the alterations attending hormonal activation (Bennett and Cuatrecasas, 1975a; Bennett *et al.*, 1975b).

Toxin stimulation converts the enzyme of toad erythrocytes to a purine nucleotide-sensitive state such that 5′GTP, 5′ITP and, with much less affinity, 5′ATP, stimulate the activity assayed in the absence of hormones by 3- to 5-fold (Fig. 1.13). At low concentrations of ATP (0.2 mM), the activation by toxin is almost entirely dependent on the presence of GTP or ITP (Bennett and Cuatrecasas, 1975a). High concentrations of ATP (> 4 mM), however, obviate the requirement for other nucleotides for expression of toxin stimulation. Similar findings have been reported for stimulation of cyclase by glucagon in rat liver (Birnbaumer *et al.*, 1972) and by oxytocin in toad bladder (Bockaert *et al.*, 1972). As a result of the dual role of ATP as a substrate and activator of the toxin-treated cyclase, the substrate-velocity relationship shows evidence of positive cooperativity. The Hill coefficient in the absence of GTP is increased by toxin while in the presence of GTP the coefficient of the toxin-treated enzyme is slightly less than that of the control (Fig. 1.14).

An enhanced response to GTP is also observed for the adenylate cyclase of toxin-treated rat fat cells (Bennett *et al.*, 1975b), although the effects of GTP are more complicated in this system. The changes in substrate-velocity relationships observed with the amphibian cells, however, are not reproduced in mammalian systems (unpublished data; Sharp *et al.*, 1973; Chen *et al.*, 1972; Kimberg *et al.*, 1971).

Table 1.8 Effect of GTP and Gpp(NH)p on control and toxin-treated adenylate cyclase of toad erythrocyte membranes

Membrane preparation	Adenylate cyclase activity[a]	
	Basal	(−) isoproterenol[b]
Control		
+ no addition	26	32
+ GTP[2]	26	58
+ Gpp(NH)p[2]	98	556
Toxin-treated[3]		
+ no addition	62	104
+ GTP[2]	155	468
+ Gpp(NH)p[2]	405	1292

[a]Picomoles of cyclic AMP formed per min per mg of protein; activity was measured (18 min at 30°C) in the presence of 0.6 MM ATP, 6.2 mM $MgCl_2$, 5 mM aminophylline, 5 mM phospho(enol) pyruvate, 50 μg ml^{-1} of pyruvate kinase, and 55 μg membrane protein in Tris-HCl buffer (50 mM, pH 8) with additions as described above.

[b]20 μM

[3]150 minutes at 37°C with 0.2 μg ml^{-1} toxin.

Purine nucleotides preferentially stimulate adenylate cyclase activity assayed in the presence of catecholamines in toad erythrocytes (Fig. 1.13) and rat fat cells (Siegel and Cuatrecasas, 1974; Bennett *et al.*, 1975b), as well as in a variety of other tissues and with various hormones (Krishna *et al.*, 1972; Wolff and Cook, 1973; Rodbell *et al.*, 1971; Leray *et al.*, 1972; Bilizekian and Aurbach, 1974; Bockaert *et al.*, 1972; Birnbaumer *et al.*, 1972; Deery and Howell, 1973). Interestingly, GTP also enhances the response of the choleragen-activated cyclase of toad erythrocytes to catecholamines (Fig. 1.13) (see below). The GTP analog, 5′-guanylylimidodiphosphate (Gpp(NH)p), which has recently been demonstrated to stimulate adenylate cyclase irreversibly in a variety of tissues (Lefkowitz, 1974; Schramm and Rodbell, 1975; Cuatrecasas *et al.*, 1975a, b; Salomon *et al.*, 1975; Pfeuffer and Helmreich, 1975; Jacobs *et al.*, 1976; Bennett and Cuatrecasas, 1976b) exerts qualitatively similar effects on the toxin-activated enzyme of toad red cells as GTP (Table 1.8).

Adenylate cyclase activity is also modulated by divalent cations such as Mg^{++}, Ca^{++}, and Mn^{++} (Birnbaumer *et al.*, 1969; Drummond and Duncan, 1970; Severson *et al.*, 1972; Bockaert *et al.*, 1972; Flawia and Torres, 1972; Siegel and Cuatrecasas, 1974). Cholera toxin treatment of toad erythrocyte adenylate cyclase decreases the apparent K_a for Mg^{++} from about 25 mM to 5 mM; the unstimulated enzyme assayed in the presence of epinephrine exhibits a similar increase in affinity for this

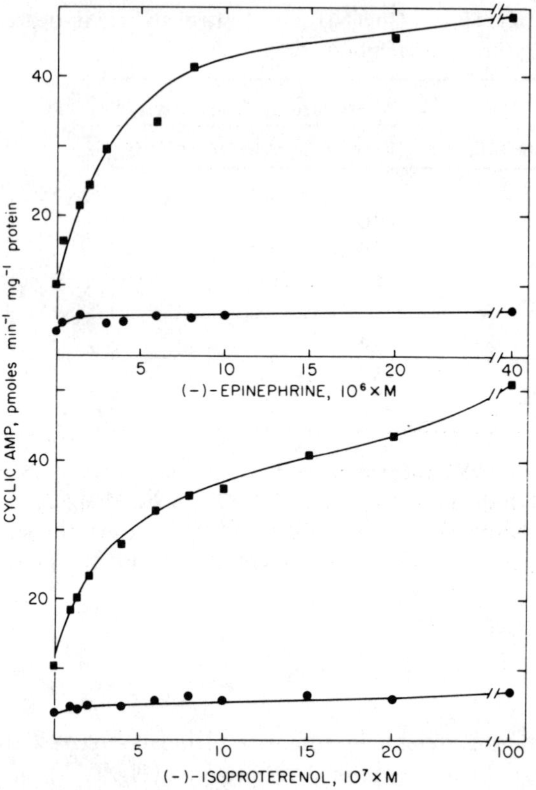

Fig. 1.15 Effect of increasing the concentration of (−)-epinephrine (top) and (−)-isoproterenol (bottom) on the adenylate cyclase activity of control (●) and cholera toxin-treated (■) toad erythrocyte plasma membranes. Cells were incubated in the presence and absence of cholera toxin (0.15 $\mu g\ ml^{-1}$) as described in Fig. 1.13. Plasma membrane adenylate cyclase activity was measured (15 min, 30°C) in the presence of $MgCl_2$ (6 mM), GTP (0.2 mM), [α-^{32}P]-ATP (0.6 mM, 92 cpm $mole^{-1}$), aminophylline (5 mM), phosphoenolpyruvate (5 mM), pyruvate kinase (50 $\mu g\ ml^{-1}$), Tris-HCl (50 mM, pH 8.0), 80–90 μg of membrane protein, and increasing concentrations of either (−)-epinephrine or (−)-isoproterenol. From Bennett and Cuatrecasas, 1975a.

metal (Bennett and Cuatrecasas, 1975a). The basis for stimulation by Mg^{++} is not clear, although a kinetic analysis (deHaen, 1974) indicates that the metal may complex with an inhibitory form of ATP to form the substrate. The specific effects of toxin and hormones on the affinity for Mg^{++} are not generally observed and may be a special feature of the toad erythrocyte.

In addition to mimicking certain effects of hormones, choleragen also increases the sensitivity of adenylate cyclase to stimulation by these agents. The maximal response of toad erythrocyte adenylate cyclase to catecholamines is increased by

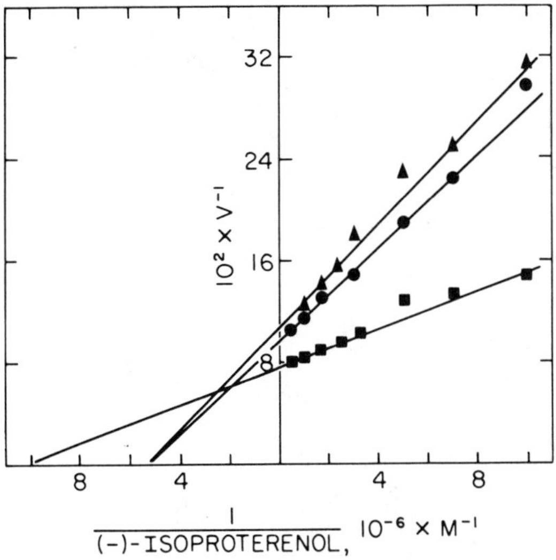

Fig. 1.16 Effect of increasing the concentration of (–)-isoproterenol on the adenylate cyclase activity of control (●), choleragenoid-treated (▲), and cholera toxin-treated (■) rat erythrocyte plasma membranes. Rat erythrocytes were washed in Krebs-Ringer-bicarbonate pH 7.4, and then incubated in the same buffer for 3 h at 37°C, either with toxin (0.28 μg ml^{-1}), choleragenoid (0.28 μg ml^{-1}), or with no additions. The plasma membrane adenylate cyclase activity was determined (12 min, 30°C) in the presence of $MgCl_2$ (6.5 mM), (α-^{32}P]-ATP (0.27 mM, 200 cpm $pmole^{-1}$), GTP (0.16 mM), phosphoenol-pyruvate (5.4 mM), pyruvate kinase (55 μg ml^{-1}), aminophylline (5 mM), Tris-HCL (50 mM, pH 8.0), 100 μg of membrane protein, and increasing concentrations of (–)-isoproterenol. The data are expressed as a double-reciprocal plot of isoproterenol-stimulated cyclase activity (stimulated-basal) vs. concenctration of isoproterenol. From Bennett and Cuatrecasas, 1975a.

3- to 5-fold in the presence of optimal concentrations of GTP (Fig. 1.13, 1.15), while the apparent affinity of these hormones is increased about 2-fold for the toxin-treated enzyme of rat erythrocytes (Fig. 16) and rat fat cells (Bennett *et al.*, 1975b). Turkey erythrocytes exhibit both an increased affinity and extent of activation by catecholamines (Field, 1974). Choleragen also increases by 2- to 5-fold the apparent affinity of rat fat cell adenylate cyclase for activation by glucagon, ACTH and vasoactive intestinal peptide (Bennett *et al.*, 1975b), and the affinity of the enzyme of rat liver for glucagon (Bennett *et al.*, 1975a). Toxin increases the affinity for vasoactive peptide in activation of cyclase in intestinal mucosal membranes of guinea pigs (Bennett *et al.*, 1975b), which could conceivably be a factor in the pathogenesis of clinical cholera. Choleragen most likely does not interact with hormone receptors (see above). Moreover, simple occupation of the binding site for choleragen is not sufficient to affect the cyclase; choleragenoid, which binds in an

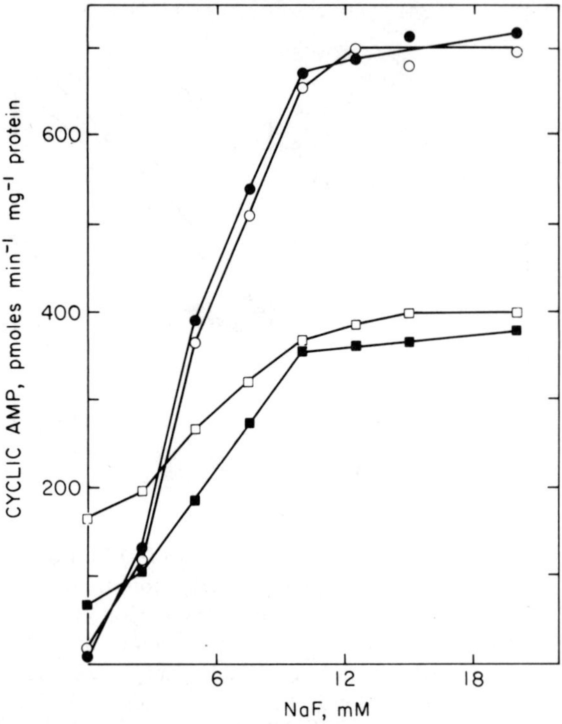

Fig. 1.17 The effect of increasing concentrations of sodium fluoride on the adenylate cyclase activity of control (●, ○) and cholera toxin-treated (■, □) toad erythrocyte plasma membranes assayed in the presence (○, □) and absence (●, ■) of 10 μM (−)-epinephrine. Erythrocytes were incubated in the presence and absence of cholera toxin (0.3 μg ml^{-1}). Plasma membrane adenylate cyclase activity was determined (15 min, 31°C) in the presence of $MgCl_2$ (6 mM), GTP (0.2 mM), [α-^{32}P]-ATP (0.6 mM, 67 cpm pmole^{-1}), aminophylline (5 mM), phosphoenolpyruvate (5 mM), pyruvate kinase (50 μg ml^{-1}), Tris-HCl (50 mM, pH 8.0), 110–130 μg of membrane protein, and various concentrations of NaF. From Bennett and Cuatrecasas, 1975a.

indentical manner to the same ganglioside receptors as choleragen (see above), has no effect on the activity of adenylate cyclase or its sensitivity to hormones (Fig. 1.16) (Bennett *et al.*, 1975b).

The stimulation by fluoride and choleragen has been distinguished by use of pyrophosphate which selectively inhibits the fluoride-stimulated enzyme (Sharp *et al.*, 1973). Toxin and fluoride ion activate adenylate cyclase of toad erythrocytes in a mutually exclusive manner. NaF prevents the activation of the toxin-enzyme by GTP (Fig. 1.13) and by catecholamines (Fig. 1.17), and also blocks the increased K_a for Mg^{++} induced by toxin (Bennett and Cuatrecasas, 1975a). Conversely, toxin depresses the stimulation by NaF (Figs. 1.13, 1.17), a phenomenon also reported for turkey erythrocytes (Field, 1974). It is of interest that an analogous inhibitory relationship between fluoride and hormone stimulation has been demonstrated in fat cell ghosts (Harwood and Rodbell, 1973).

Features of the activation of adenylate cyclase by choleragen, including conversion of the enzyme to a GTP-sensitive state and the alteration in the apparent affinities for activation by hormones, suggest analogies between the basic mechanism of action of choleragen and the events following binding of hormones to their receptors*. Cholera toxin may thus provide an experimentally accessible model for the interaction of hormone receptors with adenylate cyclase. The possibility is discussed below that the action of choleragen involves mobility of the initial toxin-ganglioside complex in the plane of the membrane with the ultimate formation of an active complex between toxin and the cyclase. The implications for the mechanism of hormone-receptor-cyclase interactions are then considered within the context of a recent proposal (Cuatrecasas, 1974, 1975) that hormone receptors are free to diffuse within the plane of the membrane, and that, when complexed with hormone, they encounter and modify membrane enzymes by analogy with protein-protein associations in solutions.

1.4.2 Kinetics and properties of the process of activation of adenylate cyclase by choleragen

The kinetics and properties of the process of activation of adenylate cyclase by choleragen have recently been described in detail (Bennett and Cuatrecasas, 1975b). When choleragen is incubated with intact cells, it stimulates adenylate cyclase in the subsequently isolated plasma membrane according to a triphasis time course. This consists of a true lag period of approximately 30 min, followed by a stage of exponentially increasing activity which continues for 110–130 min and finally a period of slow activation (Fig. 1.18). The influence of choleragen on basal activity, catecholamine sensitivity and the activation by fluoride occur in parallel (Fig. 1.18). The progressive development of the toxin effects can be arrested at any time by chiling or hypotonic lysis of the cells (Bennett and Cuatrecasa, 1975b).

A semi-log of adenylate cyclase activity versus time of incubation during the initial period of exponential change permits, by interpolation, an estimate of the length of the lag phase (Fig. 1.19). The fact that the semi-log plots do not extrapolate through zero time indicates that the delay is absolute in nature rather than a period of slowly accelerating activation. Thus, the initial toxin-membrane complex in incapable of activation adenylate cyclase, as had been proposed previously (Cuatrecasas, 1973d). It is of interest that such different cells as rat adipocytes and toad erythrocytes exhibit nearly identical delays of about 30 min (Fig. 1.19).

* Additional evidence of an important analogy between the function of choleragen and hormones is the recent observation of a peptide sequence common to the binding subunit of toxin and the β chains of thyrotropin, luteinizing hormone and human chorionic gonadatropin (Mullin *et al.*, 1976). Furthermore, gangliosides were demonstrated to inhibit the binding of ^{125}I-thyrotropin with the following order of potency: $G_{D1b} > G_{T1} > G_{M1} > G_{M2} >= G_{M3} > G_{D1a}$ (Mullin *et al.*, 1976).

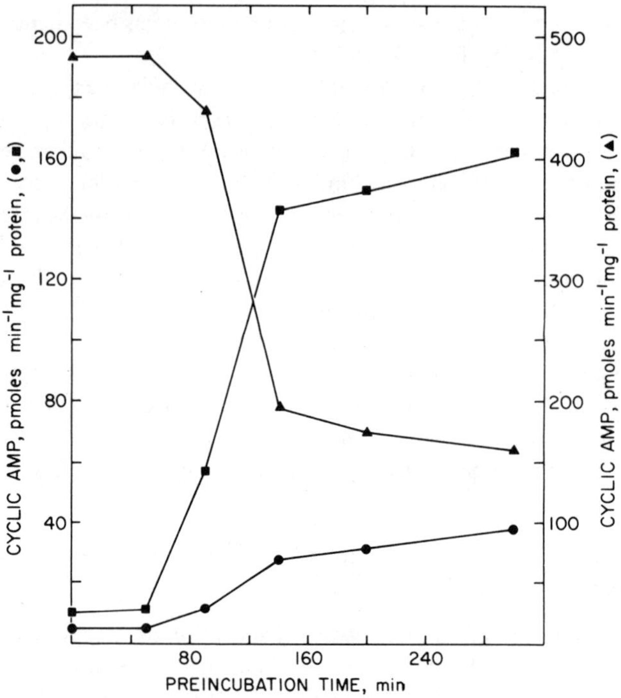

Fig. 1.18 Effect of increasing the time of incubation of cholera toxin with intact toad erythrocytes on the adenylate cyclase activity of plasma membranes assayed with 20 μM (–)-epinephrine (■), 20 mM sodium fluoride (▲), or with no additions (●). The cells were thoroughly washed with amphibian-Ringer's, pH 7.5, suspended in this buffer, and divided into six 30 ml portions, each containing about 2 x 10^8 cells. The cell samples were incubated in a 30°C water bath and cholera toxin (5.6 x 10^{-11} moles) was added at various times. One sample (zero incubation time) received no toxin. After 5 h, the cells were rapidly chilled, and the plasma membrane adenylate cyclase activity determined. All values were determined in triplicate. From Bennett and Cuatrecasas, 1975b.

The lag period is not due to a slow conversion of toxin into active, soluble form released into the medium, or by secretion of active cellular metabolites (Fig. 1.11) (see above). The duration of the delay is independent of the number of toxin-membrane complexes (Bennett and Cuatrecasas, 1975b; Gill and Ging, 1975).

The lag period is not due to a slow conversion of toxin into active, soluble form released into the medium, or by secretion of active cellular metabolites (Fig. 1.11) (see above). The duration of the delay is independent of the number of toxin-membrane complexes (Bennett and Cuatrecasas, 1975b; Gill and King, 1975). Furthermore, the events occurring this time apparently proceed in the presence of inhibitors of RNA and protein synthesis (including diptheria toxin) and of agents which disrupt microtubules and microfilaments (Bennett and Cuatrecasas, 1975b;

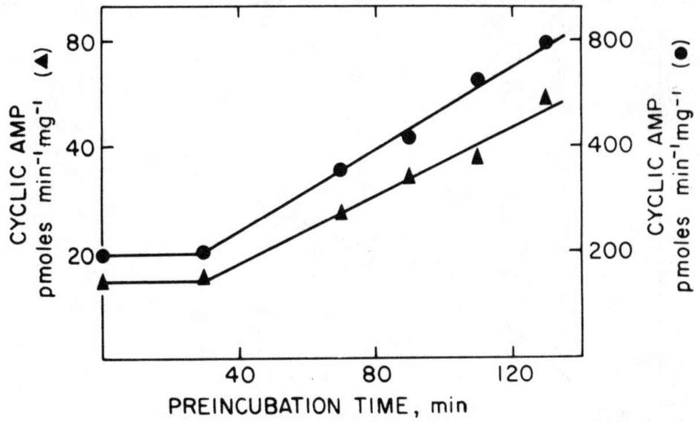

Fig. 1.19 Effect of increasing the time of cholera toxin pre-incubation with intact toad erythrocytes (▲) and rat adipocytes (●) on the plasma membrane adenylate cyclase activity. The experiment with toad erythrocytes was performed as described in Fig. 1.18. Adipocytes were isolated from the epididymal fat pads of eight 100–200 g rats by the method of Rodbell (1964), suspended in oxygenated Krebs-Ringer's bicarbonate, 2% BSA (w/v), pH 7.4, and divided into six 15 ml portions. These were incubated at 37°C under an atmosphere of 95% O_2, 5% CO_2. Cholera toxin (5.6 μg) was added at various times. After 130 min, the cells were chilled and homogenized in cold Tris-HCl (50 mM, pH 7.6) with a Brinkman Polytron (30 s at a setting of 3.0). The suspensions were centrifuged at 40 000 x *g* for 30 min at 0°C, and the membrane pellets were resuspended in Tris-HCl (50 mM, pH 8.0). Adenylate cyclase activity was determined under identical conditions as with the toad enzyme except that the concentration of ATP was 0.25 mM and the assay was conducted at 33°C. The enzyme activity of toad and rat membranes was determined in the presence of 20 μM (−)-epinephrine. The values were determined in triplicate, and are plotted as the log of enzyme activity vs. time. From Bennett and Cuatrecasas, 1975b.

Bennett *et al.*, 1975b). The delay is probably not due to slow penetration of choleragen across the membrane since direct measurements of entry of iodotoxin into the cytoplasm demonstrate that at most 10–20 molecules enter a cell out of 6000 bound during a 90 min incubation (Table 1.3) (see above).

It is unlikely that the latent period represents the time required for choleragen to remove or inactivate some cell component which is initially present in excess and is therefore rate-limiting. Such a hypothesis would predict that once the putative molecule is lost subsequently added toxin should experience no delay in activation of cyclase. It has been shown, however, that the lag is still present after adding a second dose of toxin to partially stimulated cells (Bennett and Cuatrecasas, 1975b). These results suggest that each toxin-ganglioside complex must independently undergo some slow process before activation of adenylate cyclase can proceed.

Considerable insight into the basis of the lag phase has been provided by the demonstration that the active subunit of choleragen stimulates adenylate cyclase in fat cells without the characteristic delay (Fig. 1.20) (Sahyoun and Cuatrecasas, 1975).

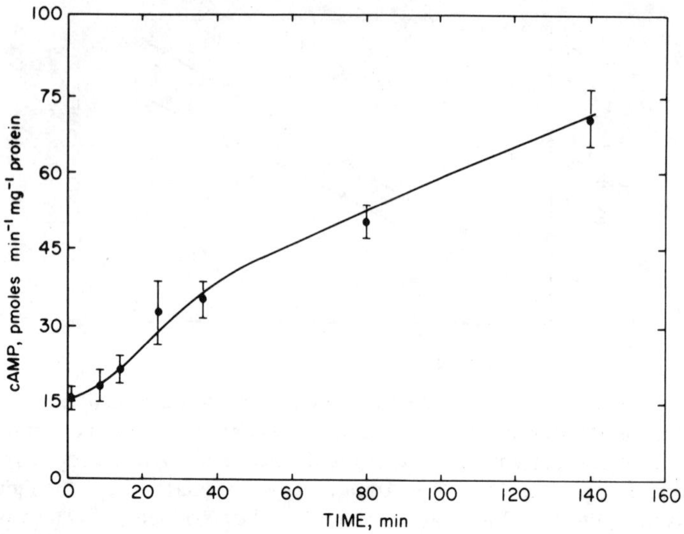

Fig. 1.20 Time course of direct activation of adenylate cyclase by choleragen in the particulate membrane fraction of fat cells. The membranes (5–15 mg proteinsml^{-1}) were pre-incubated for various times at 33°C in 25 mM Tris-HCl, pH 7.7, containing 2 mM ATP and the ATP-regenerating system with or without choleragen (0.5 μg ml^{-1}). Before assay, the suspension was diluted with 30 volumes of ice-cold buffer and centrifuged at 40 000 x *g* for 30 min at 0°C. Basal activity decreased from about 20 pmoles cAMP produced/min/mg protein to about 20% of this value after 130 min. Assays were in triplicate and the 1% stimulation (relative to the basal activity at each time) was calculated ± S.E.M. (From Sahyoun and Cuatrecasas, 1975).

The 36 000 MW subunit apparently does not interact with the initial binding site for choleragen since its effects are not blocked by gangliosides or choleragenoid (van Heyningen and King, 1975; Sahyoun and Cuatrecasas, 1975). It has also been possible to demonstrate activation of the enzyme by intact choleragen in washed membranes partially purified on sucrose gradients, although the lag is still evident in this system (Fig. 1.21) (Sahyoun and Cuatrecasas, 1975). These findings exclude an intracellular metabolic event in the action of toxin and suggest rather that the delay represents the time required for conversion of choleragen to an active form, a process the active subunit can bypass.

It has been reported recently that, in certain cell systems, activation by choleragen begins immediately, provided the toxin is added to concentrated cell homogenates (Gill, 1975; Gill and King, 1975; Bitensky *et al.*, 1975) or washed membrane preparations (van Heyningen and King, 1975; Bitensky *et al.*, 1975). Gill (1975) has reported that such activation requires a cytoplasmic factor which he believes to be NAD, although Bitensky *et al.*, (1975) have demonstrated that DTT, reduced glutathione and NADH function equally well. Bitensky *et al.*, (1975) have shown that incubation of choleragen with either DTT, reduced glutathione, NAD, NADH or a partially purified 700 MW cytoplasmic factor, followed by exposure to washed

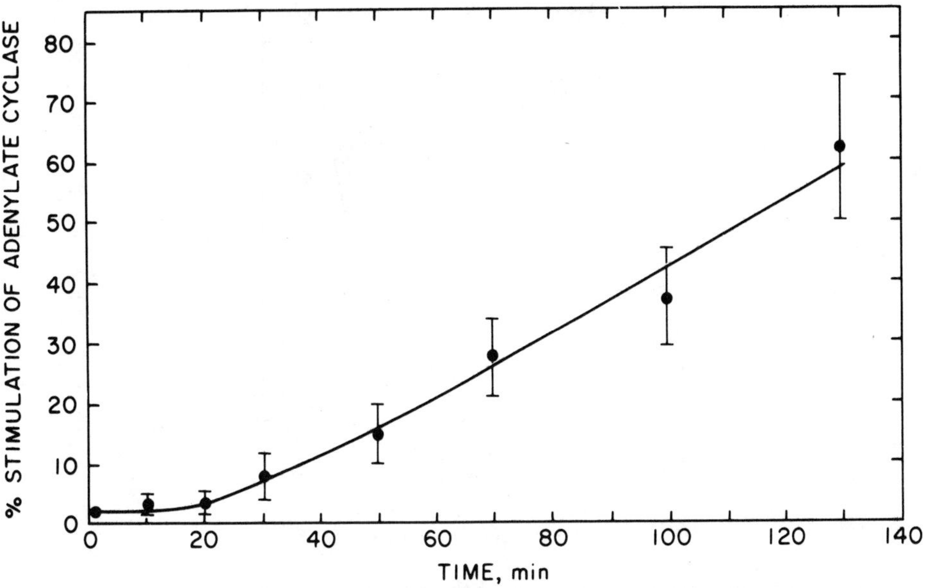

Fig. 1.21 Direct activation of adenylate cyclase by the 36 000 MW subunit of choleragen. Cells from 10 fat pads, suspended in 20 ml of Krebs-Ringer bicarbonate (pH 7.4), 2% bovine serum albumin, were incubated with the subunit (5 μg ml^{-1}) at 37°C. At various times the cells were washed twice with the same buffer and once with 25 mM Tris-HCl, pH 7.7, 2% albumin, followed by homogenization and preparation of the particulate fraction for assay. The 36 000 MW subunit contained 0.5% Lubrol PX (in 50 mM phosphate buffer, pH 7.5) to maintain its solubility; the final concentration of detergent in the pre-incubation mixture was less than 0.0001%. Equal concentrations of Lubrol and phosphate were added to the control cells; a slight time-dependent increase in activity resulted which was substracted from the activity in the presence of the subunit. From Sahyoun and Cuatrecasas, 1975.

membranes from certain cancer cells, leads to partial conversion of choleragen to a 26 000 MW fragment. This molecule, following partial purification on either sucrose gradients or by gel filtration, is capable of activating cyclase in washed membrane preparations of the same select cells without a lag period. The fragment is inactive in membranes from mouse liver, heart, kidney, and lymphocytes as well as turkey red blood cells. Furthermore, the fragment is produced selectively and is not observed in turkey erythrocytes or mouse liver.

These observations suggest that under the conditions used to obtain direct, immediate activation in subcellular systems, choleragen may dissociate into its component subunits, and that some form of the free active subunit is then capable of stimulating cyclase activity without a delay as has been demonstrated in intact cells (Sahyoun and Cuatrecasas) (Fig. 1.20). The fragmentation of choleragen apparently requires cytoplasmic factor(s) present in only certain cells (which may be replaced by thiol reducing agents) and, moreover, the 26 000 MW portion is

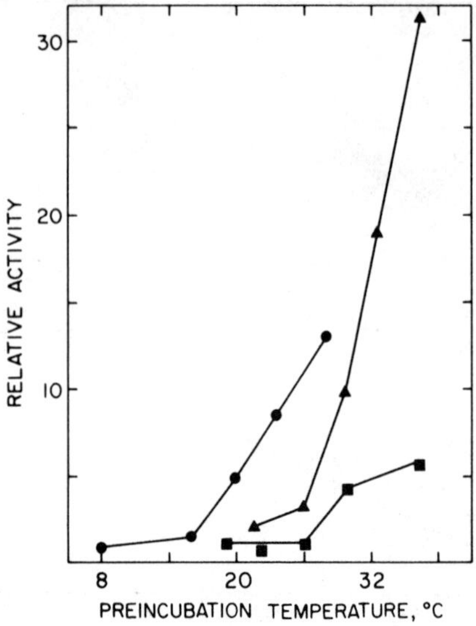

Fig. 1.22 Temperature dependence of the activation by cholera toxin of the adenylate cyclase activity of toad erythrocytes (●), turkey erythrocytes (■), and rat adipocytes (▲). Toad and turkey erythrocytes were washed in amphibian Ringer and Krebs-Ringer bicarbonate, respectively, and incubated at various temperatures with cholera toxin (0.3 $\mu g\ ml^{-1}$) for 4 h. The plasma membrane adenylate cyclase activity was determined as described (Bennett and Cuatrecasas, 1975b). Rat fat cells were prepared and incubated for 135 min at various temperatures with cholera toxin (0.2 $\mu g\ ml^{-1}$), as described in Fig. 1.17. Adenylate cyclase activity was determined in triplicate, and the values are expressed as the enzyme activity relative to control values (cells incubated under identical conditions but in the absence of cholera toxin). From Bennett and Cuatrecasas, 1975b.

effective in a limited number of cell types. Thus, these findings cannot be interpreted as reflecting a general mechanism of action of choleragen, and it is particularly misleading to invoke a special role for NAD based on such data (Gill, 1975).

The activation of adenylate cyclase activity by choleragen occurs at a very slow rate below certain critical temperatures which depend on the animal species studied (Fig. 1.21) (Bennett and Cuatrecasas, 1975b). The transition in toad erythrocytes occurs at 15–17°C, while rat fat cells and turkey erythrocytes exhibit a discontinuity at 26–30°C. The temperature effect is observed not only for the lag period, but also for the later, exponential phase of activation. These data suggest that membrane phospholipids may be involved in the action of cholera toxin, and that some degree of lipid 'fluidity' may be required throughout the process of enzyme activation.

The degree of stimulation of enzyme depends on the number of membrane-bound toxin molecules. During the initial phase of activation, the predominant effect of

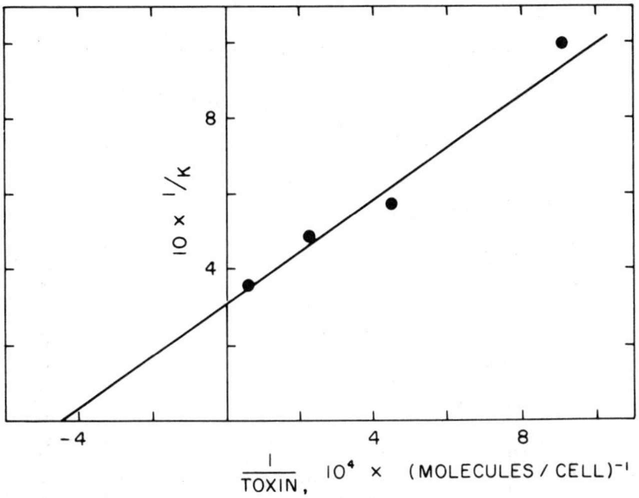

Fig. 1.23 Double-reciprocal plot of the slope (K) of semi-log plots of time courses of toad erythrocyte adenylate cyclase activation vs. the number of molecules of cholera toxin bound per cell. Washed erythrocytes were suspended in amphibian-Ringer and divided into twenty-four 10 ml portions, each containing 6 x 10^8 cells. The time course was examined as described in Fig. 7, at fixed concentrations of cholera toxin: 15 400 molecules/cell, 4440 molecules/cell, 2200 molecules/cell and 1100 molecules/cell. Adenylate cyclase activity was measured in the presence of 20 μM (——-epinephrine. The values were determined in triplicate. From Bennett and Cuatrecasas, 1975b.

increasing the amount of cell-bound toxin is an elevation in the rate of exponential increase of enzyme activity with no detectable change in the length of the lag phase (Bennett and Cuatrecasas, 1975b). The rate constant for the exponential process can be estimated from the slope of the semilog plot of activity versus time of incubation (Fig. 1.19), and this value is related to the number of toxin-receptor complexes by Michaelis-Menton saturation kinetics (Fig. 1.23). The half-maximal increase in the rate of exponential activation with toad erythrocytes occurs with about 2200 toxin molecules bound per cell (Fig. 1.23). The extent of enzyme activation at near-equilibrium conditions (four hours at 30°C) also depends in a hyperbolic manner on the amount of bound choleragen, with an apparent K_a of about 1500 toxin molecules per cell (Fig. 1.24).

The quantitative similarity of the dependence of the rate constant and extent of enzyme activation on the amount of cell-bound toxin suggests that catalytic processes are not involved in the sense that one toxin molecule does not lead progressively to stimulation of many enzyme molecules. If such were the case, half-maximal activation would require fewer toxin molecules with increasing length of incubation. Furthermore, if cholera toxin acts as a 'catalyst', the rate of activation should continue to increase with higher doses of choleragen, at least until the membrane binding sites have been saturated. The data in Fig. 1.23 demonstrate, however, that the rate

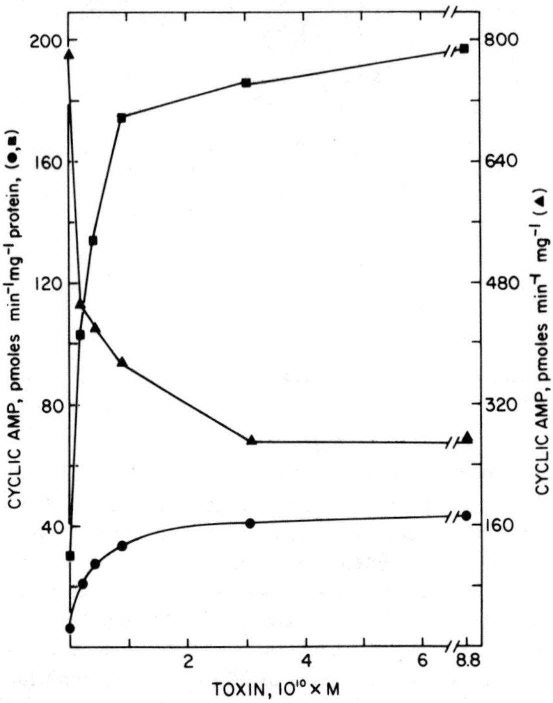

Fig. 1.24 (a) Effect of pre-incubating toad erythrocytes with increasing concentrations of cholera toxin on the adenylate cyclase activity of plasma membranes assayed with 20 μM (–)-epinephrine (■), 20 mM sodium fluoride (▲), or with no additions (●). Cells were washed with amphibian-Ringer, pH 7.5, suspended in this buffer, and divided into six 45 ml portions, each containing 6 x 10^8 cells. Various concentrations of cholera toxin were added, and the cell suspensions were incubated for 4 h at 30°C before preparation of the plasma membranes for assays of adenylate cyclase activity. Nearly all of the choleragen is membrane bound under the conditions employed in this experiment ($> 10^7$ cells ml^{-1}, 10^{-11} to 10^{-9} M choleragen). The amount of cell-bound toxin can be estimated by dividing the total number of toxin molecules by the number of erythrocytes: 10^{-10} M toxin = 2.7 x 10^{12} molecules/45 ml = 4500 molecules/cell. From Bennett and Cuatrecasas, 1975b.

of activation is saturable with respect to bound toxin, and the half-maximal value (2200 molecules per cell) is quite small compared to the total binding capacity of these cells (about 50 000 sites per erythrocyte) (Bennett and Cuatrecasas, 1975b). Similar considerations argue against other mechanisms based on catalytic processes initiated by interaction of toxin with the cell.

Diptheria toxin (Gill *et al.*, 1969; Honjo *et al.*, 1971) and the bacterial colicin E_3

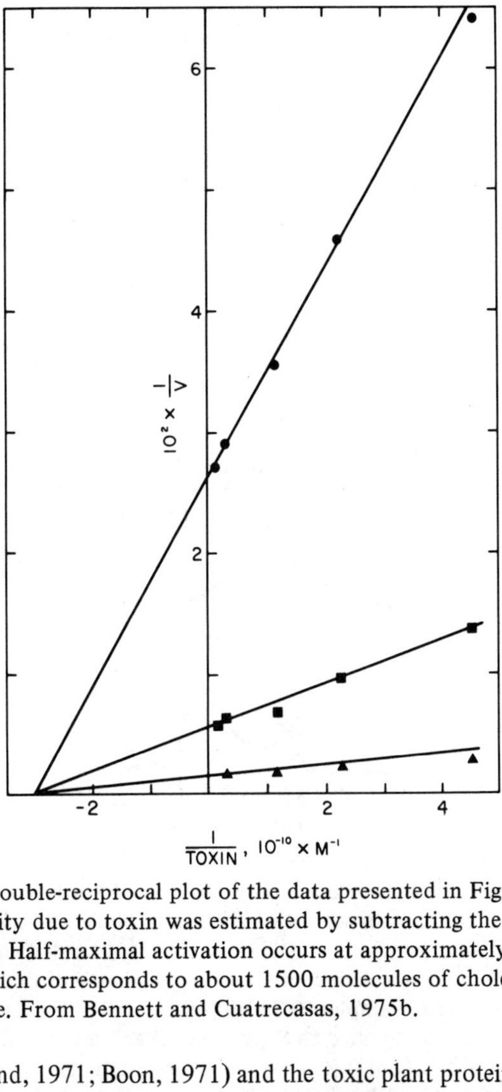

Fig. 1.24 (b) Double-reciprocal plot of the data presented in Fig. 1.20 (a). The change in activity due to toxin was estimated by subtracting the control (untreated) values. Half-maximal activation occurs at approximately 3×10^{-11} M choleragen, which corresponds to about 1500 molecules of choleragen bound per erythrocyte. From Bennett and Cuatrecasas, 1975b.

(Senior and Holland, 1971; Boon, 1971) and the toxic plant proteins, abrin and ricin (Refsnes *et al.*, 1974) provide examples of molecule which initially interact with cell membranes and ultimately function intracellularly by enzymatic mechanisms. In these cases the concentration-response relationships obey Poisson distributions for 'single hit' kinetics (Jacob *et al.*, 1952; Nomura, 1963). The data described in Fig. 1.24 clearly deviate from the predicted linear relationship (not shown).

These considerations suggest that each molecule of toxin leads to the alteration of a limited number of adenylate cyclase enzyme molecules and imply that at some point these molecules may interact 'directly'. The inactive nature of the initial toxin-ganglioside complex implies that the biological effects of toxin are due to

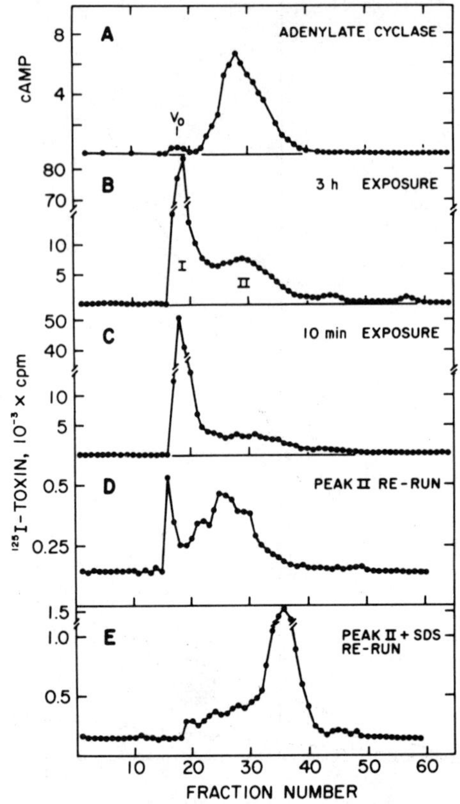

Fig. 1.25 Sepharose 6B chromatography of Lubrol PX-solubilized adenylate cyclase from fat cells incubated with ^{125}I-labeled cholera toxin. Cells from 25 rats (200 *g*) were incubated at 37°C for 3 h. ^{125}I-labeled cholera toxin (2.4 μg, 9.4 μCi μg^{-1}) was added either at the beginning of the incubation or for the final 10 min. Samples (0.75 ml) of the 250 000 x *g* membrane detergent extracts were applied to Sepharose 6B columns (1 cm x 57 cm; flow rate 10 ml h^{-1}; 0.6 ml fractions) and cyclase activity (panel A) was determined. Panels A and B refer to cells exposed to toxin for 3 h while panel C is for the 10 min exposure to ^{125}I-labeled toxin. The peak of cyclase in panel A (fractions 28–30) was rechromatographed with (panel E) or without (panel D) heating the sample in 1% sodium dodecyl sulfate (50 min at 40°C). From Bennett, O'Keefe, and Cuatrecasas, 1975.

interaction with other, secondary 'receptor' sites. The simplest possibility, in view of the lack of evidence for a catalytic mechanism, is that the 'receptor' is the adenylate cyclase molecule itself. Thus, activation would occur by a relatively stable association between the active form of choleragen and adenylate cyclase. It is

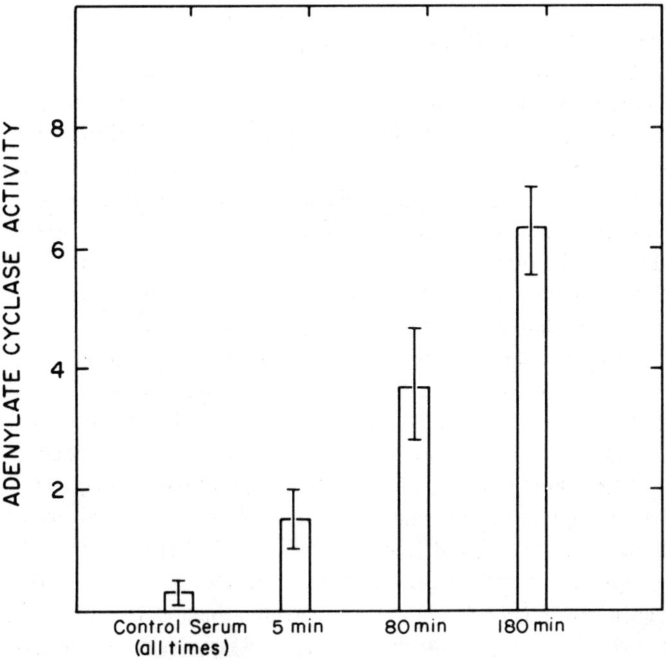

Fig. 1.26 Immunoprecipitation of adenylate cyclase with antiserum to the 36 000 MW subunit of choleragen. Fat cells were incubated with choleragen at 37°C for various periods, the particulate cell fractions were solubilized, and 200 to 300 μl were incubated with 50 μl of immune or normal rabbit serum for 12 min at 30°C followed by 2 h in ice. Goat anti-rabbit gamma globulin (200 μl) were added and the mixture incubated at 30°C for 5 min and in ice for 1 h. The precipitate was washed three times with 1 ml of 25 mM Tris-HCl, pH 7.7, suspended in 200 μl, and assayed (50 μl). 15–20% of the ^{125}I-label was precipitated when ^{125}I-choleragen was used. Activity is in pmoles cAMP produced per 50 μl per 12 min at 37°C ± S.E.M. From Sahyoun and Cuatrecasas, 1975.

pertinent that the concentration-response relationships for the toxin-membrane complex can be approximated by a Langmuir adsorption isotherm. This type of relationship is consistent with the occurrence of a bimolecular interaction between bound choleragen and another molecule in the rate-limiting step of adenylate cyclase activation. The absolute nature of the latency phase between binding of choleragen and the onset of activation of clyclase indicates that some transformation process must be occurring during this time which requires at least 20–30 min for even the most advanced choleragen molecules. Thus, the proposed association of choleragen with cyclase would not begin until *after* the lag period. It is pertinent to this view that the isolated active subunit acts without a delay (Fig. 1.20). Data will be presented below suggesting that membrane-bound choleragen may dissociate into its component subunits which remain attached at the cell surface (Sahyoun and Cuatrecasas, 1975).

Studies with detergent-solubilized adenylate cyclase

Evidence has been reported recently suggesting that following the initial binding interaction, choleragen or some portion of the molecule forms a detergent-stable complex directly with adenylate cyclase (Bennett *et al.*, 1975; Sahyoun and Cuatrecasas, 1975). Activation of adenylate cyclase by choleragen is retained following conversion of the membranes with Lubrol PX to a nonsedimentable form (30 min at 250 000 x *g*) (Beckman *et al.*, 1974; Bennett *et al.*) and after gel filtration on agarose columns in the presence of detergent (Bennett *et al.*, 1975a). Furthermore, a small peak of ^{125}I-choleragen corresponding to about 1000 molecules per cell is associated with adenylate cyclase activity on gel filtration provided the enzyme has been activated by labeled choleragen (Fig. 1.25). The direct association of at least a portion of the small radioactive peak with adenylate cyclase is supported by the finding that antisera directed against choleragen, and specifically against the active subunit, will immunoprecipitate solubilized choleragen-stimulated adenylate cyclase activity (Fig. 1.26) (Bennet *et al.*, 1975; Sahyoun and Cuatrecasas, 1975). It is pertinent that agarose derivatives containing the 'active' subunit can specifically adsorb adenylate cyclase activity (Bennett *et al.*, 1975a). The fact that about 90 percent of the membrane-bound choleragen is not associated with adenylate cyclase on gel filtration (Fig. 1.25) is not surprising; the apparent K_m for stimulation of adenylate cyclase in toad erythrocytes (Fig. 1.24) and cultured melanocytes (O'Keefe and Cuatrecasas, 1974) and fibroblasts (Hollenberg *et al.*, 1974) occurs with 1500–3000 molecules of toxin bound per cell, whereas specific binding is half-maximal at 20 000 to 500 000 molecules per cell. Studies using choleragenoid to 'decrease' the number of toxin-binding sites suggest that these excess sites represent equivalent 'spare' receptors (Bennett *et al.*, 1975a).

It is of interest to consider the mechanism by which active, membrane-bound toxin may encounter and complex with adenylate cyclase. The fact that equivalent receptors are present in excess over adenylate cyclase molecules implies that a specific mechanism exists for translocating the toxin to adenylate cyclase following formation of the initial toxin-membrane complex. Otherwise virtually all the ganglioside receptors would have to exist in direct contiguity with adenylate cyclase. Since some cells, such as cultured human fibroblasts, may bind as many as 10^6 toxin molecules per cell, this situation would pose serious geometrical problems. Furthermore, direct visualization of the initial binding of fluorescent choleragen to fibroblasts indicates a random, nonclustered distribution of toxin molecules (Craig and Cuatrecasas, 1975). Relocation of the bound toxin would most likely occur by lateral diffusion of the toxin-ganglioside complex within the plane of the membrane, since rotation through the membrane bilayer to the inner membrane surface may be energetically unfeasible for larger molecules (Singer and Nicolson, (1972). It is unlikely that toxin enters the cytoplasm in a soluble form; direct measurements of the intracellular level of ^{125}I-labeled choleragen in toad erythrocytes indicate that considerably less than 1 percent of bound toxin is present in the cell lysate after a 90 minute incubation at 30°C (Table 1.3). It is also improbable that pinocytosis

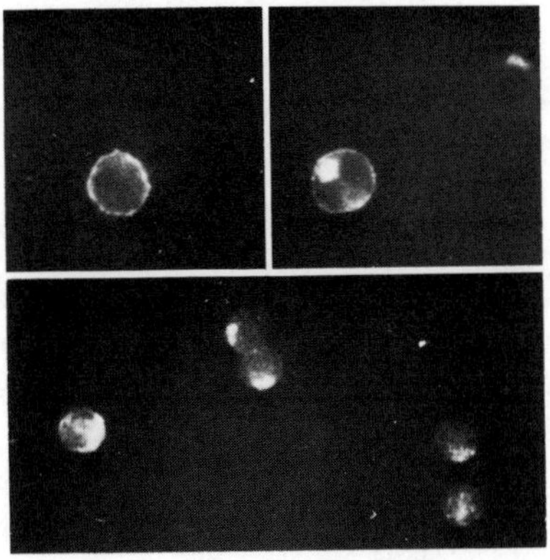

Fig. 1.27 Patterns of fluorescence on rat lymphocytes incubated in F-CT. (a) Lymphocytes (10^7) were incubated (0°C, 30 min) in 1 ml of Hanks containing 1 μg of F-CT. The cells were washed in cold Hanks before fixation and mounting. (b) and (c) Lymphocytes strained at 0°C were washed and incubated at 37°C for 20–30 min before fixing. All 37°C incubations were carried out in a humidified incubator equilibrated with 95% O_2-5% CO_2. From Craig and Cuatrecasas, 1975.

of bound choleragen is an obligatory step since anucleate rat erythrocytes, which are incapable of pinocytosis (Hirsh *et al.*, 1973) respond well to choleragen (Fig. 1.15).

Studies with immunofluorescent staining (Revesz and Greaves, 1975) and with fluorescein-labeled choleragen (Craig and Cuatrecasas, 1975) provide direct evidence for the possibility of redistribution of choleragen-receptor complexes in cell surfaces. Viable lymphocytes which are membrane-stained with fluorescent cholera toxin at 0°C and then fixed, exhibit a smooth ring fluorescence (Fig. 1.27c) and/or random micropatches when viewed under the fluorescence microscope. If the cells stained at 0°C are subsequently incubated for 20 min at 37°C, the smooth ring fluorescence redistributes into macro-patches, polar caps with residual surface fluorescence, or small tight caps without detectable residual surface fluorescence (Fig. 1.27b, c). The multivalency of cholera toxin (see above) appears to be essential to its ability to redistribute membrane toxin receptors since agents which act to reduce the effective valence of membrane-bound cholera toxin, e.g., anti-cholera toxin antibody, inhibit patching and redistribution very effectively. These observations suggest that cross liking of toxin receptors may be a necessary step in the process of redistribution of membrane-bound choleragen. The fact that anti-toxin prevents activation of adenylate cyclase in parallel to inhibition of surface redistribution is consistent with the requirement of the latter in toxin action (Craig and Cuatrecasas, 1975).

Binding and redistribution of fluorescent cholera toxin can be observed using concentrations of fluorescent cholera toxin as low at 10^{-10} M, which is within the physiological range for cholera toxin activation of adenylate cyclase in rat mesenteric lymph node lymphocytes. Thus, the events described probably reflect the properties of the biologically relevant toxin receptors. Although it is believed that lateral mobility or redistribution a necessary step for biological activation, it is important to emphasize that no evidence exists that '*capping*' of membrane-bound toxin *per se* is an obligatory step in activation of adenylate cyclase. Formation of polar aggregates is also observed with choleragenoid (Craig and Cuatrecasas, 1975; Revesz and Greaves, 1975). Also, capping presumably requires the integrity of the cell cytoskeletal system whereas choleragen activation occurs in isolated membranes (Fig. 1.21). The capping of choleragen is interpreted instead as a dramatic visual demonstration of the possibility of lateral mobility of the toxin-ganglioside complex and of the capacity of toxin to cross-link membrane gangliosides.

1.4.4 Proposed mechanism of action of choleragen

The proposed sequence of events in activation of adenylate cyclase by choleragen involves first the formation of an inert toxin-ganglioside complex which is converted to an active state by a time and temperature dependent process which involves lateral redistribution of the complex in the plane of the membrane. The activated membrane-bound choleragen, which probably is some form of the 36 000 MW 'active' subunit, diffuses laterally in the plasma membrane, and thus encounters complexes with adenylate cyclase, resulting in activation of the enzyme. The possible interaction of these two molecules would depend on their diffusion properties within the matrix of the membrane. The collision frequency can be estimated once the diffusion constants and average intermolecular distances are known. Such calculations, for example, have been made for the rate of collision of rhodopsin molecules in the photoreceptor membrane (Poo and Cone, 1974). These steps may be viewed schematically (see Fig. 1.28).

1. $\text{Choleragen} + \text{Membrane} \overset{\text{fast}}{\rightleftharpoons} \text{Choleragen}_{\text{I}}\text{-Membrane (inactive)}$

2. $\text{Choleragen}_{\text{I}}\text{-Membrane} \xrightarrow[\text{(lag)}]{\text{slow}} \text{Choleragen}_{\text{II}}\text{-Membrane (active)}$

3. $\text{Choleragen}_{\text{II}}\text{-Membrane} + \text{Adenylate Cyclase-Membrane}$

   $\xrightleftharpoons{\text{(Exponential Phase)}} \text{Choleragen}_{\text{II}}\text{-Adenylate Cyclase-Membrane}$

3′. $\text{'Active' Subunit} + \text{Membrane} \xrightarrow{\text{fast}} \text{Choleragen}_{\text{II}}\text{-Membrane} \rightleftharpoons \text{Choleragen-Cyclase}_{\text{II}}\text{-Membrane}$

This model is consistent with the kinetics and concentration-response relationships for the action of choleragen, and would explain the acute dependence on temperature

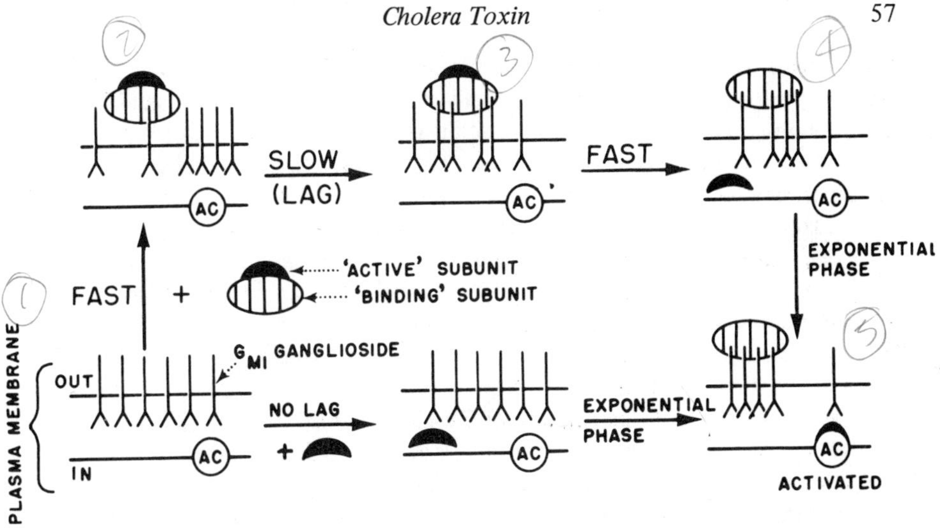

Fig. 1.28 Diagrammatic summary of postulated sequence of events in the activation of adenylate cyclase (AC) by native choleragen and by the 'active' subunit[1] (see text). The specific mechanism by which the 'active' subunit modifies the cyclase complex is unknown and is therefore unspecified. Although the 'binding' subunit is fundamentally not an essential component, it serves to give specificity and orientation, to enhance the affinity for the toxin, as well as to provide a specialized water-soluble vehicle for direct delivery of the active molecular species. From Sahyoun and Cuatrecasas, 1975.

during the lag period and for the later stages in the activation process. The ultimate formation of a complex between choleragen and cyclase is supported by studies with the detergent-solubilized enzyme, and especially by the demonstration of specific immunoprecipitation of soluble toxin-activated adenylate cyclase activity with antisera directed against the active subunit (Fig. 1.26) (Sahyoun and Cuatrecasas, 1975). The possibility of lateral motion of the initial complex (step 2) is suggested by the presence of huge numbers of functionally equivalent spare receptors in some cells (Bennett *et al.*, 1975) and has been demonstrated directly by morphological studies (Craig and Cuatrecasas, 1975; Revesz and Greaves, 1975). No evidence exists as yet, however, that the 'active' subunit itself is capable of diffusion in the plane of the membrane.

The capability for lateral relocation of membrane-bound choleragen, at some stage, could explain the apparent paradox in the intestine, where toxin binding occurs on the brush borders (Peterson *et al.*, 1972) while adenylate cyclase activity is presumably localized on the lateral and basal surfaces of the mucosal cells (Parkinson *et al.*, 1972). Bound toxin may simply move in the membrane directly to the cyclase without having to penetrate these cells and traverse the cytoplasm. Redistribution of fluorescent toxin occurs within 20 min of 37°C, which is sufficiently rapid to explain the time course of cyclase activation.

The basis for the absolute nature of the lag period is not understood. The process occurring during this time is approximately zero-order with respect to toxin concentration or membrane-bound toxin, and is still observed after adding a second dose of toxin to partially stimulated cells. This behavior suggests that each toxin molecule may *independently* experience a similar interval between contact with the cell surface and conversion to an active state. On the basis of the extreme persistence of the toxin's biological effects and the irreversible binding of ^{125}I-labeled choleragen to cell membranes (Fig. 1.5), it has been proposed that during the lag phase the active subunit of choleragen becomes incorporated into the phospholipid bilayer as an 'integral' membrane protein (Singer and Nicolson, 1972) and that this is the active form of the molecule (Bennett and Cuatrecasas, 1975a,b; Bennett *et al.*, 1975a, b; Sahyoun and Cuatrecasas, 1975). The demonstration that the isolated 'active' sububit stimulates cyclase without a delay (Fig. 1.20) (Sahyoun and Cuatrecasas, 1975) provides direct support for these speculations. It is also pertinent that this molecule has strongly hydrophobic properties such that it partitions into organic solvents and must be maintained in the presence of detergents to prevent aggregation and precipitation.

Evidence in support of the separation of the binding and active subunits of choleragen on the cell surface is provided by the finding that an increasing proportion of the binding subunit will exchange between separate cells with increasing length of incubation, while the active subunit is retained (Sahyoun and Cuatrecasas, 1975). It is of interest that measurement of intrinsic fluorescence of choleragen indicate that a conformational change occurs in this molecule attending the binding of GM1 ganglioside in solution, with evidence of positive cooperativity with increasing ratios of glycolipid to protein (unpublished data). These data and the demonstrated capacity of choleragen to bind gangliosides multivalently and cross-link these molecules suggest a possible explanation for the absolute delay. The toxin may require interaction with more than one ganglioside in order for dissociation of subunits to occur, and this multivalent binding would be obtained only as the result of lateral mobility of the initial, monovalently bound cell complex (represented schematically in Fig. 1.28). Thus, no active forms of toxin would exist until the first complex had contacted and cross-linked with at least one other ganglioside. Such multivalent binding might then induce a conformation change causing the active subunit to dissociate, 'dissolve' or partition in the lipid matrix of membrane. Subsequently, this subunit might encounter and associate with adenylate cyclase molecules.

1.4.5 Analogies between the action of choleragen and the mode of activation of adenylate cyclase by hormones.

The postulated lateral movement of membrane-bound choleragen and subsequent interaction with adenylate cyclase is similar to general features of the mobile receptor model recently proposed for the modulation of adenylate cyclase by hormones* (Cuatrecasas, 1974, 1975; DeHaen, 1975; Bennett *et al.*, 1975a; Jacobs and

* *See footnote on page 43*

Cuatrecasas, 1976) with emphasizes the fluidity of cell membranes. It was suggested that hormone receptors, which in their free state are considered to exist uncomplexed with the enzyme, are free to diffuse laterally along the plane of the membrane, as has been observed with other 'integral' membrane proteins. Modulation of adenylate cyclase activity would occur as the result of brief, direct encounters between the enzyme and hormone receptors which are embedded within the membrane matrix. The dynamic interaction between cyclase and receptors would be analogous to the associations between molecules in three dimensional aqueous solution, and could be described in terms of rates of association and dissociation. Productive collisions would presumably occur only if the receptors were complexed with the appropriate hormone. The following equilibria should exist:

1. $H + R \underset{k_{-1}}{\overset{k_1}{\rightleftharpoons}} H{\cdot}R$

2. $H{\cdot}R + AC \underset{k_{-2}}{\overset{k_2}{\rightleftharpoons}} H{\cdot}R{\cdot}AC$

3. $H{\cdot}R{\cdot}AC \underset{k_{-3}}{\overset{k_3}{\rightleftharpoons}} H + R{\cdot}AC$

4. $H{\cdot}R{\cdot}AC \underset{k_{-4}}{\overset{k_4}{\rightleftharpoons}} H{\cdot}R + AC$

5. $R + AC \underset{k_{-5}}{\overset{k_5}{\rightleftharpoons}} R{\cdot}AC$

where H refers to hormone, R to hormone receptor and AC to adenylate cyclase. Choleragen may be visualized as functioning as an 'active' hormone receptor which can stimulate adenylat cyclase in the absence of complexed hormone and which dissociates very slowly.

Several unique predictions follow from this scheme which cannot be encompassed within concepts of a static receptor-adenylate cyclase complex (Cuatrecasas, 1974, 1975; Bennett *et al.*, 1975a; Jacobs and Cuatrecasas, 1976; Cuatrecasas and Hollenberg, 1976). For example, hormones may dissociate at different rates from free receptors than the cyclase-receptor complex. It should therefore be possible to modify the binding properties of hormones by affecting the proportion of receptors complexed with adenylate cyclase without directly perturbing the hormone binding site of the receptor. Since enzyme activation involves two separate and sequential steps, $H + R \rightleftharpoons H{\cdot}R$ and $H{\cdot}R + AC \rightleftharpoons H{\cdot}R{\cdot}AC$, the apparent K_a for hormones and the extent of activation may not depend in a simple way on the binding properties of the hormone.

These considerations could explain the unusual findings that choleragen increases the extent of catecholamine stimulation in toad erythrocytes (Fig. 1.15), the extent and affinity in turkey erythrocytes, and the apparent affinity for cyclase stimulation by a variety of hormones in mammalian cells (see above). This phenomenon has been proposed to be due to stabilization of hormone-receptor-cyclase complexes by choleragen (Bennett and Cuatrecasas, 1975b; Field, 1974; Bennett *et al.*, 1975a, b). The average residence time of each hormone-receptor complex would be lengthened resulting in an increased efficiency of activation, and perhaps an altered rate of dissociation of membrane-bound hormones. It is of interest in this regard that choleragen has recently been demonstrated to decrease the rate of spontaneous dissociation of specifically bound ^{125}I-labeled glucagon from liver membranes (Bennett *et al.*, 1975a). A novel aspect of the mobile receptor theory and the proposed mechanism of action of choleragen is the concept of functionally important dynamic collisions and associations between membrane proteins. This has been postulated previously on the basis of estimates of the collision frequency between rhodopsin molecules in rod outer segment membranes (Poo and Cone, 1974) and for mitochondrial membrane proteins involved in electron-transport (Strittmatter *et al.*, 1975).

REFERENCES

Agre, P., Parikh, I. and Cuatrecasas, P. (1976), in preparation.
Beckman, B., Flores, J., Witkum, P. and Sharp, G.W.G. (1974), *J. Clin. Invest.* **53**, 1202.
Bedwani, J.R. and Okpako, D.T. (1975), Prostaglandins **10**, 110.
Bennett, V. and Cuatrecasas, P. (1973), *Biochem. Biophys. Acta* **311**, 362.
Bennett, V., O'Keefe, E., and Cuatrecasas, P. (1975a), *Proc. Nat. Acad. Sci. USA* **72**, 33.
Bennett, V. and Cuatrecasas, P. (1975a), *J. Membrane Biol.* **22**, 1.
Bennett, V. and Cuatrecasas, P. (1975b), *J. Membrane Biol.* **22**, 29.
Bennett, V., Mong, L. and Cuatrecasas, P. (1975b), *J. Membrane Biol.* **24**, 107.
Bennett, V. and Cuatrecasas, P. (1976a), *Methods in Receptor Research*, M. Blecher ed., Marcel Dekker Inc., in press.
Bennett, V. and Cuatrecasas, P. (1976b), *J. Membrane Biol.*, in press.
Benyajati, C. (1966), *Brit. Med. J.* **1**, 140.
Bourne, H.R., Lehrer, R.I., Lichtenstein, L.M., Weissman, G. and Zurier, R. (1973), *J. Clin. Invest.* **52**, 698.
Bilezikian, J.P. and Aurbach, G.D. (1974), *J. Biol. Chem.* **249**, 157.
Birnbaumer, L., Pohl, S.L. and Rodbell, M. (1969), *J. Biol. Chem.* **244**, 3468.
Birnbaumer, L., Pohl, S.L. and Rodbell, M. (1972), *J. Biol. Chem.* **247**, 2038.
Birnbaumer, L. (1973), *Biochim. Biophys. Acta* **300**, 129.
Bitensky, M.W., Wheeler, M.A., Mehta, H. and Miki, N. (1975), *Proc. Nat. Acad. Sci. USA* **72**, 2572.
Bockaert, J., Roy, C. and Jard, S. (1972), *J. Biol. Chem.* **247**, 7073.

Boon, T. (1971), *Proc. Nat. Acad. Sci. USA* **68**, 2421.
Boyle, J.M. and Gardner, J.D. (1974), *J. Clin. Invest.* **53**, 1149.
Brady, R.O. and Mora, P.T. (1970), *Biophys. Acta* **218**, 308.
Burnet, F.M. and Stone, J.P. (1947), *Australian J. Exp. Biol. Med. Sci.* **25**, 219.
Case, R.M. and Smith, P.A. (1975), *J. Phys.* (Lond.) **244**, 28.
Chen, L.C., Rohde, J.E. and Sharp, G.W.G. (1971), *Lancet* **1**, 939.
Chen, L.C., Rohde, J.E. and Sharp, G.W.G. (1972), *J. Clin. Invest.* **51**, 731.
Chisari, F.V., Northrup, R.S. and Chen, L.C. (1974), *J. Immunol.* **113**, 729.
Chisari, F.V. and Northrup, R.S. (1974), *J. Immunol.* **113**, 740
Cook, R.G., Stavitsky, A.B. and Schoenberg, M.D. (1975), *J. Immunol.* **114**, 426.
Craig, S. and Cuatrecasas, P. (1975), *Proc. Nat. Acad. Sci. USA* **72**, 3844.
Chang, K.-J., Bennett, V. and Cuatrecasas, P. (1975), *J. Biol. Chem.* **250**, 488.
Cuatrecasas, P. (1971a), *Proc. Nat. Acad. Sci. USA* **68**, 1264.
Cuatrecasas, P. (1971b), *J. Biol. Chem.* **246**, 6522.
Cuatrecasas, P. (1971c), *J. Biol. Chem.* **246**, 6532.
Cuatrecasas, P. (1971d), *J. Biol. Chem.* **246**, 7265.
Cuatrecasas, P. (1972), *Proc. Nat. Acad. Sci. USA* **69**, 318.
Cuatrecasas, P. (1973a), *Biochemistry* **12**, 1312.
Cuatrecasas, P. (1973b), *Biochemistry* **12**, 3547.
Cuatrecasas, P. (1973c), *Biochemistry* **12**, 3558.
Cuatrecasas, P. (1973d), *Biochemistry* **12**, 3567.
Cuatrecasas, P. (1973e), *Biochemistry* **12**, 3577.
Cuatrecasas, P. (1974), *Ann. Rev. Biochem.* **43**, 169.
Cuatrecasas, P. (1975), in *Advances in Cyclic Nucleotide Research,* Vol. **5**, eds. G.I. Drummond, P. Greengard and G.A. Robison, Raven Press, N.Y., 79.
Cuatrecasas, P., Parikh, I. and Hollenberg, M.D. (1973), *Biochemistry* **12**, 4253.
Cuatrecasas, P., Jacobs, S. and Bennett, V. (1975), *Proc. Nat. Acad. Sci. USA* **72**, 1739.
Cuatrecasas, P., Bennett, V. and Jacobs, S. (1975), *J. Membrane Biol.* **23**, 249.
Cuatrecasas, P. and Hollenberg, M. (1976), *Adv. Prot. Chem.* **30**, in press.
Cumar, F.A., Brady, R.O., Kolodny, E.H., McFarland, V.W. and Mora, P.T. (1970), *Proc. Nat. Acad. Sci. USA* **67**, 757.
Davoren, P.R. and Sutherland, E.W. (1963), *J. Biol. Chem.* **238**, 3016.
De, S.N. and Chatterje (1953), *J. Pathol. Bacteriol.* **66**, 559.
De, S.N. (1959), *Nature* (London) **183**, 1533.
De, S.N., Ghose, M.L. and Sen, A. (1960), *J. Pathol. and Bacter.* **79**, 373.
de Haen, L. (1974), *J. Biol. Chem.* **249**, 2756.
de Haen, (1975), submitted for publication.
Deery, D.J. and Howell, S.L. (1973), *Biochim. Biophys. Acta* **329**, 17.
Den, H., Sela, B.-A., Roseman, S. and Sachs, L. (1974), *J. Biol. Chem.* **249**, 659.
DeRobertis, F.R., Zenser, T.V., Adler, W.H. and Hudson, T. (1974), *J. Immunol.* **113**, 151.
Donta, S.T., King, M. and Sloper, K. (1973), *Nature New Biol.* **243**, 246.
Donta, S.T. (1974), *J. Infect. Dis.* **129**, 728.
Drummond, G.I. and Duncan, L. (1970), *J. Biol. Chem.* **245**, 976.
Dutta, N.K. and Habu, M.K. (1955), *Brit. J. Pharmacol Chemotherapy* **10**, 153.
Dutta, N.K., Panse, M.V. and Kulkarni, D.R. (1959), *J. Bacteriol.* **78**, 594.

Evans, D.J., Chen, L.C., Curlin, G.T. and Evans, D.G. (1972), *Nature New Biol.* **236**, 137.
Field, M., Plotkin, G. and Silen, W. (1968b), *Gastroenterology* **54**, 1233.
Field, M., Plotkin, G.R. and Silen, W. (1968a), *Nature* (London) **217**, 469.
Field, M., Fromm, D., Wallace, C.K. and Greenough, W.B., III (1969), *J. Clin. Invest.* **48**, 24a.
Field, M. (1971), *N. Engl. J. Med.* **284**, 1137.
Field, M. (1974), *Proc. Nat. Acad. Sci. USA* **71**, No. 8, 3299.
Finkelstein, R.A., Norris, H.T. and Dutta, N.K. (1964), *J. Infect. Dis.* **114**, 203.
Finkelstein, R.A., Atthasampunna, P., Chulosamaya, M. and Charunmethee, P. (1966a), *J. Immun.* **96**, 440.
Finkelstein, R.A. Sobocinski, P.Z., Atthasampunna, P. and Charunmethee, P. (1966b), *J. Immun.* **97**, 25.
Finkelstein, R.A. and LoSpalluto, J.J. (1969), *J. Exp. Med.* **130**, 185.
Finkelstein, R.A. and LoSpalluto, J.J. (1970), *J. Infect. Dis.* **121**, 563.
Finkelstein, R.A., Peterson, J.W. and LoSpalluto, J.J. (1971), *J. Immunology* **106**, 868.
Finkelstein, R.A. and LoSpalluto, J.J. (1972), *Science* **175**, 529.
Finkelstein, R.A., LaRue, M.K. and LoSpalluto, J.J. (1972), *J. Infect. Immun.* **6**, 934.
Finkelstein, R.A. (1973), *CRC Critical Reviews in Microbiology* **2**, 553.
Finkelstein, R.A., Boesman, M., Neoh, S.H., LaRue, M., and Delaney (1974), *J. Immunology* **113**, 145.
Fishman, P.H., Brady, R.O., Bradley, R.M., Aaaronson, S.A. and Todaro, G.J. (1974), *Proc. Nat. Acad. Sci. USA* **71**, 298.
Fishman, P.H., Brady, R.O. and Mora, P.T., (1973), in *Tumor Lipids: Biochemistry and Metabolism* (R. Wood, ed.), American Oil Chemists' Society Press, Champaign, III., 250.
Flawia, M.M. and Torres, H.N. (1972), *J. Biol. Chem.* **247**, 6873.
Formal, S.B., Kundel, D., Schneider, H., Kunev, N. and Spring, H. (1961), *Brit. J. Exp. Pathol.* **42**, 504.
Fresh, J.W., Versage, P.M. and Reyes, V. (1964), *Arch. Pathol.* **77**, 529.
Gangarosa, E.J., Beisel, W.R., Benyajati, C. and Spring, H. (1960), *Am. J. Trop. Med. Hyg.* **9**, 125.
Gill, D.M., Pappenheimer, A.M., Jr., Brown, R. and Kurnick, J.T. (1969), *J. Exp. Med.* **129**, 1.
Gill, D.M., Pappernheimer, A.M. and Uchida, T. (1973), *Fed. Proc.* **32**, 15.
Gill, D.M. and King, C.A. (1975), *J. Biol. Chem.* **250**, 6424.
Gill, D.M. (1975), *Proc. Acad. Sci. USA* **72**, 2064.
Gilman, A.G. and Nurenberg, M. (1971), *Nature* **234**, 356.
Goodpasture, E.W. (1923), *Philipp. J. Sci.* **22**, 413.
Goodgame, R.W. and Greenough, W.B., III (1975), *Ann. Int. Med.* **82**, 101.
Gordon, R.S. (1962), *SEATO Conference on Cholera,* Dacca, East Pakistan, Dec. 5–8, 1960, Bangkok, Thailand, 54.
Gorman, R.E. and Bitensky, M.W. (1972), *Nature* **235**, 439.
Grady, G.F., Keusch, G.T. and Deschner, E.E. (1975), *J. Infect. Dis.* **131**, 210.
Grand, R.J., Torti, F.M. and Jaksina, S. (1975), *J. Clin. Invest.* **52**, 2053.
Grayer, D.T., Serebro, H.A., Iber, F.L. and Hendrix, T.H. (1970), *Gastroenterology* **58**, 815.

Greenough, W.B., III, Pierce, N.F. and Vaughan, M. (1970), *J. Infect. Dis.* **121**, 5111.
Guerrant, R.L., Chen, L.C. and Sharp, G.W.G. (1972), *J. Infect. Dis.* **125**, 377.
Hakomori, S. (1975), *Biochim. Biophys. Acta* **417**, 55.
Hakomori, S. (1970), *Proc. Nat. Acad. Sci. USA* **67**, 1741.
Haksar, A., Mandsley, D.V. and Peron, F.G. (1975), *Biochim. Biophys. Acta* **381**, 308.
Haksar, A., Mandsley, D.V. and Peron, F.G. (1974), *Nature* **251**, 514.
Hart, D.A. and Finkelstein, R.A. (1975), *J. Immunol.* **114**, 476.
Hart, D.A. (1975), *Infect. Immun.* **11**, 742.
Hart, D.A. (1975), *J. Immunol.* **115**, 871.
Harwood, J.P. and Rodbell, M. (1973), *J. Biol. Chem.* **248**, 4901.
Haywood, A.M. (1974), *J. Mol. Biol.* **83**, 427.
Henney, C.S. Lichtenstein, L.M., Gillespie, E. and Rolley, R.T. (1973), *J. Clin. Invest.* **52**, 2853.
Hewlett, E.L., Guerrant, R.L., Evans, D.J., Jr. and Greenough, W.B., III (1974), *Nature* **249**, 371.
Hill, M.W. and Lester, R. (1972), *Biochim. Biophys. Acta* **282**, 18.
Hirsch, G.C., Rushka, H., Sitte, P., eds. (1973), *Grudlagen der Cytologie,* p. 618, Gustov Fischer Verlag, Stuttgart.
Hochman, J., Insel, P.A., Bourne, H., Coffino, P. and Tompkins, G.M. (1975), *Proc. Nat. Acad. Sci. USA* **72**, 5051.
Hollenberg, M.D. and Cuatrecasas, P. (1973), *Proc. Nat. Acad. Sci. USA* **70**, 2964.
Hollenberg, M.D., Fishman, P.H., Bennett, V. and Cuatrecasas, P. (1974), *Proc. Nat. Acad. Sci. USA* **71**, 4224.
Hollenberg, M.D. and Cuatrecasas, P. (1975), *Fed. Proceed.* **34**, 1556.
Honjo, T., Nishizuka, Y., Kato, I. and Hayaishi, O. (1971), *J. Biol. Chem.* **246**, 4251.
Holmgren, J. (1973), *Infect. and Immun.* **8**, 851.
Holmgren, J., Lonnroth, I. and Svennerholm, L. (1973a), *Infect. and Immunity* **8**, 208.
Holmgren, J., Lonnroth, I. and Svennerholm, L. (1973b), *Scand. J. Infect. Dis.* **5**, 77.
Holmgren, J., Lindholm, L. and Lonnroth, I. (1974a), *J. Exp. Med.* **139**, 801.
Holmgren, J., Mansson, J.-E. and Svennerhelm, L. (1974b), *Medical Biology* **52**, 229.
Holmgren, J. and Lonnroth, I. (1975), *J. Gen. Microbiol.* **86**, 49.
Holmgren, J., Lonnroth, I., Mansson, J.-E. and Svennerholm, L. (1975), *Proc. Nat. Acad. Sci. USA* **72**, 2520.
Hudson, N., Poppe, L. and Wilson, D.E. (1974), *Clin. Res.* **22**, 604A.
Hugh, R. (1964), *Int. Bull. Bacteriol. Nomencl. Taxon.* **14**, 87.
Hunter, W.M. and Greenwood, F.C. (1962), *Nature* **194**, 495.
Hynie, S., Cepelick, J., Cernohorsky, M. and Wenke, M. (1974), *Toxicon* **12**, 369.
Issaeff, F. and Kolle, W. (1894), *Z. Hyg. Infekt. Fr.* **18**, 17 Citation furnished by Pollitzer (1959).
Jacob, F., Siminovitch, L. and Wollman, E.L. (1952), *Ann. Inst. Pasteur* **83**, 295.
Jacobs, J.W., Niall, H.D. and Sharp, G.W.G. (1974), *Biochem. Biophys. Res. Comm.* **61**, 391.
Jacobs, S. and Cuatrecasas, P. (1976), *Biochim. Biophys. Acta,* **433**, 482.

Jacobs, S., Bennett, V. and Cuatrecasas, P. (1976), *J. Cyclic Nucleotide Res.*, in press.
Johnson, G.S. and Pastan, I. (1971), *J. Natl. Cancer Inst.* **47**, 1357.
Kately, J.R., Kasarov, L. and Friedman, H. (1975), *J. Immunol.* **114**, 81.
Kately, J.R. and Friedman, H. (1975), *Cell. Immunol.* **18**, 239.
Kempen, H.J.M., de Pont, J.J.H.H.M. and Bonting, S.L. (1975), *Biochim. Biophys. Acta* **392**, 276.
Kimberg, D.V., Field, M., Johnson, J., Henderson, A. and Gershon, E. (1971), *J. Clin. Invest.* **50**, 1218.
Kimberg, D.V., Field, M., Gershon, E., Schooley, R.T. and Henderson, A. (1973), *J. Clin. Invest.* **52**, 1376.
Kimberg, D.V., Field, M., Gershon, E. and Henderson, A. (1974), *J. Clin. Invest.* **53**, 941.
King, C.A. and van Heyningen, W.E. (1973), *J. Infect. Dis.* **127**, 639.
King, C.A. and van Heyningen, W.E. (1975), *J. Infect. Dis.* **131**, 643.
Koch, R. (1884), *Dtsch. med. Wschr.* **10**, 499; citation furnished by Pollitzer (1959).
Koch, R. (1887), *Arbt. Kaiserl. Ges.* **3**, 156; citation furnished by Pollitzer (1959).
Krishna, G., Harwood, J., Barber, A. and Jamieson, G.A. (1972), *J. Biol. Chem.* **247**, 2253.
Kurokawa, K., Friedler, R.M. and Massry, S.G. (1975), *Biol. Int.* **7**, 137.
Lefkowitz, R.J., Roth, J., Prices, W. and Pastan, I. (1970), *Proc. Nat. Acad. Sci. USA* **65**, 745.
Lefkowitz, R.J. (1974), *J. Biol. Chem.* **249**, 6119.
Lichtenstein, L.M., Henney, C.S., Bourne, H.R. and Greenough, W.B. (1973), *J. Clin. Invest.* **52**, 691.
Leray, F., Chambaut, A.-M. and Hanoune, J. (1972), *Biochem. Biophys. Res. Commun.* **38**, 86.
Lonnroth, I. and Holmgren, J. (1973), *J. Gen. Microbiol.* **76**, 417.
LoSpalluto, J.J. and Finkelstein, R.A. (1972), *Biochim. Biophys. Acta* **257**, 158.
Marchalonis, J. and Cone, R.E. (1973), *Transplant. Rev.* **14**, 3.
Masui, H. and Garren, L.D. (1971), *Proc. Nat. Acad. Sci. USA* **68**, 3206.
Mertens, R.B., Wheeler, H.O. and Mager, S.E. (1974), *Gastroenterology* **67**, 898.
Miller, R.J. and Kelly, P.H. (1975), *Nature* **255**, 163.
Minneman, K.P. and Iversen, L.L. (1976), *Science,* **192**, 803.
Morse, S.I., Stearns, C.D. and Goldsmith, S.R. (1975), *J. Immunol.* **114**, 6.
Moss, J., Fishman, P.F., Manganiello, V.C., Vaughan, M., and Brady, R.O. (1976), *Proc. Nat. Acad. Sci. USA* **73**, 1034.
Mullin, B.R., Fishman, P.H., Lee, G., Aloj, S.M., Ledley, F.D., Winand, R.J., Kohn, L.D., and Brady, R.O. (1976), *Proc. Nat. Acad. Sci. USA,* **73**, 842.
Nomura, M. (1963), *Symp. Quant. Biol.* XXVIII, 315.
Northrup, R.S. and Fauci, A.S. (1972), *J. Infect. Dis.* **125**, 672.
O'Keefe, E. and Cuatrecasas, P. (1974), *Proc. Nat. Acad. Sci. USA* **71**, 2500-2504.
Ohki, S. and Hono, O. (1970), *J. Colloid. Interface Sci.* **32**, 270.
Oye, I. and Sutherland, E.W. (1966), *Biochim. Biophys. Acta* **127**, 347.
Olsnes, S. and Pihl, A. (1973), *Eur. J. Biochem.* **35**, 179.
Olsnes, S. and Pihl, A. (1973), *Biochemistry* **12**, 3121.
Parkinson, D.K., Ebel, H., DiBona, D. and Sharp, G.W.G. (1972), *J. Clin. Invest.* **51**, 2292.

Peterson, J.W., LoSpalluto, J.J. and Finkelstein, R.A. (1972), *J. Infect. Dis.* **126,** 617.

Peterson, J.W. and Verney, W.F. (1974), *Proc. Biol. Exp. Biol. Med.* **145**, 1187.

Pfeuffer, T. and Helmreich, E.J.M. (1975), *J. Biol. Chem.* **250**, 86.

Pierce, N.F., Carpenter, C.C.J., Elliot, H.L. and Greenough, W.B., III (1971a) *Gastroenterology* **60,** 20.

Pierce, N.F., Greenough, W.B., III and Carpenter, C.C.J. (1971b), *Bacteriol. Rev.* **35**, 1.

Pierce, N.F. (1973), *J. Exp. Med.* **137**, 1009.

Pollitzer, R. (1959), *Cholera,* World Health Organization, Geneva, Switzerland.

Poo, M.-m. and Cone, R.A. (1974), *Nature* **247**, 438.

Rappaport, R.S., Rubin, B. and Tint, H. (1974), *Infect. and Imm.* **9**, 294.

Rappaport, R.S. and Grant, N.H. (1974), *Nature* **248**, 73.

Refsnes, K., Olsnes, S. and Pihl, A. (1974), *J. Biol. Chem.* **249**, 3557.

Revesz, T. and Greaves, M. (1975), *Nature* **257**, 103.

Richardson, S.H. and Noftle, K.A. (1970), *J. Infect. Dis.* **121** Suppl., S73.

Robbins, P.W. and MacPherson, I. (1971), *Proc. Roy. Soc. Lond.* **177**, 49.

Rodbell, M., Birnbaumer, L., Pohl, S.L. and Sundby, F. (1971), *Proc. Nat. Acad. Sci. USA* **68**, 909.

Rosenberg, C.E. (1962), *The Cholera Years,* The University of Chicago Press, Chicago.

Rodbell, M., Birnbaumer, L., Pohl, S.L. and Krans, H.M. (1971), *J. Biol. Chem.* **246,** 1877.

Sahyoun, N. and Cuatrecasas, P. (1975), *Proc. Nat. Acad. Sci. USA* **72**, 3438.

Salomon, Y., Lin, M.C., Londos, C., Rendell, M. and Rodbell, M. (1975), *J. Biol. Chem.* **250**, 4239.

Sato, K., Miyachi, Y., Olsawa, N. and Kosaka, K. (1975), *Biochem. Biophys. Res. Comm.* **62**, 696.

Schramm, M. and Rodbell, M. (1975), *J. Biol. Chem.* **250,** 2232.

Shottmuller, H. (1904), *Munch med. Wschr.* **51**, 294; citation furnished by Pollitzer (1959).

Senior, B.W. and Holland, I.B. (1971), *Proc. Nat. Acad. Sci. USA* **68**, 959.

Serebro, H.A., Iber, F.L., Yardley, J.H. and Hendrix, T.H. (1969), *Gastroenterology* **56**, 506.

Severson, D.L., Drummond, G.I. and Sulakhe, P.V. (1972), *J. Biol. Chem.* **247**, 2949.

Shafer, D.E., Lust, W.D., Sircar, B. and Goldberg, N.D. (1970), *Proc. Natl. Acad. Sci. USA* **67**, 851.

Sharp, G.W.G. and Hynie, S. (1971), *Nature* **229**, 266.

Sharp, G.W.G., Hynie, S., Ebel, H., Parkinson, D.K. and Witkum, P. (1973), *Biochim. Biophys. Acta* **309**, 339.

Sharp, G.W.G. (1973), *Ann. Review of Medicine* **24**, 19.

Siegel, M.I. and Cuatrecasas, P. (1974), *Molec. Cell. Endocrinol.* **1**, 89.

Singer, S.J. and Nicolson, G.L. (1972), *Science* **175**, 720.

Spyrides, G.J. and Feeley, J.C. (1970), *J. Infect. Dis.* **121**, Suppl., S96.

Staerk, J., Ronneberger, H.J., Wiegandt, H. and Ziegler, W. (1974), *Eur. J. Biochem.* **48**, 103.

Strittmatter, P. and Rogers, M.J. (1975), *Proc. Nat. Acad. Soc. USA* **72**, 2658.

Sultzer, B. and Craig, J. (1973), *Nature (New Biol.)* **244**, 178.
Sutherland, E.W., Rall, T.W. and Menon, T. (1962), *J. Biol. Chem.* **237**, 1220.
Svennerholm, L. (1963), *J. Neurochem.* **10**, 613.
Tanford, C. (1972), *J. Mol. Biol.* **67**, 59.
Tomasi, V., Koretz, S., Ray, T.K., Dunnick, J. and Marinetti, G.V. (1970), *Biochim. Biophys. Acta* **211**, 31.
Vischer, T.L. and LoSpalluto, J.J. (1975), *Immunol.* **29**, 275.
Uchida, T., Papperheimer, A.M. and Harper, A.M. (1973), *J. Biol. Chem.* **248**, 38.
Vaughan, M., Pierce, N.F., Greenough, W.B., II (1970), *Nature* **226**, 659.
van Heyningen, W.E., Carpenter, W.B., Pierce, N.F. and Greenough, W.B., III (1971), *J. Infect. Dis.* **124**, 415.
van Heyningen, S. (1974), Science **183**, 656.
van Heyningen, S. and King, C.A. (1975), *Biochem. J.* **146**, 269.
Viole, H. and Crendiropoulo (1915), *C.R. Soc. Biol.* (Paris) **78**, 331; citation furnished by Pollitzer (1959).
Walker, W.A., Field, M. and Isselbacher, K.J. (1974), *Proc. Nat. Acad. Sci. USA* **71**, 320.
Wishnow, R.M. and Feist, P. (1974), *J. Infect. Dis.* **129**, 690.
Wolff, J. and Cook, G.H. (1973), *J. Biol. Chem.* **248**, 350.
Wolff, J., Temple, R. and Cook, G.H. (1973), *Proc. Nat. Acad. Sci. USA* **70**, 2741.
Wolff, J. and Cook, G.H. (1975), *Biochim. Biophys. Acta* **413**, 283-290.
Zenser, T.V. and Metzger, J.F. (1974), *Infect. Immun.* **10**, 503.
Zieve, P.D., Pierce, N.F. and Greenough, W.B., III (1971), *Johns Hopkins Med. J.* **129**, 299.
Zuckerman, S.H. and Douglas, S.D. (1975), *Nature* **255**, 410.

2 Inhibition of Protein Synthesis by Exotoxins from *Corynebacterium diphtheriae* and *Pseudomonas aeruginosa*

R. J. COLLIER

Acknowledgements
The work in my laboratory has been supported by USPHS grant AI-07877 from the National Institute of Allergy and Infectious Diseases, and by NSF grant GB-32406.

The Specificity and Action of Animal, Bacterial and Plant Toxins
(Receptors and Recognition, series B, volume 1)

Edited by P. Cuatrecasas

Published in 1976 by Chapman and Hall, 11 Fetter Lane, London EC4P 4EE

INTRODUCTION

Among the organisms known to produce toxins which inhibit protein synthesis specifically are two pathogenic bacteria, *Corynebacterium diphtheriae* and *Pseudomonas aeruginosa.* Although the two pathogens differ markedly in gross characteristics, habitat, transmissibility, and infectious characterisitics, it now seems clear that the major toxic proteins produced by the two have almost identical modes of action at the molecular level. Both catalyze an unusual covalent modification of elongation factor 2 (EF-2), an essential factor in protein synthesis. The factor is thereby inactivated, and protein synthesis is inhibited. This inhibition is believed to be largely or partly responsible for the major toxemic symptoms observed in infections by the two pathogens.

2.1 DIPHTHERIA TOXIN

2.1.1 Diphtheria, corynebacteria, and corynephage

Diphtheria is a potentially fatal disease of man, in which the causative organism, *C. diphtheriae,* is normally localized in the upper respiratory tract. The most characteristic diagnostic feature of the disease is the leathery *pseudomembrane* , formed on infected epithelial surfaces and composed largely of necrotic epithelial cells and diphtheria bacilli enmeshed in a fibrous exudate from underlying tissue. Infections by diphtheria bacilli occur less commonly at other sites in the body, including the skin and mucous membranes, and in wounds.

A noteworthy feature of infections by the diphtheria bacillus is the difference in distributions of bacilli and tissue damage in the body of the host. Whereas the bacilli almost always remain confined to the localized area of the primary infection, necrosis and other manifestations of damage are observed in diverse organs, including the heart, adrenals, kidneys, lungs, and others (Andrewes *et al.*, 1923). In 1884 Loeffler suggested that this pattern might be caused by a soluble toxic substance, produced by the bacilli at a localized site and transported throughout the body by the bloodstream. Subsequently it was shown that diphtheria bacilli in culture did indeed produce a highly toxic substance and that immunization against the toxin provided protection against severe forms of the disease.

By the late 1930s the toxin was obtained in sufficiently pure form to determine with some confidence that it was protein in nature (Eaton, 1936; Pappenheimer, 1937). It has since been purified, crystallized, and partially characterized in many laboratories. Various numbers for the molecular weight have been published, the most recent values being slightly above 60 000 (62 000–63 000) (Collier and Kandel, 1971; Gill and Dinius, 1971). No non-protein moieties or unusual amino acids have been found, and the toxin has no gross features which might give a clue to its toxicity.

Most animals, including guinea pigs, rabbits, monkeys, and various fowl, have about the same sensitivity as man on a per body weight basis. A standard measure of toxicity is the guinea pig minimum lethal dose (MLD), which is the amount of toxin required (when injected subcutaneously) to cause death of a 250 g guinea pig within 4–5 days. The most potent preparations of the toxin contain about 40 MLD per microgram of protein, but lower figures are frequently found. A few micrograms is known to be sufficient to cause death of a human. The only mammals known with grossly different sensitivities to the toxin are rats and mice. In these animals about a thousand-fold greater amount must be injected (per unit body weight) for a comparable response.

Gross manifestations of the toxin's action are observed in animals only after a delay of several hours, and even with a massive dose of the toxin (several thousand MLD) the animal never dies before about 10 hours. On the other hand the toxin's ultimate effect becomes irreversible much sooner. A large excess of antitoxin injected 1 hour after administration of 10 MLD of toxin will not prevent death.

Analysis of the gross morphological and physiological changes in animals after injection of lethal quantities of toxin gives no clear indication of how the toxin might act at a cellular or molecular level. As indicated above, many organs show evidence of damage, and the physiological changes are so complex as to defy analysis. The gross picture is consistent with the view that the toxin has little tissue specificity and affects a common metabolic function vital for most tissues.

The production of diphtheria toxin by *C. diphtheriae* has been studied in detail. Among the interesting facts which have come to light are that lysis of cells is not required for toxin release, that toxin production is greatly accelerated when the bacteria approach stationary phase, and that the concentration of iron in the medium is critical. High concentrations of iron suppress toxin production.

Perhaps the most intriguing fact to emerge about the production of the toxin, however, is its relationship to infection of the bacteria by certain temperate bacteriophage. Following from an original observation by Freeman (1951), it has become clear that only strains of *C. diphteriae* which are lysogenic for, or vegetatively infected with, certain bacteriophage are capable of producing the toxin. More recently evidence has been published indicating virtually beyond doubt that the structural gene for diphtheria toxin resides on the genome of these phage. Murphy *et al.*, (1974) have shown that toxin may be synthesized in a cell-free system from *Escherichia coli* to which DNA from corynephage β has been added, and other work with mutants of the same phage also supports this conclusion (Uchida *et al.*, 1971).

The evolutionary and functional implications of the fact that diphtheria toxin is formally a phage protein are not yet clear. We do not yet know the role of diphtheria toxin in phage replication, if any, and have only the vaguest clues to

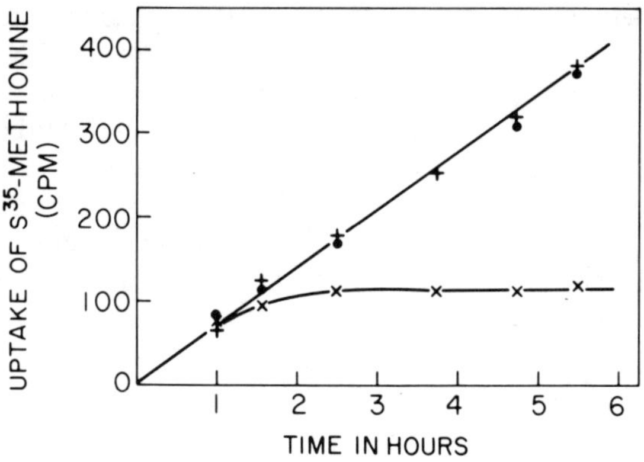

Fig. 2.1 Kinetics of inhibition of protein synthesis in HeLa cells treated with a saturating concentration of toxin. ^{35}S-Methionine and toxin (x), toxin plus excess antitoxin (+), or buffer as a control (●), were added at zero time, and the incorporation of label into trichloroacetic acid precipitable material was measured at the times indicated.

what might be the evolutionary precursor of the toxin. The basic relationship between the toxin and its parent phage is still one of the intriguing features remaining to be explained.

2.1.2 Action of diphtheria toxin

(a) *Inhibition of protein synthesis in tissue culture*

The use of mammalian cells in culture has proved to be extremely valuable for studying the mode of action of diphtheria toxin. Two groups reported in 1957 that the toxin was lethal for cultures from sensitive animals and that only very low concentrations were required (on the order of 2 ng/ml) (Lennox and Kaplan, 1957; Placido-Sousa and Evans, 1957). Later it was found that cell cultures from rats and mice exhibited a level of resistance corresponding to that of the parent animals (Gabliks and Solotorovsky, 1962; Solotorovsky and Gabliks, 1965).

The effects of diphtheria toxin on various metabolic functions of sensitive cells were first explored by Strauss and Hendee (1959) who used HeLa cells grown in monolayer culture. The earliest and most striking effect found was an inhibition of protein synthesis, which occurred from one to several hours after addition of the toxin (Fig. 2.1). The latent period before amino acid incorporation was affected was shorter with high concentrations of the toxin but was never less than about 1–1.5 hours. In certain other types of cells the latent period may be as short as 15 minutes, but an immediate inhibition of protein synthesis has never been observed (Moehring *et al.*, 1967; Pappenheimer *et al.*, 1963).

The fact that morphological changes in the cell are observed only after the inhibition of protein synthesis rules out cell lysis as an explanation for the inhibition. Moreover, among a variety of other metabolic parameters studied (RNA synthesis, DNA synthesis, glycolysis, respiration, transport of potassium ions, intracellular levels of ATP, GTP, and hexose mono- and di-phosphates), none has been found to be altered before the inhibition of protein synthesis, and many remain unaltered for long periods following the inhibition (Kato and Pappenheimer, 1960; Strauss, 1960; Collier and Pappenheimer, 1964a). These results are consistent with the notion that the toxin might be acting directly on protein synthesis. In view of the central role of protein synthesis, subsequent events might well be explained by inhibition of this process.

(b) *Inhibition of protein synthesis in cell-free systems*
Cell-free systems provided a means of testing for a possible direct effect of the toxin on protein synthesis. Kato and co-workers reported that such a system from guinea pig liver could be inhibited by the toxin (Kato, 1962; Kato and Sato, 1962). However, the extremely high concentrations required (on the order of 0.5 mg/ml) caused some doubt about the specificity of the effect.

Subsequent work by Collier and Pappenheimer (1964b) showed that cell-free systems from HeLa cells could be inhibited by concentrations of toxin on the order of a thousand-fold lower than those employed in the earlier work in the guinea pig system. Also it was shown that, if HeLa cells were treated for several hours with toxin before extraction, cell-free amino acid incorporation was reduced to about 20 per cent of that in extracts from normal cells.

Work on HeLa cell extracts also revealed an interesting and unexpected co-factor requirement for the action of the toxin on protein synthesis. If the crude extracts which were routinely used in these studies had been dialyzed or passed through a column of G-25 Sephadex they were no longer sensitive to the toxin. Furthermore, the system became sensitive once again when the low molecular weight, heat-stable fraction of the cells was added back. The fact that this fraction had been removed from the liver system used by Kato and co-workers may explain the low sensitivity to toxin observed.

Empirical tests revealed the nature of the essential co-factor removed by dialysis as nicotinamide adenine dinucleaotide (NAD). Among a variety of co-factors and co-factor mixtures tested, only those containing NAD were capable of sensitizing dialyzed cell extracts to the toxin. Furthermore, treatment of crude extract with streptococcal NAD glycohydrolase rendered them virtually insensitive to the toxin. Later it was shown that the oxidized but not the reduced form of NAD was active in promoting the toxin's action.

(c) *Identification of the toxin-sensitive factor*
Further work on HeLa cell extracts and cell-free systems from rabbit reticulocytes elucidated the target of the toxin's action as elongation factor 2 (EF-2), a protein of molecular weight about 100 000, which is required for the translocation step of protein synthesis (Collier, 1967; Goor and Pappenheimer, 1967a). In this step

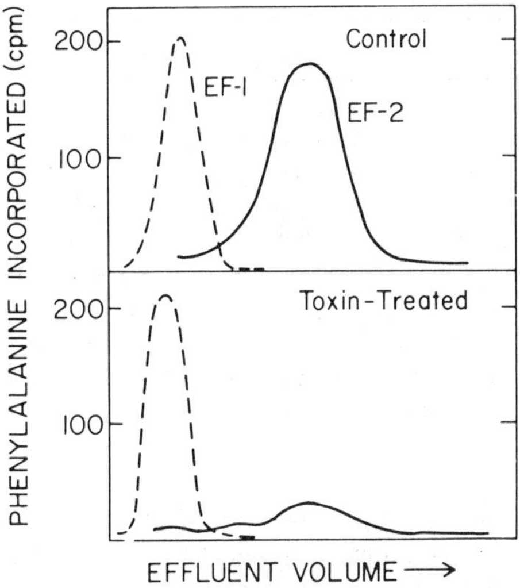

Fig. 2.2 Chromatography on Sephadex G-100 of normal and toxin-treated high-speed supernatant fraction from rabbit reticulocytes. Upper panel, control supernatant fraction incubated with NAD in buffer before chromatography. Lower panel, supernatant fraction incubated with toxin and NAD before fractionation.

peptidyl-tRNA is shifted from the acceptor to the donor site on the ribosome, messenger RNA is translocated by one trinucleotide codon on the ribosome, and GTP is hydrolyzed to GDP and inorganic phosphate. Although neither the relationships among these events nor the role of EF-2 have been precisely defined, there is a clear requirement of this factor under physiological conditions. The complementary elongation factor, EF-1, functions in the binding of substrate aminoacyl-tRNA to ribosomes.

When an ammonium sulfate precipitated fraction from cell supernatant fraction was treated with toxin in the presence of NAD and then fractionated by molecular exclusion chromatography, it was found that EF-2 had been almost completely inactivated by the treatment, whereas EF-1 retained fully activity (Fig. 2.2). Other experiments to test for effects on other components of the protein synthesizing system, for example, ribosomes, aminoacyl-tRNA synthetases, etc., were negative, and it was concluded that the effect on EF-2 probably explained the inhibition of protein synthesis by the toxin (Collier, 1967).

(d) *Covalent attachment of ADP-Ribose to EF-2*

Goor, Pappenheimer, and Ames (1967) reported an interesting observation relevant to the mechanism of inactivation of EF-2 by the toxin. They noted that nicotinamide

Table 2.1 Incorporation of radioactivity from NAD into acid-insoluble fraction*

NAD^+ employed	Incorporation (pmol)
NAD^+-(adenine)-^{14}C	51.2
NAD^+-(adenosine)-^{3}H	50.6
NAD^+-(both phosphates)-^{32}P	50.0
NAD^+-(ribose in NMN)-^{14}C	50.0
NAD^+-(nicotinamide)-^{14}C	0.3

* Reaction mixtures contained tris (hydroxymethyl) aminomethane buffer, EF-2 from rat liver, and diphtheria toxin, in addition to the radioactive NAD indicated. After incubation for 10 minutes at 37°C, the trichloroacetic acid insoluble material was collected on filters and counted. From Honjo *et al.*, (1969).

at high concentrations (30 mM) promoted *reactivation* of the toxin-inactivated EF-2. They interpreted this finding in terms of a model involving inactivation of EF-2 by its incorporation into a ternary complex together with the toxin and EF-2. It was hypothesized that nicotinamide might dissociate the complex by competing with NAD, thus releasing active EF-2.

In 1968 Honjo and co-workers suggested a different mechanism of inactivation of EF-2 by the toxin, on the basis of compelling data (Honjo *et al.*, 1968). They had incubated mixtures of EF-2 and toxin with each of various preparations of radioactive NAD, with label located in various parts of the molecule. The incorporation of label into the trichloroacetic acid precipitable fraction was then measured. As shown in Table 2.1, equivalent amounts of label were incroporated from all parts of the molecule contained within the adenosine diphosphate ribose (ADP-Rib) moiety, whereas no label was incorporated from the nicotinamide moiety. The protein-bound label could be shown to be covalently attached to EF-2, and the kinetics of inactivation of EF-2 coincided with the incorporation of label (Fig. 2.3). It was suggested, on the basis of these results, that toxin inactivated EF-2 by promoting the covalent attachment of ADP-Ribose (Fig. 2.4):

$$\text{EF-2} + \text{NAD} \rightleftharpoons \text{ADP-Rib–EF-2} + \text{Nicotinamide} + H^+$$

This mechanism also provides a clear explanation of the action of nicotinamide in reversing the inactivation of EF-2. Because nicotinamide is a product of the forward reaction, high concentrations should shift the equilibrium toward the left, thus releasing free EF-2. Thermodynamic measurements are consistent with this mechanism.

(e) *Mode and site of attachment of ADP-Ribose*

The fact that ADP-Ribose is covalently attached to EF-2 may be deduced either from chemical stability studies or data on the reversibility of the reaction. The stability of the product is demonstrated by the fact that it resists treatment in

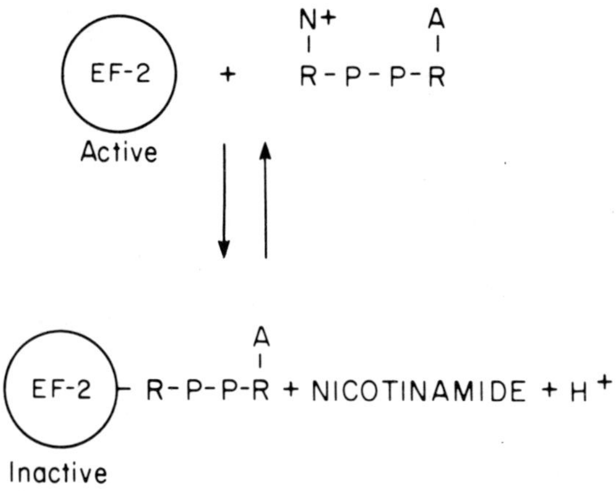

Fig. 2.3 Transfer of ADP-Ribose to EF-2.

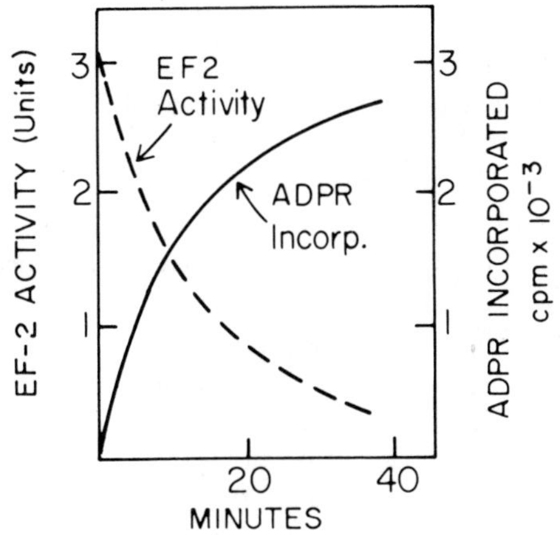

Fig. 2.4 Comparison of the kinetics of inactivation of EF-2 and of transfer of ADP-Ribose into acid insoluble fractions. From Honjo *et al.*, (1969).

0.1 N-HCl or 0.1 N-NaOH at 95°C for 6 minutes, and is quite stable in 5% trichloroacetic acid for longer periods (Honjo *et al.*, 1968, 1971).

Other studies show that the equilibrium position of the reaction is far to the right under physiological conditions (Honjo *et al.*, 1971). A value of 6.3×10^{-4} M has been determined for the equilibrium constant, from which a standard free energy change of –5.2 kcal/mole at pH 7 and 25°C may be calculated. The free energy of hydrolysis of the nicotinamide–ribose linkage of NAD is known to be about –9.2 kcal/mole, and thus, by subtraction, the value for the ADP-Ribosyl–EF-2 linkage must be about –4.0 kcal/mole at pH 7 and 25°C. The facts that the reaction is clearly reversible and that the equilibrium position is controlable by varying the concentration of nicotinamide confirm that the linkage of ADP-Ribose to EF-2 is covalent. The data also explain the fact that protein synthesis is virtually completely inhibited by the toxin *in vivo.*

Although it has been shown that there is only a single attachment site for ADP-Ribose per molecule of EF-2, the precise nature of the site remains uncertain (Raeburn *et al.*, 1971; Robinson and Maxwell, 1972). Maxwell and co-workers have isolated a fifteen-residue peptide containing the attachment site, and the amino acid sequence of this has been determined, except for the residue to which ADP-Ribose is attached (Robinson *et al.*, 1974a, b). This residue, termed X for the moment, does not correspond to any of the commonly occurring amino acids in proteins. It is known to be weakly basic. Inasmuch as X is present in EF-2 which has not been exposed to diphtheria toxin, it is presumably a product of a post-translational event. It has been shown that ribose 5′-phosphate remains attached to EF-2 after removal of the adenosine 5′-phosphate portion of ADP-Ribose with snake venom phosphodiesterase (Honjo *et al.*, 1968; Honjo and Hayaishi, 1973). This implies that ADP-Ribose is linked by its nicotinamide ribose moiety.

(f) *Catalytic role of toxin*

From the fact that the molar ratio of ADP-Ribosyl–EF-2 formed, to toxin, can be at least 50, it is clear that the role of the toxin is catalytic and not stoichiometric (Honjo *et al.*, 1968). In addition, it can be shown that the amount of product formed at completion of the reaction depends only on the concentration of EF-2, whereas the initial rate of the reaction is dependent on the concentrations of both toxin and EF-2.

Later work described below indicates that the whole toxin molecule itself is not enzymically active, but rather is a proenzyme (Collier and Cole, 1969; Collier and Kandel, 1971; Gill and Pappenheimer, 1971). The capacity to ADP-Ribosylate EF-2 seems to be a property only of certain proteolytic fragments of the toxin. The fact that enzymic activity is observed when whole toxin is used reflects the fact that essentially all toxin preparations appear to contain some active fragments produced by bacterial proteases.

(g) *Substrate specificity*

Specificities for the protein and pyridine dinucleotide substrates are high, but not absolute. Under normal conditions of assay, EF-2 from all eukaryotic organisms tested (including various vertebrates, invertebrates, wheat, and yeast) is active as substrate, whereas other proteins are not to any significant degree (Pappenheimer and Gill, 1972). Experiments to detect differences in K_M or rate constants among EF-2 from various organisms have not been conducted. At extraordinarily high concentrations of active toxin fragment (Fragment A) it appears that ADP-Ribosylation of other proteins besides EF-2 may be forced. C. Goff has shown (personal communication) that, when bovine serum albumin or RNA polymerase core enzyme is incubated with adenine-labeled NAD and a concentration of Fragment A (20 mM) about 10^6 times higher than that normally used in our assays, limited amounts of label are incorporated into the albumin, the β and β' subunits of RNA polymerase, and Fragment A itself. In those chains which were labeled, only about one ADP-Ribose was incorporated per thousand molecules of protein. Although these findings are interesting, they probably have little bearing on the physiological effects of toxin.

It is interesting that EF-2 from rats and mice is apparently fully active as substrate (Goor *et al.,* 1967; Moehring and Moehring, 1968). This implies that the resistance of these animals to the toxin results from a blockage at an earlier step in the intoxication process. It is also noteworthy that EF-G, the bacterial elongation factor corresponding in function to EF-2, and the homologous factor from mitochondria, are inactive as substrates (Johnson *et al.*, 1968; Richter and Lipmann, 1970). There are reports that toxin inhibits protein synthesis in bacterial extracts under appropriate conditions, but NAD is not required, and the mechanism appears to be different from that in eukaryotic extracts (Goto *et al.*, 1968; Tsugawa *et al.*, 1970). The significance of this effect is not understood.

Measurements of the substrate specificity of the pyridine dinucleotide involved in the ADP-Ribosylation of EF-2 indicate that probably only NAD is utilized *in vivo.* Of other pyridine nucleotides tested, only certain unnatural analogs of NAD with minor structural changes can substitute for NAD (Goor and Pappenheimer, 1967b; Honjo *et al.,* 1971; Kandel *et al.*, 1974). Oxidized and reduced NADP, and NADH, are inactive as substrates. NMN may have low activity, but this is uncertain at present.

2.1.3 Structure–activity relationships in the toxin

(a) *Activation of the toxin by proteolysis and reduction*

After the reaction responsible for the inactivation of EF-2 was identified, studies were undertaken to characterize the enzymic properties of the toxin in greater detail. A key finding was the discovery that crude or partially purified preparations of toxin contained a protein, less than half the size of whole toxin, with very high specific enzymic activity in catalyzing the ADP-Ribosylation of EF-2 (Collier and Cole, 1969). Further studies revealed that this active protein is in fact a proteolytic fragment (Fragment A) from the toxin (Drazin *et al.,* 1971; Gill and Pappenheimer, 1971).

Diphtheria toxin appears to be synthesized as a single, intact polypeptide chain (molecular weight 60 000–63 000), containing two disulfide bridges and no free sulfhydryls (Collier and Kandel, 1971; Gill and Dinius, 1971). In this form it is toxic for animals and cell cultures, but is devoid of enzymic activity. *In vitro,* ADP-Ribosylation activity is expressed after treatment of the toxin with a protease, such as trypsin, followed by reduction of disulfide bridges. Tryspin preferentially cleaves the toxin into two large fragments, A (21 145 daltons) and B (about 39 000 daltons). Fragment A, free in solution, is active in catalyzing the ADP-Ribosylation of EF-2. Fragment B has no known enzymic activity.

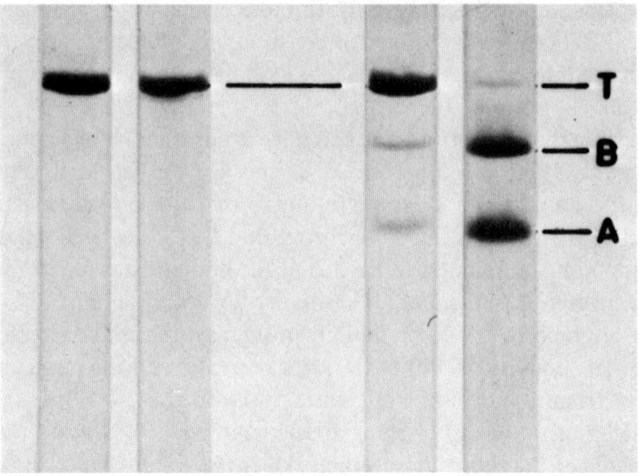

Fig. 2.5 Electrophoresis of untreated and trypsin-treated toxin on SDS polyacrylamide gels. Identical samples of toxin were incubated for 45 minutes with or without 1 μg/ml trypsin. Duplicate portions of each were then heated in 1% SDS at 100°C with or without mercaptoethanol, and an aliquot of each was electrophoresed on an SDS gel. The gels show samples treated (from left to right) without trypsin or thiol, with trypsin only, with thiol only, and with both trypsin and thiol. The positions of Fragments A and B and of intact toxin are indicated in the figure.

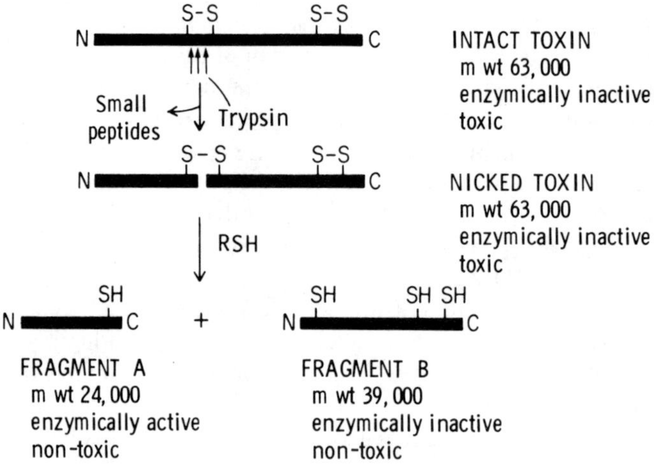

Fig. 2.6 Diagrammatic representation of the activation of diphtheria toxin by tryptic digestion and reduction.

Fragment A, together with traces of related fragments, probably accounts for all the enzymic activity previously attributed to whole toxin. The free Fragment A present in impure toxin undoubtedly results from cleavage by bacterial proteases followed by reduction or disulfide interchange promoted by thiols in the medium.

The cleavage and reduction processes may be demonstrated by polyacrylamide gel electrophoresis in the presence of sodium dodecyl sulfate (SDS) as shown in Fig. 2.5. About 20 per cent of the toxin in this preparation had been nicked by bacterial proteases. The A and B fragments formed by dissociation of this 'naturally nicked' toxin are indistinguishable at a gross level from those produced by trypsin nicking.

The activation process is illustrated diagrammatically in Fig. 2.6. Amino acid sequence studies have shown that Fragment A corresponds to the amino terminus of the toxin (Michel *et al.*, 1972; DeLange *et al.*, 1976). Trypsin attacks intact toxin preferentially in a region spanned by a single disulfide, creating Fragments A and B linked by the disulfide. The nicked form of toxin is indistinguishable (from the intact form) by SDS polyacrylamide gel electrophoresis and, like the intact form, is toxic and lacks enzymic activity. After reduction of the disulfide linking A and B, the two fragments tend to remain associated through non-covalent forces. It is clear that Fragment A is enzymically active in the free form, but it is unclear if dissociation from B is absolutely necessary for this activity to be expressed. It is not known if either or both of the substrates of the ADP-Ribosylation reaction may facilitate dissociation of the fragments.

Despite its enzymic activity, Fragment A is essentially devoid of toxicity when tested in guinea pigs or tissue culture. No toxic effects were observed in guinea pigs injected with 8 nmol per kg body weight, whereas 4 pmol of whole toxin per kg body weight is sufficient to cause death. Similarly, a concentration of 4μM of Fragment A causes no inhibition of protein synthesis in HeLa cells. However, Moehring has detected very low levels of toxicity of Fragment A in certain cells (personal communications). It is probable that the inability of free Fragment A to attach to and penetrate into the cells explains the virtual absence of toxic activity. Fragment B probably provides these functions in the whole toxin molecule. There is good evidence that Fragment B alone is also non-toxic.

(b) *Other active fragments besides A*

As far as is known, scission of the 60 000 dalton chain of toxin is absolutely required for ADP-Ribosylation activity to be expressed. Because Fragment A seems to be the predominant active fragment found in most toxin preparations, as well as the easiest to generate *in vitro*, we have generally assumed that it is also produced *in vivo*. However, there is no evidence concerning the nature of putative cleavage products formed in cultured cells or in animals injected with toxin, and it is clear that certain, longer amino-terminal fragments of the toxin which contain Fragment A have ADP-Ribosylation activity.

One such active fragment, termed Fragment F (mol.wt. 28 000), is found in certain preparations of toxin. It contains the entire disulfide bridge which normally connects Fragments A and B, and is bound to its complementary fragment, E (mol. wt. 34 000), only by non-covalent forces. The cleavage of intact toxin into Fragments E and F is apparently due to a bacterial protease of unknown specificity.

Besides the active fragments generated by proteolytic cleavage, others are known which are apparently due to chain-termination mutations in the *tox*$^+$ gene. Various non-toxic or partially toxic, immunologically cross-reacting forms of toxin have been isolated (Uchida *et al.*, 1971, 1973a). Some of these are identical in size to normal toxin and apparently represent missense mutations. Two others (CRM 30 and 45) are N-terminal fragments, presumably resulting from chain-termination mutations. CRM 30 and 45 (30 000 and 45 000 daltons, respectively) are both enzymically active and, after trypsin treatment, yield Fragment A with the normal level of activity. Their lack of toxicity results from the absence of large C-terminal sections of Fragment B.

It is clear, then, that intact diphtheria toxin may be fragmented at more than one point in its structure, to yield enzymically active N-terminal fragments. Fragmen A is at present the shortest of the known active fragments, but other N-terminal fragments up to at least twice the length of Fragment A contain enzymic activity. These facts should be borne in mind in studying possible fragmentation patterns *in vivo.*

2.1.4 Fragment A

(a) *Chemical and physical properties*

The amino acid composition of Fragment A is shown in Table 2.2 (Delange *et al.*, 1976). The numbers of residues found by sequence analysis are presented for comparison with those determined by amino acid analysis. The lower value for arginine obtained by amino acid analysis is probably due to the shorter forms of Fragment A (190 and 192 residues) which are present and contain one or two fewer arginine residues than the 193-residue molecule. Undercorrection for destruction of threonine may account for the fact that one additional residue was found in the sequence. The overall composition is distinctive in that there is only one residue each of histidine and cysteine, two of tryptophan, and a large number of aromatic residues (19 out of a total of 193 residues).

The complete amino acid sequence of Fragment A is shown in Fig. 2.7 (Delange *et al.*, 1976). The N-terminal sequence of Fragment A is identical with the N-terminal sequence of whole toxin. The C-terminal sequence is interesting because of its heterogeneity. From CNBr digests it is possible to isolate three C-terminal peptides, with sequences – Asn-Arg-COOH, – Asn-Arg-Val-Arg-COOH, and – Asn-Arg-Val-Arg-Arg-COOH. These apparently derive from three forms of Fragment A containing the corresponding C-terminal sequences. Each additional arginine should add an extra positive charge, which probably explains the electrophoretic heterogeneity of the fragment. Under non-denaturing conditions,

Table 2.2 Amino acid composition of fragment A from diphtheria toxin

	Analysis[a]	Sequence
Lys	15.9 (16)	16
His	0.93 (1)	1
Arg	6.03 (6)[b]	7[b]
Asx[c]	22.6 (23)	
Asp		13
Asn		10
Thr	8.25 (8)	9
Ser	17.0 (17)	17
Glx[c]	22.8 (23)	
Glu		15
Gln		7
Pro	6.16 (6)	6
Gly	20.0 (20)	20
Ala	14.2 (14)	14
Val	16.2 (16)	16
Met	4.24 (4)	4
Ile	5.78 (6)	6
Leu	12.4 (12)	12
Tyr	9.80 (10)	10
Phe	6.96 (7)	7
Cys	1.08 (1)[d]	1
Trp	2.1 (2)[e]	2
Total	192	193

[a] Average values (or extrapolated values for serine and threonine) from two 24-hour and two 72-hour hydrolyzates in 6N-HCl at 110°C.

[b] Seven arginyl residues are present in the 193-residue Fragment A, but the 190- and 192-residue molecules contain only 5 and 6 residues of arginine, respectively.

[c] Asx includes aspartic acid and asparagine; Glx includes glutamic acid and glutamine.

[d] Determined as S-carboxymethylcysteine.

[e] Determined spectrophotometrically.

three major, closely spaced bands are observed when Fragment A is electrophoresed under non-denaturing conditions on polyacrylamide gels (Kandel *et al.*, 1974; Michel *et al.*, 1972). All three forms are enzymically active and do not differ greatly in specific activity (Fig. 2.8).

The C-terminal heterogeneity of Fragment A implies that trypsin attacks intact

10
H_2N–Gly–Ala–Asp–Asp–Val–Val–Asp–Ser–Ser–Lys–Ser–Phe–Val–Met–Glu–
20 30
Asn–Phe–Ser–Ser–Tyr–His–Gly–Thr–Lys–Pro–Gly–Tyr–Val–Asp–Ser–
40
Ile–Gln–Lys–Gly–Ile–Gln–Lys–Pro–Lys–Ser–Gly–Thr–Gln–Gly–Asn–Tyr–
50 60
Asp–Asp–Asp–Trp–Lys–Gly–Phe–Tyr–Ser–Thr–Asp–Asn–Lys–Tyr–Asp–
70
Ala–Ala–Gly–Tyr–Ser–Val–Asp–Asn–Gly–Asn–Pro–Leu–Ser–Gly–Lys–
80 90
Ala–Gly–Gly–Val–Val–Lys–Val–Thr–Pro–Gly–Leu–Thr–Lys–Val–Leu–
100
Ala–Leu–Lys–Val–Asp–Asn–Ala–Glu–Thr–Ile–Lys–Lys–Glu–Leu–Gly–
110 120
Leu–Ser–Leu–Thr–Glu–Pro–Leu–Met–Glu–Gln–Val–Gly–Thr–Glu–Glu–
130
Phe–Ile–Lys–Arg–Phe–Gly–Asp–Gly–Ala–Ser–Arg–Val–Val–Leu–Ser–
140 150
Leu–Pro–Phe–Ala–Glu–Gly–Ser–Ser–Val–Glu–Ser–Tyr–Ile–Asn–Asn–
160
Trp–Glu–Gln–Ala–Lys–Ala–Leu–Ser–Val–Glu–Leu–Glu–Ile–Asn–Phe–
170 180
Glu–Thr–Arg–Gly–Lys–Arg–Gly–Gln–Asp–Ala–Met–Tyr–Glu–Tyr–Met–
190 193
Ala–Gln–Ala–Cys–Ala–Gly–Asn–Arg–Val–Arg–Arg–COOH

Fig. 2.7 Primary structure of Fragment A from diphtheria toxin.

toxin at any of the three arginine residues (residues 190, 192, and 193) with similar frequencies. It is not known whether the longest form of Fragment A is attached directly to the N-terminus of Fragment B, or whether the two fragments are linked by one or more small peptides split out by trypsin.

A prominent feature of Fragment A is its stability. It may be exposed to a temperature of 100°C at neutral pH or to pH of 2 or 12 at room temperature for brief periods with negligible loss of activity. In addition, it is markedly resistant to trypsin or chymotrypsin, particularly in the presence of saturating levels of NAD. The sedimentation coefficient of Fragment A is 2.19S as determined by the method of differential sedimentation, using ribonuclease A as a standard (1.84S).

Our knowledge of Fragment B is very limited at present, in part because of its instability. Fragment B denatures rapidly and precipitates after dissociation from A. In addition, preliminary studies indicate that B is heterogeneous at its N-terminus, which prevents N-terminal amino acid sequence analysis.

(b) *Enzymic properties*

Much of what we know about the catalysis of ADP–Ribosylation of EF-2 comes from studies of interactions of Fragment A with NAD (Kandel *et al.*, 1974; Michel *et al.*, 1974). It has been shown that the fragment contains a single, highly specific binding site for NAD. This may be demonstrated by dynamic dialysis or by quenching of native protein fluorescence. Bound NAD quenches the total fluorescence from the two tryptophan residues by up to 80 per cent.

Binding of NAD may also be demonstrated by a change in the absorbance properties of the protein. The complex of Fragment A with NAD has a weak, broad

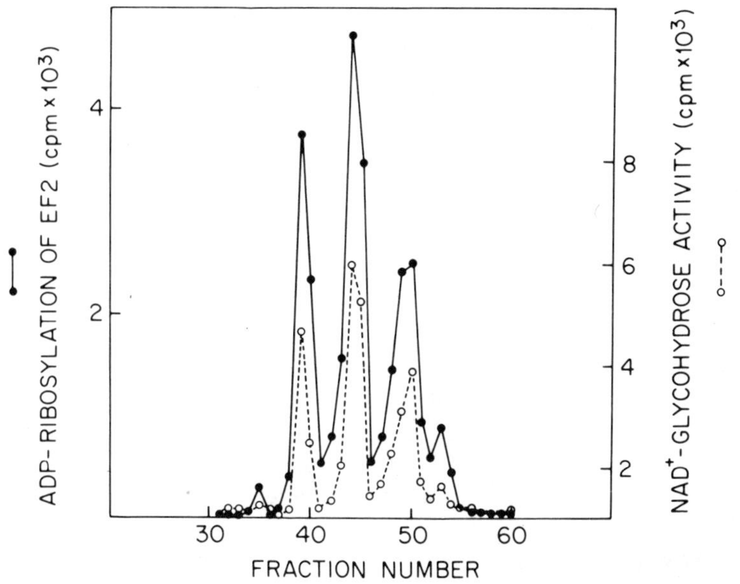

Fig. 2.8 Profiles of NAD-glycohydrolase and ADP-Ribose transferase activities of Fragment A electrophoresed on polyacrylamide gels under non-denaturing conditions. Fragment A was electrophoresed, and the gel was then sliced into 2 mm fractions. Each of these was eluted overnight in 0.1 ml buffer, and the activities were measured in each eluate.

absorbance band with a maximum at about 360 nm and ϵ of about 500, which is not present in either Fragment A or NAD alone (Kandel *et al.*, 1974; Collier *et al.*, 1974a, b).

The appearance of this absorbance band and the quenching of the protein fluorescence can both be shown to result from the interaction of the protein with the nicotinamide moiety of NAD. This conclusion follows from the fact that N-methylnicotinamide and NMN produce both 360 nm absorbance and fluorescence quenching, whereas adenine does not, even under conditions in which it binds competitively with NAD. These effects may result from a charge-transfer complex formed by the interaction of the positively charged nicotinamide ring with the indole side chain of one (or perhaps both) of the tryptophans of Fragment A. Nicotinamide, being uncharged, neither quenches fluorescence nor generates 360 nm absorbance.

When NAD is incubated with high concentrations of Fragment A in the absence of EF-2, the dinucleotide is slowly hydrolyzed (Kandel *et al.*, 1974). The products have been identified as nicotinamide and ADP-Ribose, which defines the hydrolysis as due to NAD glycohydrolase activity (EC 3.2.2.5). That this slow reaction is due to Fragment A and not to traces of contaminating NAD glycohydrolase is shown by the correlation of the profiles of this activity with ADP-Ribosylation activity eluted from polyacrylamide gels run on native Fragment A. Both profiles have three peaks corresponding to the three major electrophoretic forms of Fragment A. The turnover number of the fragment in this reaction is about 0.05 mole NAD minute per mole of the fragment at the optimal temperature of 37°C.

The binding constants determined for NAD and for various inhibitors of NAD binding are consistent with the idea that the single NAD binding site detected by dialysis or optical methods is responsible for both the NAD glycohydrolase reaction and the ADP-Ribosylation of EF-2. For example, both adenine and nicotinamide inhibit both reactions and the binding of NAD with K_i values of about 30 and 200 μM, respectively. It is clear that the binding of NAD to Fragment A involves both the adenine and nicotinamide moieties of NAD. NAD is apparently bound in such a manner as to facilitate nucleophilic attack on the nicotinamide–ribose linkage, by water, or presumably, the X side chain of EF-2.

Much less is known about the interaction of Fragment A with its other substrate, EF-2. This is in large part due to the difficulty in obtaining significant quantities of EF-2 in highly purified form. The purification procedures that have been devised so far are lengthy and laborious.

With the limited quantities of EF-2 obtained, it has been possible to determine apparent K_M as about 0.2 μM (Chung and Collier, unpublished results). Similar figures have been obtained with relatively crude preparations of EF-2. Attempts to demonstrate a direct interaction of EF-2 with Fragment A in the absence of NAD have failed. Experiments have been conducted in which mixtures of pure EF-2 and radioactively labeled Fragment A are chromatographed on Sephadex G-100 under conditions in which binding should have been detected if the affinity were similar if the K_D were comparable to the value of K_M.

(c) *Mechanism of ADP-Ribosyl transfer*

All data obtained to date indicate that the interaction between Fragment A and NAD involves only non-covalent forces (Goor and Maxwell, 1970; Kandel *et al.*, 1974). Attempts have been made to find a covalent ADP-Ribosyl-Fragment A intermediate, but without success. Labeled nicotinamide has been shown not to exchange into NAD in the presence of Fragment A, and when mixtures of Fragment A and adenosine-labeled NAD are fractionated on Sephadex no label is bound to the protein.

These results imply that the ADP-Ribosylation of EF-2 proceeds through a ternary intermediate complex, containing Fragment A, NAD, and EF-2. The major unknown is the pathway by which this complex is formed. One would like to know if there is an obligatory order in the binding of the dinucleotide and protein substrates to Fragment A. The fact that binding of NAD to Fragment A clearly occurs in the absence of EF-2, whereas binding of EF-2 in the absence of NAD could not be detected, suggests that the dinucleotide must bind first. We have attempted to determine if adenine or nicotinamide might facilitate Fragment A EF-2 interaction, but without success. We therefore tentatively assume that there is an obligatory, sequential binding of substrates, NAD being the first.

2.1.5 Events at the cell surface

(a) *Attachment to specific receptors*

Before EF-2 can be inactivated the Fragment A moiety of toxin must gain access to the factor. At a minimum this must involve projection of the active site of Fragment A through the internal surface of the plasma membrane, and it is likely that the entire Fragment A peptide, if not the whole toxin, penetrates to the cytosol. Experimental evidence concerning the mechanisms of attachment and penetration is meager, however, largely because there appears to be only a relatively small number of attachment sites for the toxin on sensitive cells.

One of the few facts which seems relatively firm in this area is that there are specific receptors for the toxin on the cell. The evidence for this comes largely from work with one of the non-toxic, cross-reacting forms of toxin, CRM 197 (Ittelson and Gill, 1973; Uchida *et al.*, 1972, 1973b, c). This protein, produced by *C. diphtheriae* lysogenized with a mutant form of phage β, is known to contain a lesion which renders its Fragment A portion enzymically inactive; its B portion is fully functional as demonstrated by molecular hybridization experiments. CRM 197 alone has no effect on cultured cells, but it has the interesting property of inhibiting the action of toxin on such cells (Fig. 2.9). Inasmuch as CRM 197 does not inhibit the ADP-Ribosylation of EF-2 in cell-free systems, this evidence implies that it competes for specific receptors at the cell surface. The fact that CRM 45, a non-toxic form lacking the C-terminal 17 000 daltons of the toxin, does not compete in a similar manner is consistent with the idea that Fragment B contains the receptor binding site.

The number and nature of the receptors remains uncertain. Pappenheimer and Brown (1968) estimated from studies of the uptake of ^{125}I-labeled toxin by HeLa cells that no more than 25–50 molecules per cell were taken up in 6 hours in the presence of a virtually saturating concentration of toxin. Because of the apparent reversibility of the initial binding event and the necessity of extensive washing to remove excess, unbound toxin, it is uncertain how these numbers relate to the number of receptors per cell. Recent estimates of 10 000–20 000 receptors per cell have been reported, based on unpublished results (Pappenheimer and Gill, 1973). Evidence regarding the chemical nature of the receptors is limited to the findings that treatment of cells with trypsin, pronase, or phospholipase C seems to desensitize them, whereas neuraminidase, lysozyme, and hyaluronidase seem not to affect intoxication (Duncan and Groman, 1969; Moehring and Crispell, 1974).

(b) *Internalization and processing*

Although models postulating direct penetration of the lipid bilayer by the toxin cannot be excluded, it seems likely that diphtheria toxin bound to cell surface receptors may be internalized by an endocytotic mechanism similar to that

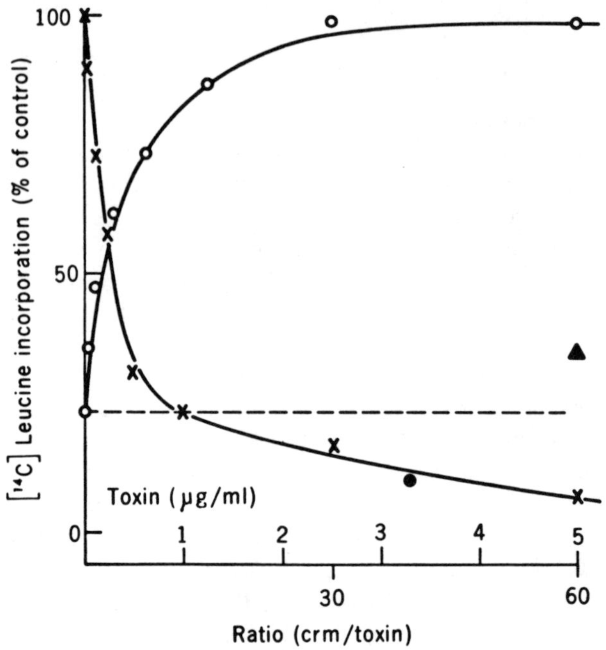

Fig. 2.9 Competition between CRM 197 and toxin for binding sites on HeLa cells. Samples of cells in roller tubes were incubated with the concentration of toxin indicated for 3.5 hours, and ^{14}C-leucine was then added. After an additional 3.5 hour incubation, the cells were harvested and leucine incorporation measured (x). The open circles indicate the incorporation observed when the cells were treated with a constant concentration of toxin (1 μg/ml) plus a variable concentration of CRM 197. Incorporation of leucine in the presence of toxin alone at 1 μg/ml is indicated by the dashed line. The closed circle and triangle show leucine uptake in the presence of CRM 45 and purified toxin in ratios to toxin of 40 : 1 and 60 : 1, respectively. From Uchida *et al.*, (1972).

described for RCA_{II}, a toxic lectin from castor beans (which is probably identical with ricin) (Nicholson 1974; Nicholson *et al.*, 1975). RCA_{II} apparently binds initially to a galactose or galactose-like moiety on the surface of cells. A lectin-induced clustering of RCA_{II} receptors is then observed, followed by endocytosis of RCA_{II} molecules bound to the receptors. RCA_{II} is apparently released from endocytotic vesicles into the cytosol, and the toxin then interacts with ribosomes, producing an inhibition of protein synthesis. Fusion of the endocytotic vesicles with lysozomes is not observed, but it cannot be ruled out. Although further evidence is needed to determine if this model represents the physiologically significant mode of entry, the kinetics of vesiculation and internalization seem to be consistent with the observed time course of inhibition of protein synthesis.

Evidence relevant to internalization of diphtheria toxin is fragmentary and indirect, but there seem to be some findings which are consistent with an endocytotic mechanism. The facts that the entry is much slower than the adsorption process, and that adsorbed toxin is maintained in an antitoxin-sensitive state at temperatures of 10°C or lower, are at least consistent, as are the findings that fluoride and arsenate strongly block the action of the toxin (Strauss and Kendel, 1959; Ivins *et al.*, 1975). Inhibitors of respiration inhibit the toxin's effects only partially.

The effects of ammonium salts and aliphatic amines on the intoxication process suggest that they somehow interfere with internalization of the toxin, but the mechanism is unclear. These compounds seem to permit adsorption to occur, but maintain it in an antitoxin sensitive state. (Kim and Groman, 1965a, b; Ivins *et al.*, 1975).

Another interesting but unexplained effect is produced by polycations, such as polyornithine or diethylaminoethyldextran (Moehring and Moehring, 1968; Ivins *et al.*, 1975). These compounds seem to increase the toxin sensitivity of mouse cells, which are normally resistant, but have the opposite effect on lines which are normally highly sensitive.

Moehring and Moehring (1972a) have taken a very interesting approach to the problems of toxin entry. They have isolated toxin-resistant variants of the KB cell line, which is normally highly sensitive. For one such resistant strain the concentration of toxin required to produce a 50 per cent reduction of incorporation of amino acids into protein was more than 2000 times higher than required by the parent strain. Tests of cell-free extracts showed that EF-2 is apparently unaltered in toxin sensitivity in the resistant strains, thus implying an alteration in the entry or processing mechanisms. Interestingly, toxin resistance seemed to be accompanied by an increased resistance to certain RNA viruses, including Mengo virus, vesicular stomatitis virus, Newcastle disease virus, and poliovirus (Moehring and Moehring, 1972b). The nature of the common blockage remains unclear.

We assume that the expression of enzymic activity by the toxin requires activation events similar to those observed in cell-free systems, but again, direct evidence is lacking. The fact that preparations containing mostly intact toxin molecules (60 000 dalton chains) differ little in toxicity from completely nicked preparations implies that nicking occurs during or after entry of toxin. One can envisage that this might occur after fusion of endocytotic vesicles with lysosomes or by other proteases associated with membranes or the cytosol. If reduction of the disulfide bridge linking Fragments A and B is necessary within cells, this may be promoted by glutathione. This compound is present in millimolar concentrations in most tissues, and such concentrations have been shown to reduce nicked toxin rapidly *in vitro.*

(c) *Kinetics of entry and turnover*

By whatever mechanism entry occurs, toxin appears to begin inactivating EF-2 within minutes, long before the end of the lag period. The total elongation factor activity in the supernatant fraction from toxin-treated cells drops markedly within 20 minutes, although protein synthesis is not affected until about 1 hour (Gill *et al.*, 1969). The fact that EF-2 does not normally limit the rate of protein synthesis may explain the apparent discrepancy. Gill and Dinius (1973) have demonstrated that EF-2 is present in a molar ratio of 1.2 molecules per ribosome in a variety of tissues. Apparently only after this ratio drops to some much lower value does EF-2 begin to limit protein synthesis.

Toxin entry continues beyond the end of the lag period, but the exact duration is unknown. Protracted entry is demonstrated by experiments with CRM 176, a partially toxic mutant form which has a Fragment A with 8–10 per cent of the normal specific activity (Uchida *et al.*, 1973c). CRM 176 inhibits protein synthesis in HeLa cells to a lesser extent than similar levels of toxin. If cells are treated with CRM 176 for 1.5 or even 3 hours, and then washed, the decline in the rate of protein synthesis halts and the cells recover. Inasmuch as the decline continues in cells further treated with CRM, entry must be occurring at these late times. In control experiments cells were treated with toxin for 1.5 hours and resuspended in toxin-free medium containing CRM 197 to prevent further entry of toxin. The decline in the rate of protein synthesis was only marginally affected by the removal of toxin, and the cells did not recover.

The recovery of CRM 176 treated cells must depend on degradation of the A-176 in the cytosol, presumably by the normal mechanism of protein turnover in cells. The higher specific activity and possibly greater stability of the normal Fragment A presumably accounts for the lack of recovery of cells treated with toxin.

After the lag period the decline in the rate of protein synthesis in toxin-treated cells seems to follow first-order kinetics. It has been suggested that this implies the presence of a constant concentration of Fragment A within cells, presumably a steady state between entry and degradation. Other explanations should be considered, however, because of the number of variables affecting the intracellular reaction.

Estimates of the specific activity of Fragment A in the intracellular milieu are highly uncertain at this point because of the large number of uncertainties about intracellular conditions. Thus, it is not possible to predict accurately the intracellular concentration of Fragment A needed to inhibit protein synthesis at a given rate. Unfortunately this concentration is too low to be measured directly.

The minimum number of toxin molecules required to kill a cell is a point of interest. With certain cell lines, several hundred molecules per cell added to the medium are sufficient to cause death within 3–4 days (Gabliks and Solotorovsky, 1962; Placido-Sousa and Evans, 1957). However, it is possible that only a small fraction of those added are taken up. From the fact that the kinetics of adsorption of a lethal dose of toxin by HeLa cells are first-order, it has been suggested that a single molecule of toxin may be lethal for a cell (Duncan and Groman, 1969).

2.1.6 Determinants of cellular sensitivity and resistance

As indicated earlier, marked resistance to diphtheria toxin is found in cells from rats and mice. In addition, it is known that at least one cell type from sensitive animals, namely reticulocytes from rabbits, is insensitive to the toxin. These variations document the existence of naturally occurring variations in sensitivity. As noted earlier, Moehring and Moehring have selected toxin-resistance variants of KB cells.

In addition, there have been reports of significant differences in sensitivity between tumor cells and normal cells from the same animal. Iglewski and co-workers have reported that Ehrlich-Lettre tumor cells are more sensitive than normal mouse spleen or thymus cells, and that certain types of human tumor cells appear to be more sensitive than normal human cells (Iglewski and Rittenberg, 1974). The same group has also found that Rous sarcoma virus-transformed chicken cells are more sensitive than normal tissue cells (Iglewski *et al.*, 1975). Pappenheimer and Randall (1975) have challenged the results on Ehrlich-Lettre cells. It should also be noted that SV40-transformed 3T3 fibroblasts are apparently more sensitive to RCA_{II} than untransformed 3T3 cells (Nicholson *et al.*, 1975). Although it is still unclear whether consistent differences exist in toxin sensitivity between tumor and normal cells, it would perhaps not be surprising that transformations which affect cell surface properties and activities should affect the intoxication process.

Our overall understanding of the factors which differentiate diphtheria toxin sensitive from insensitive cells is still poor, although the vague outlines are apparent. It seems virtually certain that resistance will most often reside in steps in the intoxication process preceding the ADP-Ribosylation of EF-2. Occasional variants may be found in which EF-2 is resistant to ADP-Ribosylation, or conceivably in which ADP-Ribosyl-EF-2 retains activity in protein synthesis. However, in examples which have been studied, EF-2 seems essentially normal, implying blockage of an earlier step.

Toxin action can probably be blocked at any of numerous steps prior to the ADP-Ribosylation reaction. For example, one can easily envisage that the absence of specific receptors might inhibit adsorption, or that cells might contain defective internalization or processing mechanisms. Unfortunately, the available data are not yet sufficient to determine, for any given toxin-resistant cell, the precise nature of the defect. Pappenheimer and Brown (1968) have reported data suggesting that mouse cells are deficient in toxin receptors, but further work is necessary before this can be considered certain. Bonventre *et al.*, (1975) have reported that mouse L cells absorb and internalize ^{125}I-labeled toxin to about the same extent as cells from sensitive animals.

Solution of these problems will probably require a large array of techniques and much effort, given the complex and dynamic nature of the cell surface in animal cells.

2.2 EXOTOXIN FROM PSEUDOMONAS AERUGINOSA

2.2.1 Infections by P. aeruginosa

P. aeruginosa contrasts strongly with *C. diphtheriae* both in its gross characteristics and in the patterns of infection which it produces (Clarke and Richmond, 1975). For example, pseudomonads are Gram-negative whereas corynebacteria are Gram-positive, and *P. aeruginosa* is capable of growing saprophytically in water containing minimal levels of nutrients whereas *C. diphtheriae* is much more fastidious in its growth requirements. The latter characteristic contributes to *P. aeruginosa's* being an important contaminant of sterile solutions in hospitals.

Except for infants, *P. aeruginosa* is virtually harmless for healthy individuals. It is an 'opportunistic' pathogen which most commonly infects patients with impaired host-defense mechanism, such as those with extensive burns, with genetic immunodeficiencies, or under treatment with immunosuppressive drugs. This property, together with the organism's ability to multiply in moist, inorganic environments, makes *P. aeruginosa* a major source of hospital infections.

P. aeruginosa is capable of causing both local and generalized infections. Localized infections occur at various sites in the body, including the respiratory and urinary tracts, the intestine, the ear, and the eye. In patients with strongly impaired resistance, generalized infections may develop from the localized foci. The mortality is very high in such cases, perhaps approaching 100 per cent. A complicating factor in treatment is that the organisms are insensitive to most antibiotics.

2.2.2 PA toxin

Although the mechanism of pathogenesis of *P. aeruginosa* is more complex and less well understood than that of *C. diphtheria,* it seems likely that a protein exotoxin (PA toxin) produced by the pseudomonad and first isolated by Liu (1969a) may prove to be an important element. Demonstration of the importance of the toxin is complicated by the fact that other toxic factors are excreted, including lecithinase, proteases, hemolysin, and various pigments. However, the purified PA toxin has been shown to cause death, and produces symptoms similar to those observed in animals infected with cultures of *P. aeruginosa* (Liu, 1969b; Callahan, 1974). Also, antiserum against the toxin prevents the lethal effects of infections in mice with live cultures (Liu and Hsieh, 1973). Data are lacking for human infections.

The PA toxin has been purified and studied in several laboratories. Liu, Yoshii, and Hsieh (1973) have reported a molecular weight of about 50 000 for the purified toxin, but more recent estimates by Iglewski and co-workers (Iglewski *et al.*, 1975a) and by Chung and Collier (unpublished data) indicate a higher figure, in the vicinity of 66 000–70 000. The toxin has been reported to have an isoelectric point of about 5.0 (Callahan, 1974) and seems to consist of a single polypeptide chain (Iglewski *et al.*, 1975a). Little else is known about the physical or chemical properties of the protein.

In studies on cultured mouse fibroblasts, PA toxin (0.02 μg/ml) has been shown to inhibit protein synthesis, beginning about 1 hour after addition (Iglewski *et al.*, 1975). RNA and DNA synthesis were not inhibited until at least 4 hours under these conditions, but may be inhibited earlier at higher toxin concentrations. These results are similar to those seen with diphtheria toxin in cells which are sensitive to it.

There is now good evidence that PA toxin inhibits protein synthesis by the same mechanisms as diphtheria toxin, i.e. by catalyzing ADP-Ribosylation of EF-2. This was first reported by Iglewski and Kabat (1975), and the work has been extended by Iglewski and co-workers (1975a) and by Dominic Chung in my laboratory (unpublished results). To summarize the evidence, it has been shown that: (1) PA toxin inhibits protein synthesis in cell-free systems, and NAD is required; (2) label is incorporated into a crude preparation of EF-2 from the adenosine or phosphate moieties of NAD, but not from the nicotinamide moiety; (3) the incorporated label is attached to EF-2 as judged by SDS gel electrophoresis, and is present in the same tryptic peptide as labeled by Fragment A from diphtheria toxin, as evidenced by thin-layer chromatography; (4) finally, Fragment A catalyzes the reversal of the reaction catalyzed by PA toxin, as determined by the release of labeled ADP-Ribose in the presence of a high concentration of nicotinamide. From these results it seems certain that both toxins catalyze the attachment of ADP-Ribose to the same site on EF-2 and with the same stereospecificity of attachment.

2.2.3 Comparisons between PA and diphtheria toxins

From the fragmentary information at present available, it appears that there are other interesting similarities, besides enzymic activity between the two toxins, but also marked differences.

(1) *Production of PA toxin.* Liu (1973) has studied certain factors affecting production of PA toxin, but much remains to be learned. We have recently found that, as in the case of diphtheria toxin, the PA toxin is produced during the period of declining growth rate as the bacteria approach stationary phase. (Chung and Collier, unpublished data). Fig. 2.10 shows a concomitant rise in ADP-Ribosylation activity and in toxicity, the latter as assayed by the inhibition of protein synthesis in HeLa cells. SDS polyacrylamide gels run on the culture supernatant show the 66 000 dalton species and several other bands arising concomitantly with the appearance of toxicity and enzymic activity. Preliminary results indicate that the concentration of iron in the medium has little effect on the level of PA toxin produced, in contrast with the case of diphtheria toxin.

(2) *Antigenic properties.* No cross reaction between PA and diphtheria toxins has been found. Antibody against diphtheria toxin Fragment A neutralizes the activity of this antigen but not that of PA toxin, and pony antiserum against PA toxin is similarly specific for the latter toxin. Iglewski *et al.*, (1975) have also reported that no cross reaction can be detected by complement fixation.

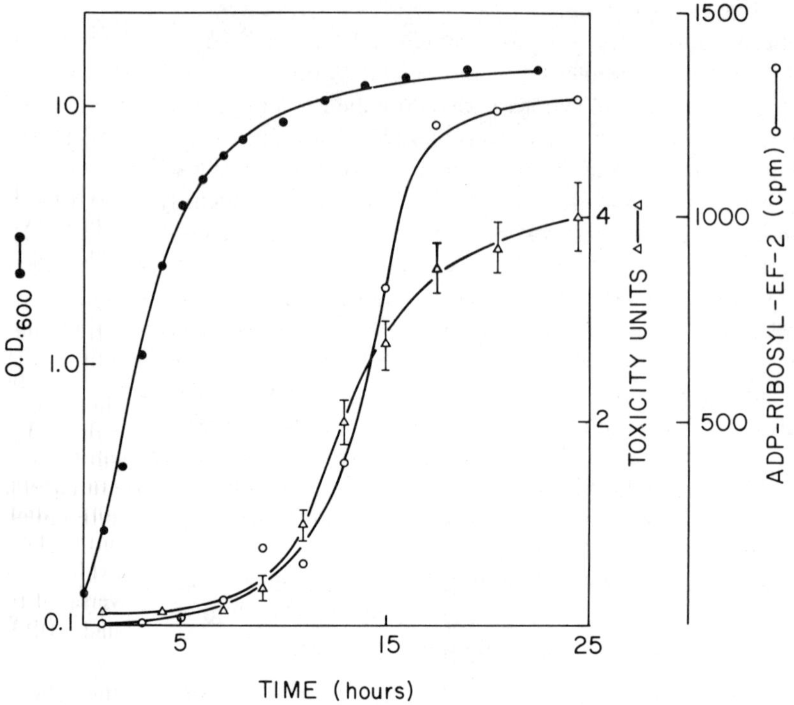

Fig. 2.10 Kinetics of growth and toxin production by *P. aeruginosa* strain PA-103. Cells were grown according to the method of Liu (1973). ADP-Ribose transferase activity in the culture supernatant was determined after desalting on Sephadex G-25 (○). Toxicity was determined by measuring the inhibition of protein synthesis in HeLa cells after a 17 hour incubation with various dilutions of toxin (△). A unit of toxin is arbitrarily defined as the amount of toxin needed to produce a 50 per cent reduction in the rate of protein synthesis as compared with untreated cells.

(3) *Cellular specificity.* Mouse cells are apparently more sensitive to PA toxin than are human cells. This strong contrast with diphtheria toxin suggests that the two toxins may bind to different receptors on the cell surface, although other explanations are possible.

(4) *Proteolysis and activation.* In contrast to diphtheria toxin, intact PA toxin appears to be enzymically active. Iglewski *et al.,* (1975) reported that this species isolated from SDS polyacrylamide gels exhibits enzymic activity, and we have conducted further experiments to demonstrate that the 66 000 dalton species is not activated by proteolytic cleavage during the course of the enzymic assay (Chung and Collier, unpublished results). In one experiment, pure PA toxin was first incubated with ^{14}C-NAD and partially purified EF-2 from rabbit reticulocytes for 30 minutes at 25°C. The mixture was then chromatographed on Sephadex G-100

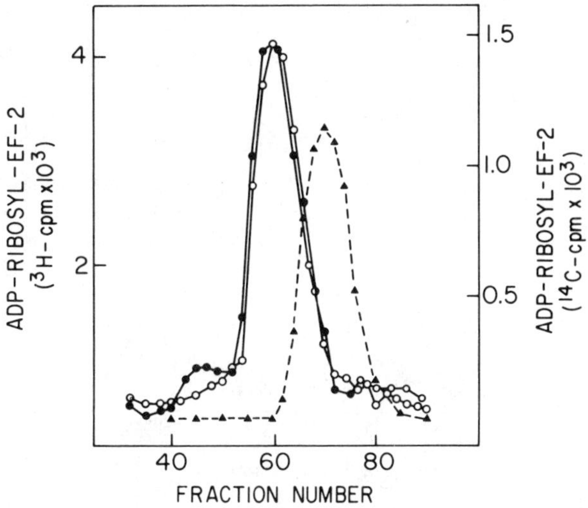

Fig. 2.11 Chromatography of PA toxin on Sephadex G-150 after incubation with EF-2 and NAD. Purified toxin (mol. wt. ~ 66 000) was incubated with crude reticulocyte EF-2 and ^{3}H-NAD under normal assay conditions, and the mixture was then chromatographed. Each fraction was assayed for ADP-Ribose transferase activity in the presence of ^{14}C-NAD and fresh EF-2. The activity profile (○) was essentially identical to that of toxin which had not been exposed to EF-2 and NAD (●). The incorporated ^{3}H is not shown in this figure. The dashed line shows the activity profile of the lower molecular weight active species chromatographed on the same column and assayed with ^{3}H-NAD (right-hand scale).

and the fractions were assayed for ADP-Ribose transferase activity, using ^{3}H-NAD and additional EF-2. As shown in Fig. 2.11, there was no change in the molecular weight of the enzymically active species during the incubation.

Also, the kinetics of ADP-Ribosyl transfer catalyzed by PA toxin are consistent with the idea that the intact form is enzymically active. The incorporation of ADP-Ribose into EF-2 is proportional to the length of incubation, and the curve extrapolates through the origin. Any conversion from an enzymically inactive into an active species in the incubation mixture would have had to be extraordinarily rapid, and we doubt that any occurred. Although proteolysis does not appear to be obligatory for enzymic activity in PA toxin, a lower molecular weight, enzymically active species has been found, and preliminary evidence indicates that it is a proteolytic fragment from the 66 000 dalton toxin. As shown in Fig. 2.12, when partially purified PA toxin is chromatographed on Sephadex G-150 two peaks of ADP-Ribosyl transferase activity are found. The larger form corresponds to a molecular weight of about 66 000, and the smaller form is about half this size. We have not yet succeeded in converting the toxin into the smaller form using trypsin,

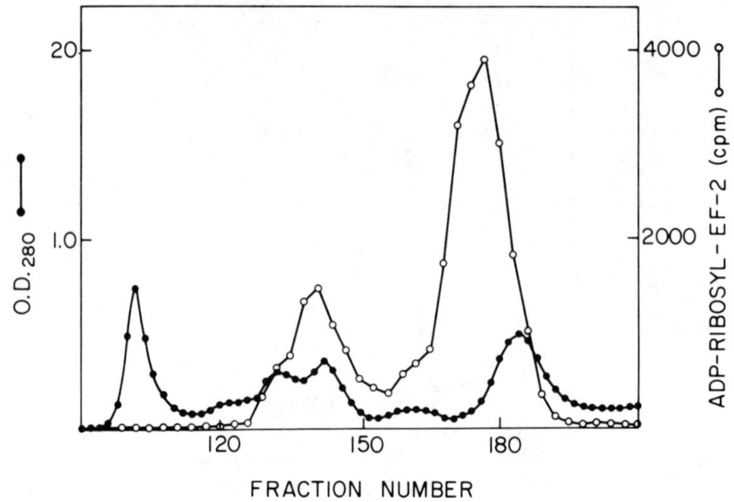

Fig. 2.12 Fractionation of partially purified PA toxin on Sephadex G-150. Enzymically active protein from DEAE-cellulose was chromatographed on Sephadex G-150 (SF) (1.5 cm x 70 cm column) equilibrated with 10 mM Tris-HCl, pH 7.6, and the fractions were assayed for enzymic activity. Two peaks of activity are apparent, the one on the left being the higher in molecular weight.

chymotrypsin, subtilisin, or pronase. However, after prolonged storage of the toxin in the frozen state, considerable degradation to the smaller form is noted. Also, if extracts from *P. aeruginosa* strain PA 103 (ribosome-free supernatant fraction) are incubated for several hours, the endogenous 66 000 dalton form of toxin which is present is converted into the lower molecular weight form, as judged by chromatography on Sephadex G-150. Presumably a bacterial protease of unknown specificity is responsible for the conversion in both cases.

We also have data suggesting that the smaller form has a higher specific activity than the intact PA toxin, but precise numbers are not yet available. The 33 000 dalton form has been shown to have NAD glycohydrolase activity, like Fragment A, but the intact PA toxin has not yet been tested.

2.3 CONCLUDING REMARKS

Much remains to be learned about the structures and activities of both the diphtheria and PA toxins. Given our basic knowledge of their enzymic activities, it will be particularly important to investigate further the mechanisms by which these proteins attach to and enter cells. As stated earlier, it is likely that these are the steps which primarily control sensitivity or resistance to the toxins. The information obtained may be relevant to much broader questions, such as the mechanisms of virus infection and of cell–cell communication in development.

Further studies on the toxins may also bring certain practical benefits. Diphtheria is essentially under complete control in countries with advanced medical care, owing to the effectiveness of immunization with diphtheria toxoid, but means for increasing immunity to infections by *P. aeruginosa* is an area for exploration. Development of effective methods of actively or passively immunizing against PA toxin could be of significance. Unfortunately there is no known mechanism for preventing or reversing the ADP-Ribosylation reaction within cells, and there seems to be little likelihood of finding one in the near future.

The production by two unrelated pathogenic bacteria of toxic proteins with identical and highly unusual activities raises many interesting academic questions. One of these is, of course, the evolutionary origins of the toxins. Divergent evolution from a common precursor seems most likely, as suggested by Iglewski and Kabat (1975). This argument rests largely on the evident improbability of convergence of separate proteins to an activity which results in transfer of the same moiety of the same co-factor to the same amino acid side chain of only one among at least several thousand different proteins in eukaryotic cells. The improbability of this is further enhanced by the apparent rarity of similar reactions in nature. The only know instance of ADP-Ribosyl transfer from NAD to a protein in prokaryotes is the ADP-Ribosylation of the α subunit of *E. coli* RNA polymerase following infection by bacteriophage T_4 (Goff, 1974a, b). In eukaryotes, attachment of ADP-Ribose to a nuclear protein may precede the formation of poly ADP-Ribose, but there is no indication that this reaction is related to that catalyzed by the two toxins.

Hard clues to the presumed precursor of the two toxins are nil. Because the structural gene for diphtheria toxin resides on a phage genome it is tempting to infer a relationship to the T_4 enzyme which ADP-Ribosylates RNA polymerase, but the polymerase is not ADP-Ribosylated by Fragment A from diphtheria toxin, and we have no further evidence concerning the possible relationship. Nothing is known about the site of the gene for PA toxin.

REFERENCES

Andrewes, F.W., W. Bulloch, S.R. Douglas, G. Dreyer, A. D. Gardner, P. Fildes, J.C.G. Ledingham, and C.G.L. Wolf. (1923), *Diphtheria, Its Bacteriology, Pathology,* and *Immunology,* Medical Research Council, London.

Bonventre, P.E., C.B. Saelinger, B., Ivins, C. Woscinski, and M. Amorini, (1975), *Infect. Immun.* **11**, 675–684.

Callahan, L.T. (1974), *Infect. Immun.* **9**, 113–118.

Clarke, P.H., and M.H. Richmond (Eds.) (1975), *Genetics* and *Biochemistry of Pseudomonas,* Wiley, London.

Collier, R.J. (1967), *J. Mol. Biol.* **25**, 83–98.

Collier, R.J. (1975), *Bacteriol. Rev.* **39**, 54–85.

Collier, R.J., and H.A. Cole (1969), *Science* **164**, 1179–1182.
Collier, R.J., R.J. DeLange, R. Drazin, J. Kandel, and D. Chung (1974a) in M. Harris (Ed.), *Poly (ADP-Ribose)*, U.S. Govt. Printing Office, Washington, D.C.
Collier, R.J., R. Drazin, J. Kandel, R.J. DeLange, and D. W. Chung. (1974b) in E. Fischer, E. Krebs, H. Neurath, and E. Stadtman (Eds.), *Metabolic Interconversion of Enzymes*, Springer-Verlag, Berlin.
Collier, R.J., and J. Kandel (1971), *J. Biol. Chem.* **246**, 1496–1503.
Collier, R.J., and A.M. Pappenheimer, Jr. (1964a), *J. Exp. Med.* **120**, 1007–1018.
Collier, R.J., and A.M. Pappenheimer, Jr. (1964b), *J. Exp. Med.* **120**, 1019–1039.
DeLange, R.J., R.E. Drazin, and R.J. Collier (1976), *Proc. Nat. Acad. Sci.* **73**, 69–71.
Drazin, R., J. Kandel, and R.J. Collier (1971), *J. Biol. Chem.* **246**, 1504–1510.
Duncan, J.L., and N.B. Groman (1969), *J. Bacteriol.* **98**, 963–969.
Eaton, M.D. (1936), *J. Bacteriol.* **31**, 347–383.
Freeman, V.J. (1951), *J. Bacteriol.* **61**, 675–688.
Gabliks, J., and M. Solotorovsky (1962), *J. Immunol.* **88**, 505–512.
Gill, D.M., and L.L. Dinius (1971), *J. Biol. Chem.* **246**, 1485–1491.
Gill, D.M., and L.L. Dinius (1973), *J. Biol. Chem.* **248**, 654–658.
Gill, D.M., and A.M. Pappenheimer, Jr. (1971), *J. Biol. Chem.* **246**, 1492–1495.
Gill, D.M., A.M. Pappenheimer, Jr., R. Brown, and J.J. Kurnick (1969), *J. Exp. Med.* **129**, 1–21.
Goff, C.G. (1974a), *J. Biol. Chem.* **249**, 6181–6190.
Goff, C.G. (1974b) in E. Fischer, E. Krebs, H. Neurath, and E. Stadtman (Eds.), *Metabolic Conversion of Enzymes*, Springer-Verlag, Berlin.
Goor, R.S., and E.S. Maxwell (1970), *J. Biol. Chem.* **245**, 616–623.
Goor, R.S., and A.M. Pappenheimer, Jr. (1967a), *J. Exp. Med.* **126**, 899–912.
Goor, R.S., and A.M. Pappenheimer, Jr. (1967b), *J. Exp. Med.* **126**, 913–921.
Goor, R.S., A.M. Pappenheimer, Jr., and E. Ames (1967), *J. Exp. Med.* **126**, 923–939.
Goto, N., I. Kato, and H. Sato (1968), *Jap. J. Exp. Med.* **38**, 185–192.
Honjo, T., and O. Hayaishi (1973), *Curr. Topics Cell Regul.* **7**, 87–127.
Honjo, T., Y. Nishizuka, O. Hayaishi, and I. Kato (1968), *J. Biol. Chem.* **243**, 3553–3555.
Honjo, T., Y. Nishizuka, I. Kato, and O. Yayaishi (1971), *J. Biol. Chem.* **246**, 4251–4260.
Iglewski, B.H., L.P. Elwell, P.V. Liu, and D. Kabat (1975a) in *Proc. Fourth Int. Symp. on Interconversion of Enzymes*, S. Shaltiel (Ed.), Springer-Verlag, Berlin and New York.
Iglewski, B.H., and D. Kabat (1975), *Proc. Nat. Acad. Sci.* **72**, 2284–2288.
Iglewski, B.H., M. Rittenberg, and W.J. Iglewski (1975b), *Virology* **65**, 272–275.
Ittelson, T.R., and D.M. Gill (1973), *Nature* **242**, 330–332.
Ivins, B., C.B. Saelinger, P.F. Bonventre, and C. Woscinski (1975), *Infect. Immun.* **11**, 665–674.
Johnson, W., R.J. Kuchler, and M. Solotorovsky (1968), *J. Bacteriol.* **96**, 1089–1098.
Kandel, J., R.J. Collier, and D.W. Chung. (1974), *J. Biol. Chem.* **249**, 2088–2097.
Kato, I. (1962), *Jap. J. Exp. Med.* **32**, 335–343.
Kato, I., and A.M. Pappenheimer, Jr. (1960), *J. Exp. Med.* **112**, 329–349.

Kato, I., and H. Sato (1962), *Jap. J. Exp. Med.* **32**, 495–504.
Kim, K., and N.B. Groman (1965a), *J. Bacteriol.* **90**, 1557–1562.
Kim, K., and N.B. Groman (1965b), *J. Bacteriol.* **90**, 1552–1556.
Lennox, E.S., and A.S. Kaplan (1957), *Proc. Soc. Exp. Biol. Med.* **95**, 700–702.
Liu, P.V. (1966a), *J. Infect. Dis.* **116**, 112–116.
Liu, P.V. (1966b), *J. Infect. Dis.* **116**, 481–489.
Liu, P.V. (1973), *J. Infect. Dis.* **128**, 506–513.
Liu, P.V., and H. Hsieh (1973), *J. Infect. Dis.* **128**, 520–526.
Liu, P.V., S. Yoshii, and H. Hsieh (1973), *J. Infect. Dis.* **128**, 514–519.
Loeffler, F. (1884), *Mitt. Klin. Gesundh.* **2**, 421–499.
Michel, A., and J. Dirkx (1974), *Biochim. Biophys. Acta* **365**, 15–24.
Michel, A., J. Zanen, C. Monier, C. Crispeels, and J. Dirkx (1972), *Biochim. Biophys. Acta* **257**, 249–256.
Moehring, T.J., and J.P. Crispell (1974), *Biochem. Biophys. Res. Commun.* **60**, 1446–1452.
Moehring, J.M., and T.J. Moehring (1968), *J. Exp. Med.* **127**, 541–553.
Moehring, T.J., and J.M. Moehring (1972a), *Infect. Immun.* **6**, 487–492.
Moehring, T.J., and J.M. Moehring (1972b), *Infect. Immun.* **6**, 493–500.
Moehring, T.J., J.M. Moehring, R.J. Kuchler, and M. Solotorovsky (1967), *J. Exp. Med.* **126**, 407–422.
Murphy, J.R., A.M. Pappenheimer, Jr., and S.T. deBorms (1974), *Proc. Nat. Acad. Sci.* **71**, 11–15.
Nicholson, G.L. (1974), *Nature* **251**, 628–630.
Nicholson, G.L., M. Lacorbier, and T.R. Hunter (1975), *Cancer Res.* **35**, 144–155.
Pappenheimer, A.M., Jr. (1937), *J. Biol. Chem.* **120**, 543–553.
Pappenheimer, A.M., Jr., and R. Brown (1968), *J. Exp. Med.* **127**, 1073–1086.
Pappenheimer, A.M., Jr., R.J. Collier, and P.A. Miller (1963), in M. Solotorovsky (Ed.), *Cell Culture in Study of Bacterial Disease,* Rutgers University Press, New Brunswick, N.J.
Pappenheimer, A.M., Jr., and D.M. Gill (1972) in E. Munoz, F. Garcia-Ferrandix, and D. Vasquez (Eds.), *Molecular Mechanisms of Antibiotic Action on Protein Synthesis,* Elsevier, Amsterdam.
Pappenheimer, A.M., Jr., and D.M. Gill (1973), *Science* **182**, 353–358.
Pappenheimer, A.M., Jr., and V. Randall (1975), *Proc. Nat. Acad. Sci.* **72**, 3149–3152.
Placido-Sousa, C., and D.G. Evans (1957), *Br. J. Exp. Pathol.* **38**, 644–649.
Raeburn, S., J.F. Collins, H.M. Moon, and E.S. Maxwell (1971), *J. Biol. Chem.* **246**, 1041–1048.
Richter, D., and F. Lipmann (1970), *Biochemistry* **9**, 5065–5070.
Robinson, E.A., O. Henriksen, and E.S. Maxwell (1974a), *J. Biol. Chem.* **249**, 5088–5093.
Robinson, E.A., O. Henriksen, and E.S. Maxwell (1974b) in *Poly (ADP-Ribose),* U.S. Govt. Printing Office, Washington, D.C.
Robinson, E.A., and E.S. Maxwell (1972), *J. Biol. Chem.* **247**, 7023–7028.
Solotorovsky, M., and J. Gabliks (1965) in M. Solotorovsky (Ed.), *Cell Culture in Study of Bacterial Disease,* Rutgers University Press, New Brunswick, New Jersey.
Strauss, N. (1960), *J. Exp. Med.* **112**, 351–359.

Strauss, N., and E.D. Hendee (1959), *J. Exp. Med.* **109**, 144–163.
Tsugawa, A., Y. Ohsumi, and I. Kato (1970), *J. Bacteriol.* **104**, 152–157.
Uchida, T., D.M. Gill, and A.M. Pappenheimer, Jr. (1971), *Nature New Biol.* **233**, 8–11.
Uchida, T., A.M. Pappenheimer, Jr., and R. Gregory (1973a), *J. Biol. Chem.* **248**, 3838–3844.
Uchida, T., A.M. Pappenheimer, Jr., and A.A. Harper (1972), *Science* **175**, 901–903.
Uchida, T., A.M. Pappenheimer, Jr., and A.A. Harper (1973b), *J. Biol. Chem.* **248**, 3851–3854.
Uchida, T., A.M. Pappenheimer, Jr., and A.A. Harper (1973c), *J. Biol. Chem.* **248**, 3845–3850.

3 Colicin E3 and Related Bacteriocins: Penetration of the Bacterial Surface and Mechanism of Ribosomal Inactivation

I. B. HOLLAND

Acknowledgements
I wish to acknowledge the many important contributions made to the work described here by numerous colleagues since work in this field was initiated in this laboratory in 1966. I also wish to thank Valerie Darby for her critical reading of, and useful comments on, the final manuscript, and Sheila Mackley for her patience and care in typing the manuscript.

The Specificity and Action of Animal, Bacterial and Plant Toxins
(Receptors and Recognition, series B, volume 1)

Edited by P. Cuatrecasas

Published in 1976 by Chapman and Hall, 11 Fetter Lane, London EC4P 4EE

3.1 INTRODUCTION

Since this volume is concerned primarily with the action of toxins on eukaryotic cell surfaces, the inclusion of a chapter on toxic proteins, active only against certain bacteria, requires some justification at the outset. Such justification is provided on the grounds that obstacles encountered by large colicin molecules, seeking to reach specific intracellular targets in sensitive cells, are likely to be fundamentally the same as those shared by many animal and plant cell toxins. In addition, the ease of handling and the ability to apply to complex problems, powerful techniques such as genetic analysis or specialized molecular biological methods, makes the study of colicin–bacteria interactions an excellent model system for studies in more complex organisms.

Before turning to the specific properties of colicin E3, and by way of introduction to the peculiarities of the envelope of its target bacteria (the Enterobacteriaceae), the detailed nature of this structure will be carefully examined (see [1, 2]). At first sight this is an extremely complex, multifunctional structure with several features unique to bacteria. These features apart, it is nevertheless possible to identify several specific membrane structures which display the same fundamental properties as those found in the eukaryote cell surface.

3.2 ORGANIZATION OF THE *E.COLI* ENVELOPE

In Fig. 3.1 the complete cell envelope of *Escherichia coli* is presented in diagrammatic form; attention is first drawn to the murein or peptidoglycan layer, which is composed of many longitudinally arranged chains frequently crosslinked through peptide side chains to form a rigid net-like structure which is responsible for the maintenance of the characteristic rod shape of *E. coli*. Beneath this layer lies the cytoplasmic, inner or plasma membrane, identical in gross structure with the plasma membrane of higher organisms. Functionally, however, this layer is extremely complex, combining the functions of mitochondrial, endoplasmic reticulum, and plasma membranes as well as providing the equivalent of the mitotic spindle. This functional complexity of the bacterial plasma membrane is reflected in turn by the complexity of its polypeptide content (see below). External to the peptidoglycan skeleton, and tightly bound to it through numerous specific lipo-protein cross-links, is the chemically complex outer membrane. This layer, whilst not as impermeable as the plasma membrane, nevertheless appears to be a continuous structure which constitutes in this bacterial species an important barrier to many toxic molecules, including antibiotics and detergents. Several factors contribute to the

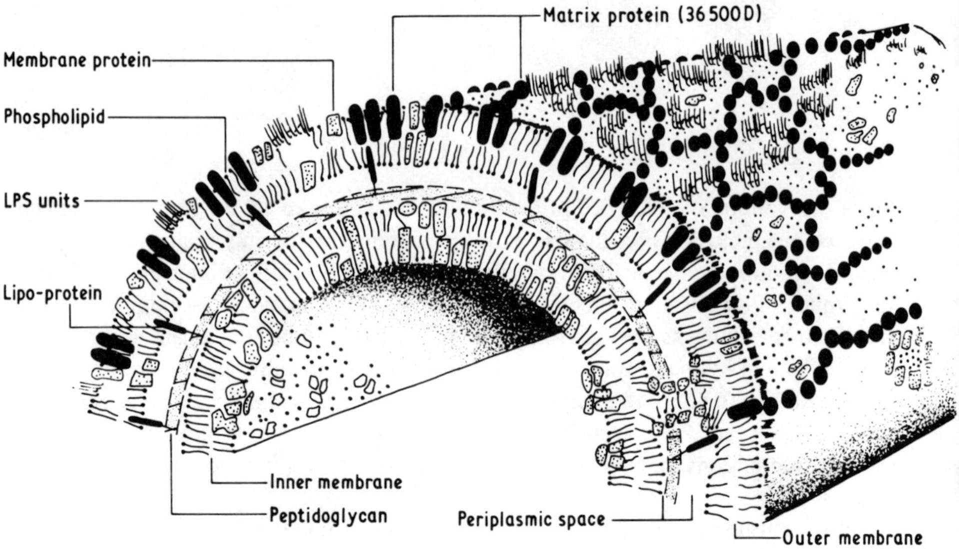

Fig. 3.1 Diagrammatic representation of the envelope of *Escherichia coli* showing inner and outer membranes separated by peptidoglycan which forms a rigid but loose mesh-like cage around the inner membrane. Also represented is a point of adhesion between the two membranes; the precise structure of these regions is not known but they appear to provide some kind of channel of communication between the inner membrane and the outer surface. Colicin receptors are located in the outer membrane. The outer membrane is dominated by one or two major species of protein which appear to form an interacting network throughout this layer.

permeability of this layer but their relative significance has not yet been fully elucidated. The basic structure appears to be a classical lipoprotein membrane, interspersed with very many lipopolysaccharide (LPS) residues, whose lipid termini are located in the external face of this layer. Considerable crosslinking may occur between LPS chains but there is no evidence to suggest covalent binding of LPS to protein molecules in the membrane, and in fact both glycoproteins and glycolipids are quite rare in bacteria. The presence of large amounts of LPS in the outer membrane of *E. coli*, and its consequent effect upon the density of this layer, provides the basis for a convenient method [3] of density gradient separation of the outer membrane from the less dense inner membrane (Fig. 3.2). As shown in Fig. 3.3 the protein content of the two membranes is quite different and the relatively simple profile of the outer membrane is dominated by one or two major or matrix proteins. Such proteins appear to form some kind of network structure involving considerable protein–protein interaction [4]. It is not yet clear, however, how this network is organized with respect to the plane of the outer membrane bilayer. If, as depicted in Fig. 3.1, these matrix proteins penetrate the bilayer to a considerable

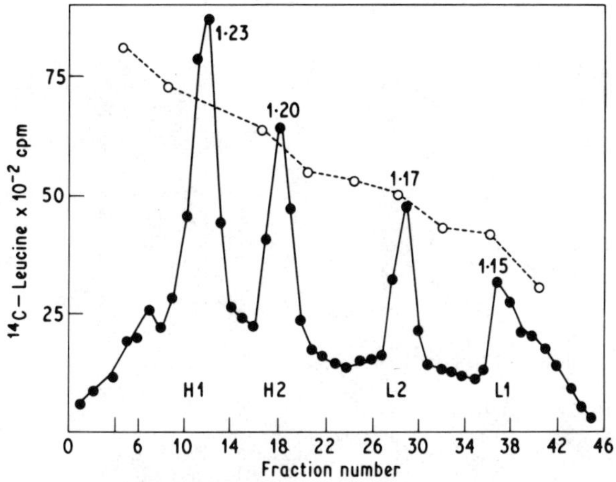

Fig. 3.2 Equilibrium density gradient separation of inner and outer membranes of *E.coli* B/r on a discontinuous sucrose gradient. Bacteria grown in the presence of [^{14}C]leucine are harvested and envelopes isolated from broken cells by differential centrifugation [26]. The outer membrane fractions H1 and H2 sediment to the heavier position because of the presence of LPS units which are absent from the inner membrane fractions L2 and L1. H1 and L1 represent the purer of the outer and inner membrane fractions respectively. Figures by the peaks are the densities of the corresponding fractions and the dotted line indicates the sucrose concentration in the gradient.

extent, then long-range lateral movements of the outer membrane components could be greatly restricted. This is an important point in this discussion since most bacterial viruses and colicin molecules, including colicin E3, bind to specific receptors located in the outer membrane as the essential first step to penetration of the surface envelope. The plasma membrane, in contrast, does not appear to contain such matrix proteins and there is no evidence at the present time of any particular mechanism for restricting free lateral movement of membrane components. An interesting feature of the envelope of *E.coli* is the presence of small tube-like structures which appear to connect inner and outer membranes in some way [5]. These areas of adhesion are distributed throughout the envelope surface and are very probably the sites of export of LPS units to the cell surface [6] and also the preferred port of entry of many viral DNA molecules [5]. Whether such structures play any specific role in the 'uptake' of colicin molecules like E3 which have specific intracellular targets is not clear at the moment.

As already indicated, the outer membrane of *E.coli* undoubtedly provides an additional, quite impermeable, barrier to numerous potentially toxic compounds. At the same time, this also constitutes an additional barrier to some nutrients essential to the cell, and whilst the role of numerous energy-dependent transport systems,

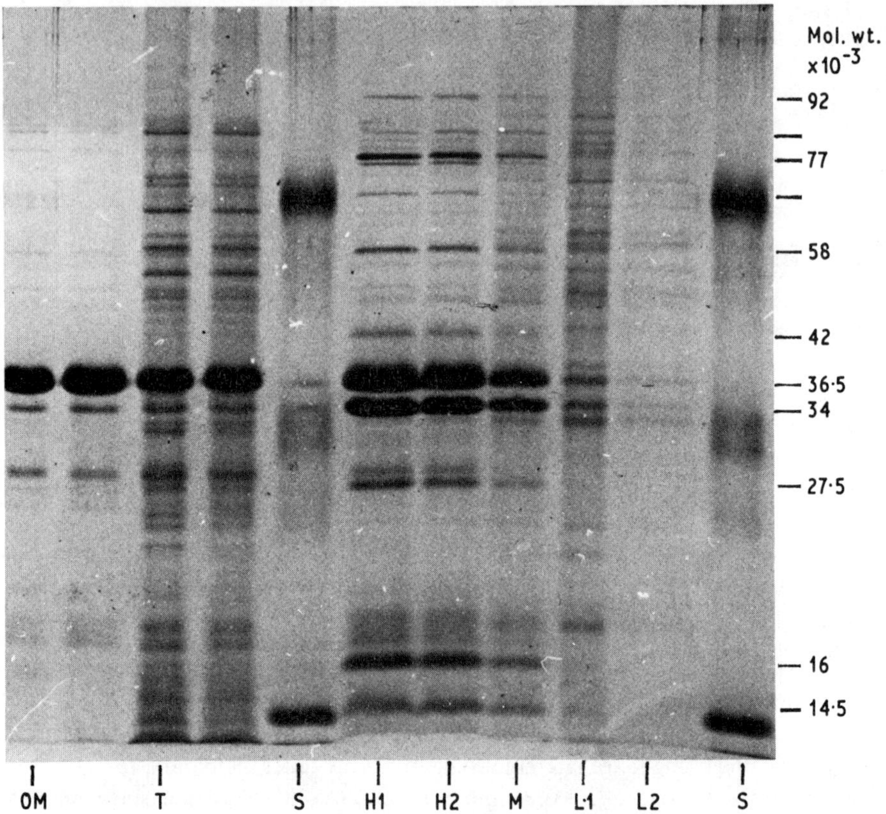

Fig. 3.3 SDS-PAGE analysis of membrane fractions (H1–L2) of *E.coli* K12 isolated as indicated in Fig. 3.2. Samples are analysed on 10% acrylamide slab-gels and polypeptides stained with Coomassie Blue [22]. Note the marked differences between inner and outer membrane polypeptides, and also the major polypeptide species with molecular weights of 36 500 and 34 000 daltons which appear in the outer membrane. OM, outer membrane of *E.coli* B/r; T, total envelope of *E.coli* B/r; S, standard proteins.

based on 'permeases' in the plasma membrane, is well established in bacteria, it is becoming clear that specific binding proteins, located in the outer membrane, may also be necessary for the first step in the uptake of these nutrients. Thus, binding proteins for ferric iron complexes [7], maltose [8], and vitamin B12 [9] have recently been identified as minor components of the outer membrane. Many colicins and bacteriophages appear to have evolved specifically to fit these binding sites, and the close similarity between the B12 and colicin E3 receptors in particular will be considered in detail later.

In the first part of this chapter the properties of the colicin receptors will be considered, and some recent developments in the study of these and other membrane proteins in bacteria will be discussed. The architecture of colicin E3 and other colicins will be described, and present ideas on how such molecules may penetrate to intracellular sites will be presented. In the second part, the mechanism of E3-induced inhibition of protein synthesis, which has recently been elucidated in quite remarkable detail, will be presented together with the new insights into the functioning of ribosomes which these and subsequent studies have brought to light. Several references will be made to a number of other colicins, particularly E2 which, although having some properties in common with E3, does not affect protein synthesis but promotes DNA degradation and inhibition of cell division in sensitive bacteria. Detailed accounts of the action of E2 and of other colicins are to be found in recent reviews [10–12].

3.3 ARCHITECTURE OF THE COLICIN MOLECULE

In order to understand some aspects of colicin action it is necessary to take account of some unusual properties of the molecule. Firstly, however, concerning the production of E3, the colicin is elaborated by bacterial strains carrying a specific small plasmid – colicinogenic factor, E3 – and, as with many other colicins, E3 synthesis is induced when DNA synthesis in the host cell is inhibited. Fully induced bacteria may contain several per cent of the total protein as colicin, and this provides for a relatively simple purification procedure. Highly purified E3 has been shown to be a water-soluble protein, with molecular weight 60 000 daltons, and free of carbohydrate [13]. Interestingly, the amino acid compositions of colicins E2 and E3 are very similar, and immunochemical analysis indicates that large portions of the two molecules are in fact identical [13]. This no doubt reflects their binding to a common receptor since, as noted earlier, the two colicins ultimately have quite distinct effects on sensitive bacteria. From studies of purified colicin E3, contradictory estimates of molecular weight by SDS-PAGE, on the one hand, and from equilibrium sedimentation analysis, on the other hand, have been obtained, indicating that the colicin is an extremely asymmetric molecule. This in fact is the general rule amongst colicins, and calculations from physical studies indicate that E3 itself is a relatively needle or plate shaped molecule with a major axis perhaps 8–9 times that of the minor axis (see [14]). This feature, as shown in Fig. 3.4, should greatly facilitate penetration of the envelope in order ultimately to interact directly with a specific region of the ribosomes as will be discussed later.

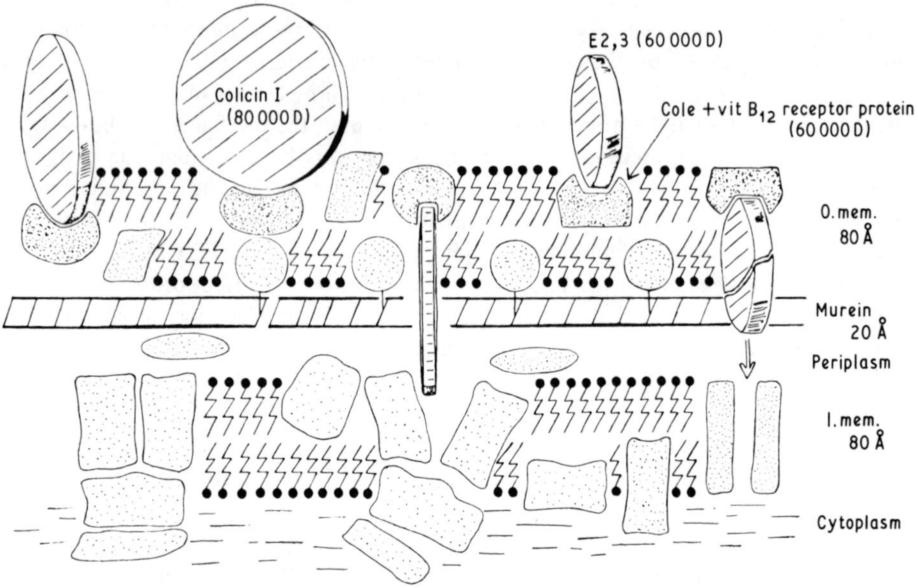

Fig. 3.4 Pictorial view of a cross-section of the *E.coli* envelope drawn approximately to scale and to shape. The LPS residues in the OM have been omitted for clarity. A specifically bound E3 (or E2) molecule is shown on the right, and one possible mechanism of entry – fragmentation and translocation through a specific energized gate – is depicted. Illustration of the mode of transition of the initial E3-receptor complex to one coupled with the inner membrane is not attempted and the diagram is not meant to imply that this proceeds via a rotational diffusion mechanism. Such a mechanism appears most unlikely on thermodynamic grounds. As discussed in the text, this transition most likely involves diffusion of colicin in the membrane and/or the formation of an intermediate complex before final coupling with the inner membrane is achieved.

3.4 NATURE AND LOCALIZATION OF E3 RECEPTORS

A combination of genetical and biochemical studies have led to the identification of a specific receptor for colicin E3 (and the related colicin E2) in the outer membrane of sensitive bacteria. Thus, colicins E2 and E3 compete for the same binding site in sensitive bacteria [15]. Wild-type strains in fact bind 2000–3000 molecules of E3 or E2 per cell [15], but E3-resistant mutants may be isolated which fail to bind detectable amounts of either protein [16]. Many such mutants have been isolated, and all mutations have been shown to map at one specific locus and to constitute a single complementation group [17]. This locus is very probably the structural gene for the major constituent of the E3 and E2 receptor; in confirmation of this, Schnaitman and his co-workers have shown that an E3 specific binding complex, extracted from the *E.coli* envelope, is absent from envelopes of resistant strains [18, 19]. This complex has been extensively purified; it consists predominantly of a single, hydrophobic protein of molecular weight 60 000 daltons and containing no more than one molecule of carbohydrate per molecule of protein [19]. This protein is located exclusively in the bacterial outer membrane and probably forms part of a larger complex which constitutes a specific receptor for another colicin, E1, and the bacteriophage BF23, as well as for colicins E2 and E3. Approximately 200–300 molecules of the E3-specific protein appear to be present per cell, indicating that perhaps several molecules of E3 bind to a single receptor in the bacterial envelope. Little is known concerning the precise location of the receptor within the outer membrane, although its resistance to trypsin treatment [15] of the intact cell suggests that it is mostly buried within that layer. Since receptors for several bacteriophages appear to be specifically clustered at the adhesion points between the inner and outer membranes [5] it is likely that the E3-receptor is similarly located.

In relation to the normal functioning of the E3-receptor De Masi *et al.* [9] have made the most interesting discovery that vitamin B12 is a competitive inhibitor of colicin E3 for binding sites in the bacterial envelope. Moreover, mutants lacking the B12 binding site are resistant to E3, and E3-resistant mutants correspondingly fail to bind and hence to transport the vitamin [20]. Normal transport of B12, however, also involves in some way the coupling of an energy-dependent, inner membrane system to the outer membrane receptor. Although, as discussed below, a late step in E3-action is also energy-dependent, this second B12 transport component is apparently not utilized by E3 since mutations blocking B12 uptake at this level have no effect on sensitivity to E3. Other transport systems involving an initial outer membrane receptor, presumably coupled to an inner membrane 'permease', have recently been identified as specific receptors for several bacterial viruses [7, 8]. In fact, utilization by viruses of elements of normal cellular transport systems is likely to be a mechanism operating throughout Nature.

3.5 ANALYSIS OF MEMBRANE PROTEINS INCLUDING RECEPTORS

Since this series is concerned with the understanding of the nature and functions of surface receptors, which in many cases are proteins, it would seem appropriate to discuss briefly some methods now in use for the routine analysis of membrane proteins in bacteria. Of necessity only a few general methods will be discussed, and these are all concerned with gel electrophoresis since this is proving particularly fruitful at the present time. The basic system involves the use of polyacrylamide gel electrophoresis in the presence of sodium dodecyl sulphate (SDS-PAGE). With appropriate supplementations to the basic technique important questions can now

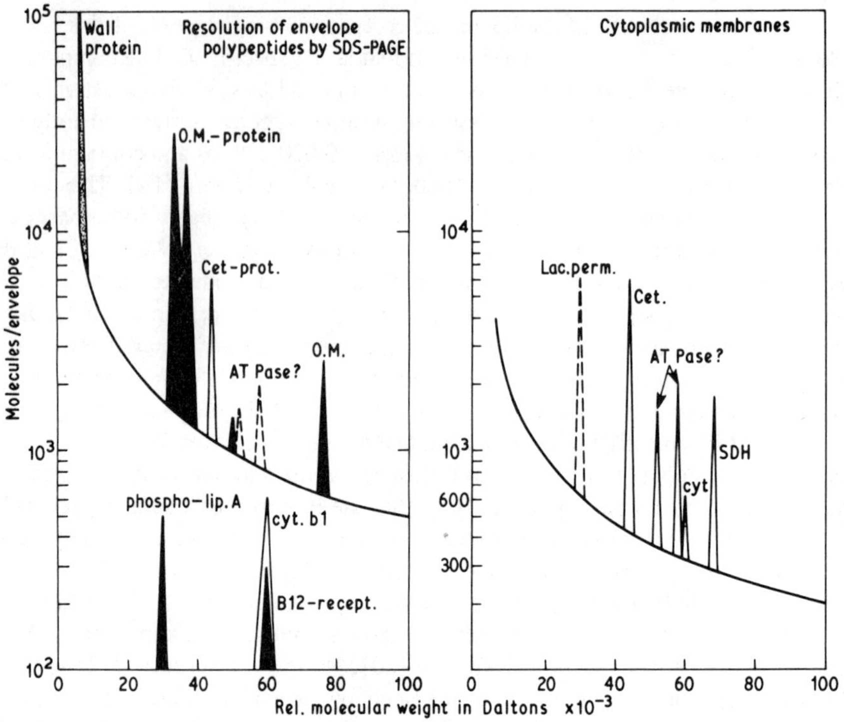

Fig. 3.5 Limits of resolution of the SDS-PAGE system. The empirical relationship is shown here between the molecular multiplicity of a bacterial membrane protein and its molecular weight. The base-line curve is fixed at a level corresponding to the minimum detection level – 0.15 μg protein per band in the SDS-PAGE system (see text). Polypeptide species sufficiently abundant to appear above this base-line will therefore be potentially detectable as a single band, whilst minor species below the base-line will not be detectable by conventional staining techniques. The specific polypeptide examples illustrated are taken from the literature and are based on reported multiplicities for the *E.coli* wall protein [25], the Cet-protein in colicin E2-tolerant mutants [26], ATPase [23], phospholipase A [27], cytochrome b1 [24], the vitamin B12 receptor [9], lactose permease [28], succinic dehydrogenase [29], and the *E.coli* major outer membrane proteins computed from Fig. 3.3. Analysis of total envelopes is shown on the left whilst on the right is shown the maximum improvement in resolution which might occur when all the outer membrane (60 per cent of the total) polypeptides have been removed and inner membrane polypeptides alone are analysed.

be asked concerning the regulation of synthesis and integration into membranes of individual polypeptides. Firstly, however, the limitations of the technique will be considered, particularly in relation to its overall sensitivity.

3.6 LIMITS TO RESOLUTION OF SDS-PAGE PROFILES

I have discussed elsewhere in this article that the colicin E3 receptor, which probably has 300 copies per bacterium, can be readily eliminated from the envelope by selecting for organisms resistant to E3. Nevertheless, whole envelope or outer membrane fractions of the mutant are indistinguishable from those of the wild-type when analysed by SDS-PAGE. This relative insensitivity of the technique derives primarily from the widely varying multiplicity (copies per envelope) of different membrane proteins in the bacterial envelpope. Thus, at the opposite extreme to the case of the E3 receptor, the major band in gel profiles of the outer membrane (Fig. 3.3) represents a single polypeptide which constitutes more than 50 per cent of the total outer membrane protein. On the basis of its molecular weight this therefore corresponds to approximately 100 000 copies per envelope in this particular bacterial species [21]. Several other envelope proteins, particularly amongst outer membrane proteins, are present in many thousands of copies per cell, and inevitably SDS-PAGE profiles are dominated by these abundant species and fail to reveal the great majority of the minor species which are also present. To some extent the ability to detect a particular level of polypeptide can be predicted in the way illustrated in Fig. 3.5. For the purposes of these estimations, calculations are based upon the favoured gel system in current use, that described by Laemmli [22] and used in combination with Coomassie Blue staining of the resolved polypeptides. The sensitivity of this system is such that 0.15 μg of protein or 0.5 per cent of the total sample analysed can be detected in a single band. On this basis, and using the relationship between molecular weight and molecular multiplicity, the base-line for detection of any particular membrane polypeptide can be deduced as shown in Fig. 3.5. Thus, individual species as abundant as ATPase [23], or the major outer membrane proteins, appear well above the base-line, whilst the E3 receptor protein or cytochrome B1 [24], in the inner membrane, are still some way below the base-line and remain undetected using conventional methods.

Somewhat improved resolutions of minor bands can of course be achieved by overloading of gels or the use, for example, of gradient gel systems [3], but real advances would need some convenient method of subfractionation of the membrane material prior to analysis. Such methods are, however, not yet available. Alternatively, a method which has recently been used most successfully to identify a variety of penicillin binding proteins in *E.coli* involves the use of a radioactive label (e.g. ^{14}C-penicillin) to tag polypeptides prior to SDS-PAGE and gel autoradiography. Using this techniques Spratt and Pardee [31] have been able to detect a number of previously unknown, but functionally extremely important, inner membrane polypeptides present in copy numbers as low as 10–50 molecules per envelope. The method of course depends upon the availability of the labelled marker and its ability to remain bound to its receptor during electrophoresis in SDS.

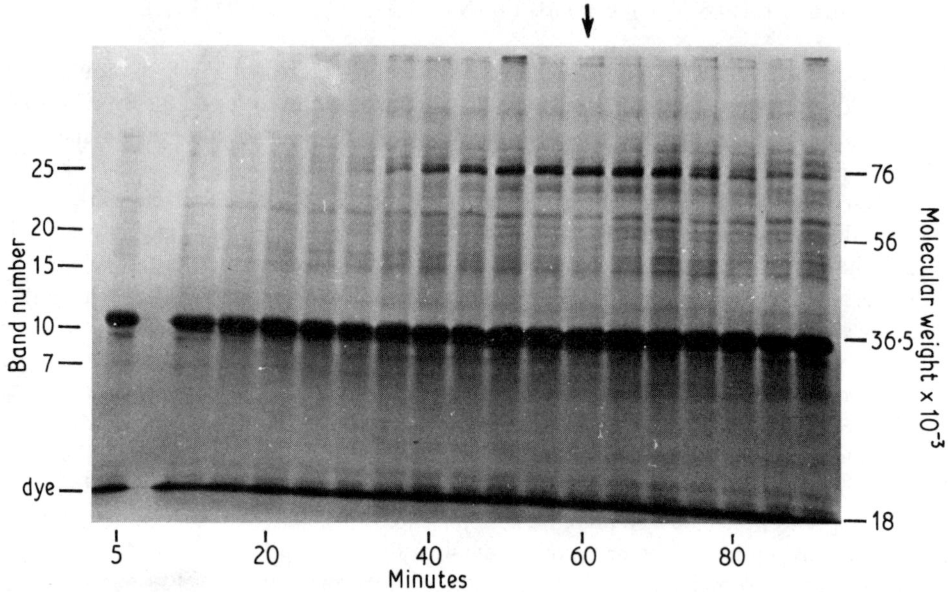

Fig. 3.6 Envelope protein synthesis in *E.coli.* A culture of *E.coli* B/r growing synchronously was pulse-labelled (5 minutes) at intervals throughout the cell cycle with [^{14}C]leucine. The isolated envelopes were then electrophoresed through a 12.5% polyacrylamide slab gel, and finally an autoradiogram was prepared [26, 32]. Note the major matrix protein at 36 500 daltons and the periodic protein at 76 000 daltons; the arrow indicates the approximate time of division.

3.7 SDS-PAGE AND THE SYNTHESIS OF MEMBRANE POLYPEPTIDES

A combination of SDS-PAGE and gel autoradiography can also be used most successfully to study the synthesis of membrane proteins under a variety of conditions by pulse-labelling cultures with ^{14}C-amino acids. This is an especially powerful technique when applied to cultures of bacteria growing synchronously, in which case the pattern of synthesis of many individual membrane polypeptides can be studied in relation to the cell cycle [32]. Such studies, as with those of Spratt and Pardee described above, have been greatly facilitated by the improvement to the autoradiography technique recently devised by Bonner and Laskey [33], which permits the ready detection of radioactive bands containing only a few radioactive counts per minute. The results of an experiment in which all these procedures are used in combination are shown in Fig. 3.6. This demonstrates the relatively simple pattern of envelope protein synthesis observed in the cell cycle of *E.coli*. The great majority of detectable polypeptides are synthesized and immediately incorporated into the surface membrane, whilst at least one polypeptide is synthesized only during a brief period, just prior to division of the bacteria. Studies of this kind, in combination with the tagging of specific polypeptides of particular functional interest, should now be applicable to both eukaryote and prokaryote systems for the study of the synthesis and the stability of membrane association of many polypeptides, including specific receptors, during the cell cycle or after modification by the action of specific effectors.

3.8 COLICIN E3 ACTION IS A TWO-STAGE PROCESS

The crucial step in the action of E3 is that which permits this large polar molecule to negotiate two impermeable barriers in order to reach its intracellular target. The precise mechanism whereby this is achieved is not clear but the outlines of the process are beginning to emerge and will now be considered.

Addition of either purified or crude preparations of E3 to exponentially growing cultures of sensitive bacteria results, after a short delay, in the complete cessation of protein synthesis (Fig. 3.7). The synthesis of other macromolecules is not affected and there is no detectable damage to the bacterial permeability barrier [34]. This strongly suggests that envelope penetration is a carefully controlled and specific process which is not accompanied, for example, by phospholipase digestion of membrane to facilitate entry. As illustrated in Fig. 3.7, a key feature of E3 action is the lag period which precedes its effect on protein synthesis. The length of the lag is dependent upon colicin multiplicity [12, 34, 35], and the bacteria may be fully rescued by trypsin digestion of the surface-bound colicin throughout this period. Thus, binding of E3 to surface receptors, which is essentially complete in 3–5 minutes at 37°C, leaves the bacterium apparently quite undamaged for several minutes before ribosome inactivation occurs. We have previously suggested [12, 36] that this delayed action effect reflects a transition from an initial inert form of bound colicin, Complex I, to the active form, Complex II, which specifically promotes ribosome disorganization. The final formation of Complex II must represent the penetration of the inner (and outer) membrane by at least a part of the colicin molecule. This rather improbable event illustrates the essence of current interest in colicin action and the unique properties which must be built into these multifunctional proteins to facilitate, first, entry into the cell and, secondly, delivery of its specific biochemical change on the ribosome. Several factors are known to influence Complex II formation and these will now be considered in turn.

3.8.1 Effect of colicin multiplicity

The primary requirement for colicin action is of course the surface receptor but unfortunately precise data about the effect of receptor concentration on Complex II

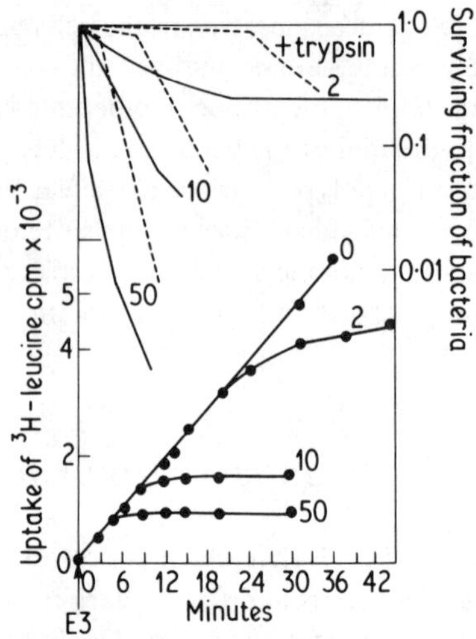

Fig. 3.7 Kinetics of Complex II formation in E3-treated cultures. *E.coli* growing exponentially in a defined medium was treated at time zero with E3 at the concentrations indicated by the curves. In the lower half of the figure protein synthesis, as measured by the incorporation of [^{3}H]leucine into acid precipitable material, is indicated. The survival of treated bacteria (continuous lines) determined by plating out for viable colonies at intervals after addition of E3 is shown in the upper half of the figure. The broken lines indicate the number of bacteria capable of forming colonies when the treated cultures are incubated with trypsin prior to plating out. The length of the trypsin rescue period defines the timing of Complex II formation for different colicin multiplicities. Note the close coincidence between the length of the trypsin rescue period and the time of onset of inhibition of protein synthesis.

formation are not available. However, as already indicated, varying the colicin multiplicity does have a marked effect on Complex II formation. Thus, it is important to emphasize that increasing the amount of E3 bound to sensitive bacteria appears to accelerate the formation of Complex II, rather than affecting the magnitude of the eventual inhibition of protein synthesis. A similar situation appears to occur with a number of colicins including E2 and also colicin K which acts to block energy metabolism [34, 36, 37]. Such relatively slow, multiplicity-dependent processes are consistent with an underlying mechanism involving lateral diffusion of colicin molecules within the bacterial surface, until interaction with a second molecule occurs which then triggers penetration [12]. Such a model is similar to that suggested to explain the early stages in the action of cholera toxin [38] as discussed in Chapter 1.

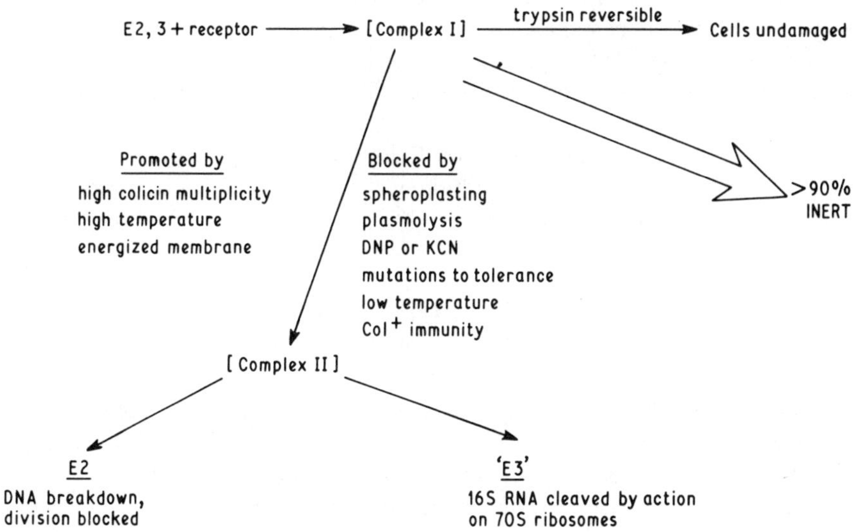

Fig. 3.8 Summary of the major factors known to influence the transition from Complex I to Complex II in colicin-treated bacteria.

3.8.2 Energy metabolism

Inhibition of protein, DNA, RNA, or peptidoglycan (cell wall) synthesis in E3-treated bacteria does not have any detectable effect on Complex II formation [12]. However, inhibition of energy metabolism during the lag period completely prevents the transition to Complex II and the bacteria remain in a trypsin resuable state [12]. Recent studies with colicins E1 and K, which specifically promote efflux of K^+ ions from bacteria, suggest that the energy required in these cases is needed to provide a suitable membrane potential, rather than a direct need for ATP [39]. This is probably true also for E3 and indicates that only a suitably energized inner membrane can be breached by the colicin molecule.

3.8.3 Other factors

The short lag period before protein synthesis is blocked by the addition of a high multiplicity of E3 may be increased 3-fold by a 10-degree reduction in temperature [12]. More specifically the effect of temperature on the action of colicins E1 and K has been correlated with the resultant variation in fluidity of the membrane [40, 41]. Unfortunately it is not possible to say whether membrane fluidity is important for penetration of the inner or of the outer membrane by the colicin. Complex II formation for E3 and E2 is also blocked by raising the osmolarity of the medium to plasmolyse the bacteria. This provides evidence that close apposition of inner and outer membranes may be required before the second step in colicin action can proceed. Finally, Complex II formation can be affected by mutations of the sensitive bacteria, and the properties of such tolerant mutants will be briefly considered in a later section.

All the various factors known to affect Complex II formation are summarized in Fig. 3.8, and a model for E3 penetration of the envelope taking account of most of these features is presented below.

3.9 INHIBITION OF PROTEIN SYNTHESIS AND SINGLE-HIT KILLING BY E3

An unusual but highly characteristic feature of all colicins is the form of the survival curves when the kinetics of killing of sensitive bacteria are determined. Since this feature must be incorporated into any model of colicin action its basis will be briefly considered. Thus, survival kinetics show that sensitive bacteria are killed by the adsorption of a single unit dose of colicin or, in other words, as a result of a single hit (see [10). Therefore, if the fraction of surviving bacteria (N) is determined at different times after treatment, the colicin multiplicity (m) in killing units (KU), can be calculated from the equation $N/N_0 = e^{-m}$. It is important to stress that single-hit killing suggests that colicin action is non-cooperative and that in theory a single molecule is capable of killing a single bacterium. Nevertheless, as already indicated, cells may adsorb considerably more than this amount of colicin before a lethal hit takes place. In fact, the number of KU, and therefore the probability of a single molecule producing a lethal hit, varies widely with the specific activity of the colicin preparation, the cultural conditions, and the inherent sensitivity of the target cells. Under the best conditions, 1 KU may contain as few as 5–50 molecules of colicin but usually the probability of achieving a lethal hit is nearer 1 in a 100 molecules. At the same time, as originally shown by Maeda and Nomura [15], more than 60 per cent of radioactive colicin remains accessible for digestion by trypsin in killed cells, and in addition the colicin can be recovered, still in the envelope fraction of sonicated cells [15]. Thus, only a small minority of adsorbed colicin molecules appear to penetrate the envelope to form Complex II and therefore achieve a lethal hit. Single-hit killing has particular implications for colicin E3 since, as will be described below, inactivation of all the 2000–3000 ribosomes of the cell can occur after formation of Complex II, although this may only involve one or two colicin molecules. Therefore, colicin E3 action at the intracellular level must be of a catalytic nature.

3.10 A MODEL FOR PENETRATION OF THE BACTERIAL ENVELOPE BY E3

In considering models for the penetration of the bacterial envelope by colicin E3, several facts must be accommodated. The most important of these are the essential role played by the receptor, the energy requirement and time dependency of Complex II formation, the low probability per adsorbed molecule of forming Complex II, the absence of any generalized membrane damage during passage of the colicin, and the elongated form of the colicin molecule. Finally, of course, in the case of colicin E3 we must assume that the E3 molecule (or an active fragment of it) penetrates not only the outer but also the inner membrane, since E3 blocks protein synthesis *in vitro* through the inactivation of ribosomes (see below). Reinforcement of this supposition has recently been obtained by Lau and Richards [42] who showed that colicin E3 bound to agarose beads, although still binding to sensitivie bacteria, no longer produced killing effects. In contrast, colicin E1, whose target appears to be the inner membrane itself, is still active under these conditions.

The importance of highly specific outer membrane receptors for later stages of colicin action is emphasized by the finding that bacterial spheroplasts, in which the close apposition of inner and outer membranes is disrupted [43], although still adsorbing colicin E3, do not undergo inhibition of protein synthesis. Similarly, in plasmolysed bacteria, in which the inner and outer membrane have become physically separated, colicin is still adsorbed but Complex II formation is blocked [12, 34]. These results indicate that specific coupling of components of both membranes is required before penetration of the envelope by the receptor-bound colicin E3 can take place. With regard to the actual mechanism of penetration of the outer membrane, the properties of this layer do not rule out the possibility that sufficiently large pores, perhaps at the sites occupied by LPS clusters or at the adhesion points between inner and outer membranes, are available to permit, in the presence of the appropriate receptor, the passage of colicin molecules of 10–20 Å in diameter. Studies by Davies and Reeves [44], with 18 different colicins differing both in their receptors and in their ultimate effects on sensitive bacteria, have shown that mutants selected for insensitivity or tolerance (see below) to an individual colicin are usually cross-tolerant to many other colicins. This is in contrast to mutations to resistance (loss of receptor), which are much more exclusive, reflecting the specificity of individual receptors. In fact, cross-tolerance patterns are limited to two groups, with cross-tolerance the rule within a group but not being observed between the groups. As discussed below, most mutations to tolerance affect the outer membrane, independently of the receptor, and therefore there may be two major pathways of penetration of the outer membrane. In either case, however, the role of the receptor may be envisaged as necessary for presenting the adsorbed colicin molecule in the correct orientation for eventual penetration of either or both outer and inner membranes.

In view of the elongated form of the colicin molecule it seems possible that the

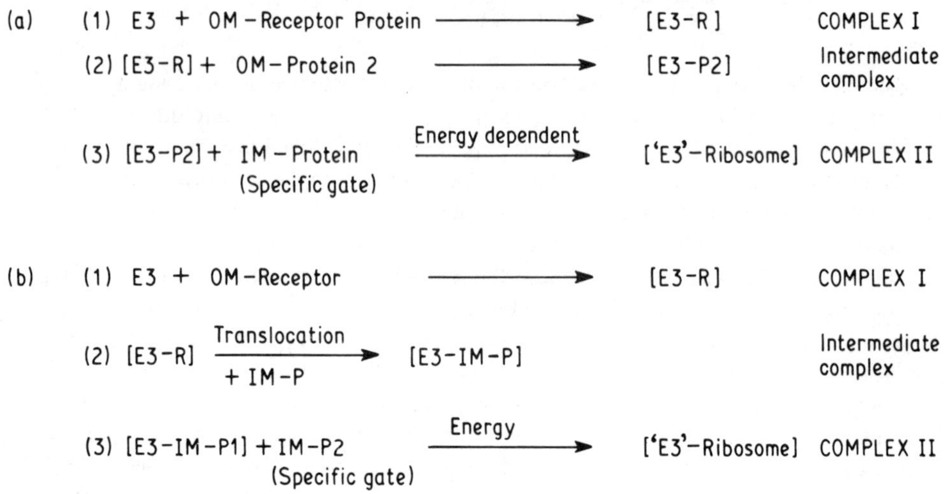

Fig. 3.9 Possible pathways of colicin action. The figure indicates the sequence of initial and intermediate complexes formed before a minority of colicin molecules penetrate the bacterial surface to form the lethal complex with ribosomes. Two basic schemes are shown but clearly others are possible including more complex ones. In (a), the colicin receptor complex is seen as diffusing within the OM until encountering a second OM protein with which it then combines to ensure its ultimate coupling with the IM gate. In (b), translocation of colicin from the OM receptor is achieved directly through binding to an IM protein. This is followed by diffusion within the IM until the intermediate complex encounters the IM gate or transport system. In both models the energy required is needed for final penetration of the IM. Both models emphasize the specificity of both the initial receptor and the IM gate for their respective colicins, whilst the intermediate complex may be relatively non-specific, there being perhaps only two alternatives for all colicins [44].

next step, interaction with the inner membrane, may take place with the receptor–colicin complex still intact. Indeed, this may be essential for the correct coupling of the colicin molecule to inner membrane constituents, which are also presumably specific for each colicin. Thus, insertion of the colicin molecule into the centre of a transmembrane cluster of proteins in the inner membrane might facilitate penetration by lateral displacement of the individual subunits of the cluster within the fluid membrane, thereby opening up a hydrophilic channel through which the large colicin molecule might then pass. Such a process would most likely require an 'energized membrane' in order to produce a suitably shaped cluster or gate for interaction and translocation of the colicin molecule (see also [45]).

The time dependence of E3 action could be envisaged to reflect the lateral diffusion of colicin within a fluid membrane, until, with a certain probability, a second interaction takes place with a membrane protein. The formation of this new complex could then in turn trigger the next step in penetration of the envelope (see Fig. 3.9). The timing of the formation of this new intermediary complex should depend upon membrane fluidity, the number of bound colicin molecules (determined both by receptor frequency and colicin multiplicity), and the multiplicity of this second membrane constituent. At the present time it is impossible to say whether these intermediate steps should take place in the outer membrane, with colicin still attached to the receptor, or within the inner membrane, perhaps after release of colicin from the receptor (see Fig. 3.9). Furthermore, in either case the possibility remains that cleavage of E3 takes place upon formation of the intermediary complex in order to simplify subsequent passage across an impermeable barrier, as depicted in Fig. 3.4. Finally, the formal model remains the same whether eventually the colicin molecule only partially penetrates the plasma membrane or whether complete penetration to the cell interior takes place. However, in view of the eventual catalytic action of an active E3 molecule upon several-thousand ribosome targets, it seems more likely that penetration of colicin is complete.

3.11 MUTATIONS AFFECTING SENSITIVITY TO COLICIN

As we have seen, the action of colicins is clearly a multistep process, with binding to the surface receptor followed by penetration of the cytoplasmic membrane and finally the specific and usually lethal modification of the target structure. Such a stepwise process, initially described in the relatively simple terms of the Complex I to Complex II transition, but now envisaged in rather more detail as shown in Fig. 3.9, should be amenable to analysis by the isolation of mutants blocked at intermediate points in the pathway, including the final target.

The isolation of mutants blocked at the first step, as already discussed, has greatly assisted the identification of the specific receptor-protein complex in the bacterial outer membrane. Elucidation of subsequent steps in the process has not however been so successful, despite the isolation of a large number of potentially interesting mutants. These mutants, which retain the capacity to bind colicin (e.g. colicin E3 or E2 or both), but remain insensitive to its effect, are termed tolerant (see Fig. 3.10). The general properties of these mutants have been summarized elsewhere [12] and will only be briefly considered here. At least nine distinct loci have been identified in *E.coli*, mutations of which confer tolerance to E3 and invariably to at least one other colicin, usually the structurally related E2. No mutants exclusively tolerant to E3 have been reported and hence none are target, i.e. ribosomal, mutants. This is perhaps not surprising since receptor negative and some E3-tolerant mutants occur at high frequency (1 per 10^6–10^7 bacteria) and are therefore likely to mask ribosomal mutations which probably occur only

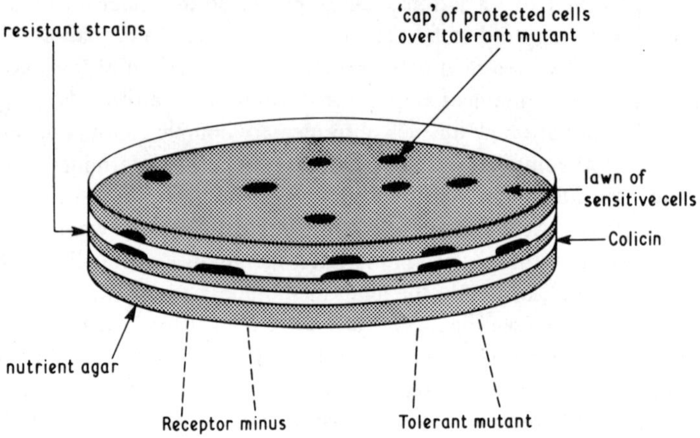

Fig. 3.10 Plate receptor test for distinguishing colicin-tolerant mutants from receptor negative mutants of *E.coli*. After isolation of strains resistant to E2 or E3, individual macrocolonies are first grown on nutrient agar over a thin layer of water agar containing colicin. The bacteria are then killed by ultraviolet irradiation or chloroform and overlayered with water agar. Finally, this last layer is itself overlayered with nutrient agar containing a low-density inoculum of colicin-sensitive bacteria and the plate is incubated overnight. Diffusion of colicin from the lower to the upper layer then kills the sensitive bacteria unless its passage is blocked by adsorption to receptors in the macrocolony of a tolerant strain. Thus, a small cap of growth of bacteria forms in the top layer over any tolerant mutant but not over receptor minus strains.

extremely rarely. A characteristic feature of all classes of E3-tolerant mutants, including some which are also conditional lethal mutants (i.e. temperature sensitive for growth), is that they have leaky outer membranes and as such they may be tolerant for relatively non-specific reasons. These mutants almost invariably show increased sensitivity to detergents or to a number of antibiotics. In one class of mutants this feature can be correlated with defective LPS synthesis but in the others the molecular basis of the envelope changes remains unknown. Nevertheless, the properties of the mutants do suggest that the overall integrity of the outer membrane is essential if the receptor–colicin complex is ultimately to be coupled, via possible intermediate complexes, with the 'energized gate' which we postulate to be located in the inner membrane and necessary for the specific entry of E3 to the cytoplasm. Unfortunately, none of the E3-tolerant mutants has been sufficiently well characterized to say whether or not any are defective in such hypothetical inner membrane proteins. Although analysis of E3-tolerant mutants have so far been relatively unsuccessful, the properties of one class of mutants (designated Cet), exclusively tolerant to the closely related colicin E2 [16, 46, 47], encourage the view that this can still be a fruitful approach. Cet mutants do not usually manifest any indication of outer membrane alterations, and membrane analysis has, on the contrary, shown that these mutants greatly over-produce a specific inner membrane protein [26, 48], when compared with the wild type strain. Genetic studies of Cet mutants [46] indicate that the synthesis of the 'Cet' protein is probably severely repressed under most conditions in the wild type but is produced constitutively in mutant bacteria. The presence of several-thousand copies of the Cet protein in the inner membrane presumably somehow blocks correct penetration of colicin E2 at low temperature where tolerance is maximally expressed in these mutants. However, final proof that the Cet protein actually interacts directly with colicin E2 in the envelope has not yet been established.

3.12 MOLECULAR BASIS OF E3 ACTION

3.12.1 Ribosome modification *in vivo*

I shall now consider the nature of the intracellular change promoted by E3 which, in combination with *in vitro* studies, clearly indicates that the colicin does indeed penetrate the inner membrane to interact directly with ribosomes. 70S or 30S ribosomal subunits extracted from sensitive bacteria treated with E3 are almost completely defective when tested in an *in vitro* protein-synthesizing system using natural messengers [49]. However, considerable activity is retained when such ribosomes are tested with the synthetic messenger poly-U [35, 49]. No other defects in the components of the protein-synthesizing machinery in extracts of E3-treated cells can be detected. Moreover, Bowman *et al.* [50] demonstrated, with ribosome reconstitution experiments, that the functional defect lay in the 16S RNA moiety and not in the proteins of the 30S subunit. In 1971, Senior and Holland [51] and Bowman *et al.* [50] showed that the 16S RNA isolated from either 30S or 70S ribosomes from E3-treated bacteria contained a single fracture about 50 nucleotides from the 3′-end of that molecule. The kinetics of the appearance of the small RNA fragment, as shown in Fig. 3.11, precisely parallels the inhibition of protein synthesis in E3-treated cultures [52]. This confirms, on the one hand, that ribosome modification does not take place during the early lag period before Complex II formation and, on the other hand, that 16S RNA fragmentation is very probably the specific and primary event which blocks protein synthesis *in vivo*. The ability of E3 to promote 16S RNA cleavage in this way appears to be independent of methylated bases in the RNA [53] and takes place whether or not the ribosomes are actually engaged in protein synthesis [35]. Finally, since from isolated 30S subunits the small E3-fragment can be extracted quantitatively without further degradation [51, 52], this strongly indicates that simple cleavage of 16S RNA is sufficient to block protein synthesis *in vivo*. The implications of this for the functional role played by 16S RNA in polypeptide bond synthesis will be considered in a subsequent section.

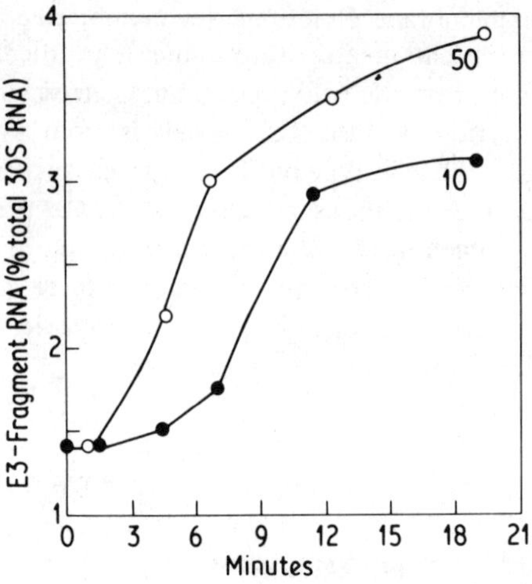

Fig. 3.11 Kinetics of appearance of the small fragment of 16S RNA (E3-RNA) produced in E3-treated cultures of *E.coli*. Samples of culture are removed at intervals, 30S ribosomes are isolated and purified, and finally the total RNA is extracted and electrophoresed through 3.5% polyacrylamide gels containing 0.5% agarose. RNA content of the individual bands produced is determined by scanning at 260 nm, and the amount of E3-RNA present at each time is expressed as the percentage of the total 30S subunit RNA. Figures by the curves are the colicin concentrations in units/ml. Comparison with Fig. 3.7 shows that the inhibition of protein synthesis in the treated cultures closely parallels the formation of the E3-RNA fragment.

3.12.2 Ribosome modification *in vitro*

Although some early stages in E3 action may be similar to those involved in the action of, for example, cholera toxin, *in vitro* studies of E3 have clearly demonstrated that the colicin molecule itself is responsible for target modification, rather than by acting through the mediation of a 'second messenger' [54, 55]. In this respect therefore E3 action is very reminiscent of that of diphtheria toxin as described in Chapter 2.

In 1971, Boon [54] and Bowman *et al.,* [55], using highly purified colicin E3, demonstrated that cleavage of 16S could be obtained *in vitro* in the absence of envelope receptors, simply by incubating E3 with 70S ribosomes. Moreover, this *in vitro* reaction appeared identical to that obtained *in vivo* and was blocked by extracts of bacteria carrying the colicinogenic E3 but not the colicinogenic E2 factor. This is significant since colicinogenic bacteria are specifically immune to the homologous colicin, and this finding is therefore consistent with the presence of a specific immunity factor in bacterial extracts which is capable of inhibiting the *in vitro* activity of E3. Using this assay the immunity protein has been purified [56, 57] and shown to interact specifically *in vitro* with E3 but not with 70S ribosomes. It is interesting that the presence of the immunity protein does not block the action of E3 on whole bacteria. This confirms that the colicin is not irreversibly inactivated by association with this protein and that the intervention of high-affinity receptors on the bacterial surface allows E3 to proceed to its ribosomal target, quite independently of the immunity protein which is presumably left behind outside the cell.

In the early *in vitro* studies with E3, 16S RNA cleavage was usually only obtained at colicin multiplicities several orders of magnitude higher than those required *in vivo*. This implied that traces of the immunity protein were still present in the colicin preparation or that normal penetration of the bacterial envelope, via the specific surface receptor, results in the production of a modified or activated colicin molecule. In fact, recent studies by Ohsumi and Imahori [58] have shown that a factor can be isolated from normal *E. coli* which can enhance E3 activity *in vitro* at least 1000-fold under certain conditions. Furthermore, this result could be simulated to a large degree by brief treatment of E3 with trypsin. Unfortunately the activated E3 obtained by either of these treatments has not yet been characterized, but these results are certainly consistent with the view that a modified E3 molecule, analogous to the active fragment of diphtheria toxin, is normally produced upon interaction with whole bacteria, presumably in association with Complex II formation.

3.12 SPECIFICITY OF THE 16S RNA CLEAVAGE REACTION

Studies *in vitro* with colicin E3 [54, 59, 60] have also revealed that the cleavage reaction is quite remarkable in its specificity (Fig. 3.12). First of all, neither 16S RNA nor the small 30S subunit alone is a substrate for E3. Complete 70S particles are required before cleavage takes place, demonstrating that a unique structural arrangement of nucleic acid and proteins, only obtained when both subunits interact, is necessary to E3 action. Moreover, alteration of ribosome structure by addition of a variety of antibiotics, acting at quite distinct sites on the 30S subunit, is also sufficient to block E3-promoted cutting activity [53], presumably through overall distrurbance of ribosome organization. More specific effects have been obtained by examination of the protective action of t-RNA molecules added to 70S particles. These studies [61] have shown that poly-U and, to a lesser extent, natural m-RNA directed binding of t-RNA at the amino-acyl site confers considerable protection against E3 action *in vitro*, but t-RNA binding, which is directed primarily to the peptidyl-tRNA site, is much less protective. This points to a possibly specific effect of E3 on the ribosomal A site. Nevertheless, fully complexed ribosomes actively engaged in protein synthesis are still inactivated by E3 *in vitro* [62] as they are *in vivo* [35]. This suggests that, if E3 does act at or near the A site, transient exposure of this ribosomal region must take place during polypeptide synthesis, and that this is sufficient for 16S RNA cleavage to occur.

In a rather different approach, Sidikaro and Nomura [63] made the most interesting

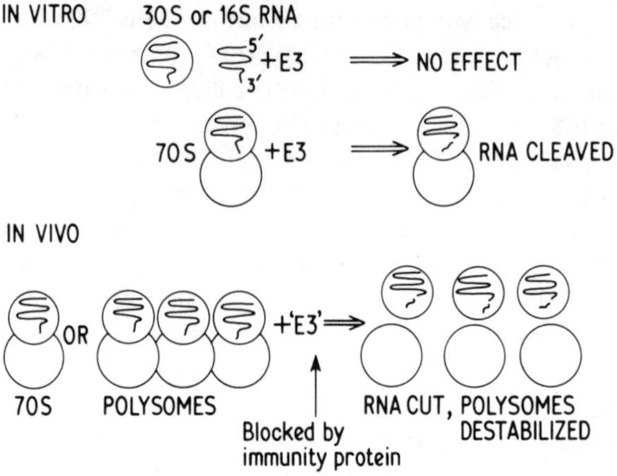

Fig. 3.12 Substrates for colicin E3 *in vitro* and *in vivo*. Cleavage *in vitro* of 16S RNA in 70S ribosomes appears identical to that which occurs *in vivo*. The low efficiency of the *in vitro* reaction may indicate that a modified form of colicin 'E3' may be the active form *in vivo*. Although polysomes are fairly stable in E3-treated cultures, they are markedly unstable compared with controls when analysed *in vitro* at low magnesium concentrations [35], i.e. conditions in which the E3-fragment is probably shed [69]. This result indicates that the 3′-terminus of 16S RNA may also be involved in the normal coupling of 30S and 50S ribosomal subunits.

observation that 70S ribosomes from *Bacillus stearothermophilus*, an organism which lacks the E3-surface receptor and has a ribosomal RNA base composition and sequence largely different from that of *E.coli*, nevertheless undergoes *in vitro* the specific E3-promoted cleavage of 16S RNA close to the 3′-terminus. This result appears to confirm that the tertiary structure of the RNA molecule in the context of ribosomal proteins, rather than specific nucleotide sequence as such, is the determining factor in the cutting reaction. Recent analysis of the base sequence at the 3′-end of 16S RNA from a number of bacteria has revealed a remarkable degree of conservation of this region [64]. Moreover, similar analysis [65] of the 18S RNA of a number of eukaryotes has also identified a conserved 3′-terminal region related in sequence to that found in 16S RNA, indicating a common functional role in protein synthesis for this region of the small subunit RNA. This raises the possibility that E3, under certain conditions at least, might be active on both eukaryote as well as prokaryote ribosomes.

Before concluding this section, the question of whether E3, or an active fragment of it, is itself an endo-ribonuclease should be considered. However, since E3-promoted cutting of RNA has only so far been observed in the presence of many ribosomal proteins, this is a difficult question to answer. Several ribonucleases which normally associate with ribosomes have been identified but the participation of any of these in E3-action appears to have been ruled out by examination of E3-induced 16S RNA cleavage in mutants lacking these various enzymes [66]. We are therefore left with the possibility that E3 itself directly causes strand cutting of a unique configuration of 16S RNA, or that E3, in conjunction with a ribosomal protein from the 30S or even the 50S subunit, actually constitutes the hydrolytic complex. A possible candidate for this role is protein 'S1' of the 30S ribosome, which, as discussed in the next section, is probably normally bound to 16S RNA close to the point of the E3-induced fracture.

3.14 E3 ACTION AND THE FUNCTIONAL ORGANIZATION OF THE RIBOSOME

Colicin E3 produces a simple yet highly specific change in its target ribosome, providing an important probe of the latter's functional organization. Studies with colicin E3 have in fact led directly to the conclusion that 16S RNA very probably plays a crucial role in polypeptide biosynthesis. As already indicated, fragmentation of 16S RNA precisely parallels inhibition of protein synthesis by E3 *in vivo*, and examination of ribosomes isolated from treated bacteria reveals no alteration other than the simple fracture in the 16S RNA. We must therefore conclude that RNA in this region, which is known to lie close to the surface of the 30S subunit [67], either participates directly in protein synthesis, e.g. as part of an mRNA recognition site, or that the functional organization of ribosomal proteins, e.g. close to the A site, requires an intact 16S RNA molecule. Some evidence for an effect of E3 on the A site has already been presented in a previous section. Now evidence will also be considered which suggests that a region of 16S RNA, between the 3′-terminus and the closely associated point of E3 induced cleavage, is directly involved in mRNA binding, and hence in the initiation of protein synthesis. Nomura and co-workers [50] first showed that ribosomes reconstituted from 16S RNA lacking the small E3-fragment failed to reassemble one protein, S1, which is located on the surface of the 30S subunit, into the reconstituted particle. Several studies have subsequently shown that S1 is normally bound close to the 3′-terminus of 16S RNA [68–70], and recently Dahlberg and Dahlberg [71] have confirmed that S1 can be specifically removed from 70S ribosomes together with the small terminal fragment after treatment with colicin E3. Moreover, *in vitro* studies showed that S1 binds reversibly to the isolated E3-fragment. Furthermore, several groups have provided evidence that both S1 and nucleotide sequences near the 3′-terminus are involved in the initiation of protein synthesis [64, 71–76]. Specifically it has been suggested that the near terminal sequence A–C–C–U–C–C, which is apparently conserved in all bacterial species so far examined, base-pairs with ribosomal recognition sequences in natural mRNA molecules on the 5′-side of the initiation codon AUG [64]. In confirmation of this, and again utilizing the specificity of the E3-nuclease, Steitz and Jakes [77] have isolated an RNA–RNA hybrid containing E3-RNA and an initiation region of bacteriophage R17-RNA after E3-treatment of ribosomes complexed with R17-RNA.

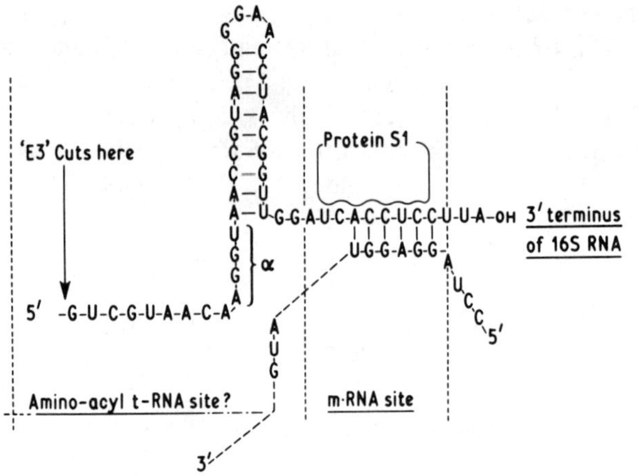

Fig. 3.13 Suggested secondary structure and functional organization of the 3′-terminal region of 16S RNA (adapted from Dahlberg and Dahlberg [71]). The complete nucleotide sequence of the E3-RNA fragment is shown, including the region of binding of the initiation specific protein S1. The presence of S1 is postulated to 'open up' the terminal RNA sequence by preventing extension of the base-paired hairpin into the α-region and thus making the sequence A–C–C–U–C–C available for base-pairing with the complementary ribosomal recognition sequences in mRNA molecules. In the example here, bacteriophage R17 RNA is the messenger. Protein S1 and part of the 16S RNA therefore constitute an important part of the mRNA binding site on the ribosome. The relative positioning and functioning of 16S RNA in the aminoacyl tRNA site are less clear. The results of studies with E3 are consistent with the model shown here but leave open the possibility that the integrity of S16 RNA is essential for direct coupling to tRNA or for the functional organization of ribosomal proteins in the A site. For the former model, note the α-region sequence which could pair with the 3′-terminal sequence, A–C–C–A, of all tRNA molecules.

A model for the role of the S1 protein in these initiation complexes and the secondary organization of the RNA in the E3-fragment has been suggested by Dahlberg and Dahlberg [71]. A modified version of this is illustrated in Fig. 3.13. This model shows that the ribosomal recognition sequence in the mRNA molecule binds to 165 RNA is close to the point of E3-fracture. Moreover, since the mRNA initiation codon, AUG, appears on the 3′-side of the ribosomal recognition sequences, the tRNA (A) site must also lie close to the point of E3-fracture. We may now consider the possible effects of E3 action on these functional centres in inhibited ribosomes. Early studies showed that poly-U directed protein synthesis with E3-ribosomes *in vitro* was still effectively carried out, demonstrating that polypeptide bond formation was still relatively normal under these conditions. On the other hand, the inability of such ribosomes to use natural messengers is consistent with a defect in the specific initiation reaction. Nevertheless, studies by Konisky and Nomura [49] appeared to show that natural mRNA binding to E3-ribosomes was relatively normal, whilst poly-U directed binding of aminoacyl tRNA was greatly reduced. This is consistent with the *in vivo* studies of Senior *et al.*, [35], which showed that pulse-labelled, nascent polypeptides are incompletely chased from polyribosomes, suggesting that at least one effect of E3 action is to block normal extension of polypeptide chains. Similar inhibition of polypeptide chain extension was also obtained by Tai and Davies [62] in an *in vitro* study of E3 action. These workers in addition obtained some evidence that phage RNA could not form functional initiation complexes with E3-treated ribosomes, suggesting an effect of E3 treatment on initiation. In order best to reconcile all these observations it seems most likely that the consequences of E3-induced 16S RNA cleavage are two-fold: first, that the colicin blocks normal mRNA binding (which might depend upon whether the cleaved 3′-terminus and its attendant S1 protein actually remain within the 70S particle), and secondly, that it also blocks the normal functioning of the aminoacyl tRNA site which is presumably close to the point of E3-induced cleavage and to the S1 binding site. Functional integrity in the former case may depend on direct binding of mRNA to the 16S RNA molecule, whilst in the latter the role of the RNA may be more indirect, providing the correct structural context for the proteins which make up the ribosomal A site. Finally, some evidence also indicates that an intact 16S RNA molecule is necessary for correct interaction of 30S and 50S ribosomal subunits. Thus, polysomes isolated from E3-treated cultures are destabilized and free 30S and 50S subunits are released when examined *in vitro* at low magnesium concentrations [35]. The reason for this phenomenon is not understood but this observation probably indicates that the 3′-terminal region of 16S RNA is located near the point of interaction of the two ribosomal subunits.

REFERENCES

1. Schnaitman, C.A. (1971), *J. Bacteriol.*, **108**, 553–563.
2. Costerton, J.W., Ingram, J.M. and Cheng, K.J. (1974), *Bacteriol. Rev.*, **38**, 87–110.
3. Osborn, M.J., Gander, J.E., Parisi, E. and Carson, J. (1972), *J. Biol. Chem.*, **247**, 3962–3972.
4. Henning, U., Höhn, B. and Sonntag, I. (1973), *Europ. J. Biochem.*, **39**, 27–36.
5. Bayer, M.E. (1968), *J. Virol.*, **2**, 346–356; *J. Gen. Microbiol.*, **53**, 395–404.
6. Mülradt, P.F., Menzel, J., Golecki, J.R. and Speth, V. (1973), *Europ. J. Biochem.*, **35**, 471–481.
7. Hantke, K. and Braun, V. (1975), *FEBS Letters*, **49**, 301–304; Braun, V. and Wolfe, H. (1973), *FEBS Letters*, **34**, 77–70.
8. Ryter, A., Shuman, H. and Schwartz, M. (1975), *J. Bacteriol.*, **122**, 295–301.
9. De Masi, D.R., White, J.C., Schnaitman, C.A. and Bradbeer, C. (1973), *J. Bacteriol.*, **115**, 506–513.

10. Reeves, P. (1973), *The Bacteriocins,* Molecular Biology Monographs II, Springer-Verlag, Berlin.
11. Luria, S.E. (1973), in *Bacterial Membranes and Walls* (L. Leive, Ed.), pp. 293–320, Dekker, New York.
12. Holland, I.B. (1975), *Advances in Microbial Physiology,* **12**, 55–139.
13. Herschmann, H.R. and Helinski, D.R. (1967), *J. Biol. Chem.,* **242**, 5360–5368.
14. Konisky, J. (1973), in *Chemistry and Functions of Colicins* (L.P. Hager, Ed.), pp. 68–80, Academic Press.
15. Maeda, A. and Nomura, M. (1966), *J. Bacteriol.,* **91**, 685–694.
16. Hill, C. and Holland, I.B. (1967), *J. Bacteriol.,* **94**, 667–686.
17. Buxton, R.S. (1971), *Molec. Gen. Genet.,* **113**, 154–156.
18. Sabet, S.F. and Schnaitman, C.A. (1971), *J. Bacteriol.,* **108**, 422–430.
19. Sabet, S.F. and Schnaitman, C.A. (1973), *J. Biol. Chem.,* **248**, 1797–1806.
20. Kadner, R.J. and Liggins, G.L. (1973), *J. Bacteriol.,* **115**, 514–521.
21. Rosenbusch, J.P. (1974), *J. Biol. Chem.,* **249**, 8019–8029.
22. Laemmli, U.K. (1970), *Nature,* **227**, 680–685.
23. Hiroshi, K. and Anraku, Y. (1972), *J. Biochem.,* **71**, 387–391.
24. Cox, G.B., Newton, N.A., Gibson, F., Snoswell, A.M. and Hamilton, J.A. (1970), *Biochem. J.,* **117**, 551–562.
25. Braun, V. and Wolff, H. (1970), *Europ. J. Biochem.,* **14**, 387–391.
26. Holland, I.B. and Tuckett, S. (1972), *J. Supramolec. Struct.,* **1**, 77–97.
27. Scandella, C.J. and Kornberg, A. (1971), *Biochem.,* **10**, 4447–4450.
28. Jones, T.H.D. and Kennedy, E.P. (1969), *J. Biol. Chem.,* **224**, 5981–5987.
29. Spencer, M.E. and Guest, J.R. (1974), *J. Bacteriol.,* **117**, 947–953.
30. Chua, N.H. and Bennoun, P. (1975), *Proc. Nat. Acad. Sci.,* **72**, 2175–2179.
31. Spratt, B.G. and Pardee, A.B. (1975), *Nature,* **254**, 516–517.
32. Churchward, G.G. and Holland, I.B. (1976), *FEBS Letters,* **62**, 347–350; and *J. Molec. Biol.,* in Press.
33. Bonner, W.M. and Laskey, R.A. (1974), *Europ. J. Biochem.,* **46**, 83–88.
34. Nomura, M. and Maeda, A. (1965), *Z. Bacteriol. Parcint. Infekt. Krank. Hygiene (Ab. I),* **196**, 216–238.
35. Senior, B.W., Kwasniak, J. and Holland, I.B. (1970), *J. Molec. Biol.,* **53**, 205–220.
36. Holland, E.M. and Holland, I.B. (1970), *J. Gen. Microbiol.,* **64**, 223–239.
37. Wendt, L.W. (1970), *J. Bacteriol.,* **104**, 1236–1241.
38. Cuatrecasas, P. (1974), *Ann. Rev. Biochem.,* **43**, 169–214.
39. Lusk, J.E. and Nelson, D.L. (1972), *J. Bacteriol.,* **112**, 148–160.
40. Cramer, W.A., Phillips, S.K. and Keenan, T.W. (1973), *Biochem.,* **12**, 1177–1181.
41. Plate, C. (1973), *Antimicrobial Agents and Chemotherapy.,* **4**, 16–24.
42. Lau, C. and Richards, F.M. (1976), *Biochem.,* **15**, 666–671.
43. Birdsell, D.C. and Cota-Robles, E.H. (1967), *J. Bacteriol.,* **93**, 427–437.
44. Davies, J.K. and Reeves, P. (1975), *J. Bacteriol.,* **123**, 96–101, 102–117.
45. Racker, E. and Hinkle, P.C. (1974), *J. Membrane Biol.,* **17**, 181–185.
46. Buxton, R.S. and Holland, I.B. (1973), *Molec. Gen. Genet.,* **127**, 69–88.
47. Holland, I.B. (1968), *J. Molec. Biol.,* **31**, 267–275.
48. Holland, I.B. and Darby, V. (1973), *FEBS Letters,* **33**, 106–108.

49. Konisky, J. and Nomura, M. (1967), *J. Molec. Biol.*, **26**, 181–195.
50. Bowman, C.M., Dahlberg, J.E., Ikemura, T., Konisky, J. and Nomura, M. (1971), *Proc. Nat. Acad. Sci.*, **68**, 964–969.
51. Senior, B.W. and Holland, I.B. (1971), *Proc. Nat. Acad. Sci.*, **68**, 959–963.
52. Samson, A.C.R., Senior, B.W. and Holland, I.B. (1972), *J. Supramolec. Struct.*, **1**, 135–144.
53. Dahlberg, A.E., Lund, E., Kjeldgaard, N.O., Bowman, C.M. and Nomura, M. (1973), *Biochem.*, **12**, 948–950.
54. Boon, T. (1971), *Proc. Nat. Acad. Sci.*, **68**, 2421–2425.
55. Bowman, C.M., Sidikaro, J. and Nomura, M. (1971), *Nature New Biology*, **234**, 133–137.
56. Jakes, K., Zinder, N.D. and Boon, T. (1974), *J. Biol. Chem.*, **249**, 438–444.
57. Sidikaro, J. and Nomura, M. (1973), *J. Biol. Chem.*, **249**, 445–453.
58. Ohsumi, Y. and Imahori, K. (1974), *Proc. Nat. Acad. Sci.*, **71**, 4062–4066.
59. Boon, T. (1972), *Proc. Nat. Acad. Sci.*, **69**, 549–552.
60. Bowman, C.M. (1972), *FEBS Letters*, **22**, 73–75.
61. Kaufmann, Y. and Zamir, A. (1975), *Europ. J. Biochem.*, **53**, 599–603.
62. Tai, P. and Davies, B.D. (1974), *Proc. Nat. Acad. Sci.*, **71**, 1021–1025.
63. Sidikaro, J. and Nomura, M. (1973), *FEBS Letters*, **29**, 15–18.
64. Shine, J. and Dalgarno, L. (1973), *Proc. Nat. Acad. Sci.* **71**, 1342–1345; *Nature*, **254**, 34–37.
65. Dalgarno, L. and Shine, J. (1973), *Nature New Biology*, **245**, 261–262.
66. Meyhack, B., Meyhack, L. and Apirion, D. (1973), *Proc. Nat. Acad. Sci.*, **70**, 156–160.
67. Santer, M. and Santer, V. (1973), *J. Bacteriol.*, **116**, 1304–1313.
68. Kenner, R.A. (1973), *Biochem. Biophys. Res. Commun.*, **51**, 932–938.
69. Dahlberg, A.E. (1974), *J. Biol. Chem.*, **249**, 7673–7678.
70. Czemilofsky, A.P., Kurland, C.G. and Stöffler, G. (1975), *FEBS Letters*, **58**, 281–284.
71. Dahlberg, A.E. and Dahlberg, J.E. (1975), *Proc. Nat. Acad. Sci.* **72**, 2940–2944.
72. Van Dieijen, G., Van der Laken, C.J., Van Kippenberg, P.H. and Van Duin, J. (1975), *J. Molec. Biol.*, **93**, 351–360.
73. Szer, W., Hermosa, J.M. and Leffler, S. (1975), *Proc. Nat. Acad. Sci.*, **72**, 2325–2329.
74. Held, W.A., Gette, W.R. and Nomura, M. (1974), *Biochem.*, **13**, 2115–2122.
75. Goldberg, M.I. and Steitz, J. (1974), *Biochem.*, **13**, 2123–2128.
76. Fiser, I., Scheit, K.H., Stöffler, G. and Kuechler, E. (1975), *FEBS Letters*, **56**, 226–229.
77. Steitz, J.A. and Jakes, K. (1975), *Proc. Nat. Acad. Sci.*, **72**, 4734–4738.

4 Abrin, Ricin, and their Associated Agglutinins

S. OLSNES *and* A. PIHL

Acknowledgement
This work was supported by the Norwegian Cancer Society.

The Specificity and Action of Animal, Bacterial and Plant Toxins
(Receptors and Recognition, series B, volume 1)

Edited by P. Cuatrecasas

Published in 1976 by Chapman and Hall, 11 Fetter Lane, London EC4P 4EE

4.1 HISTORICAL INTRODUCTION

It has been known since ancient times that the seeds of *Abrus precatorius* L. (*Semen Jequiriti*)* and *Ricinus communis* L.† (castor bean) are highly toxic, and the seeds have been used in folk medicine against a wide variety of diseases and also for various non-medical purposes. Castor beans were used in the classical Greek medicine, and both plants are described in the Sanskrit work on medicine *Susruta Ayurveda* from the sixth century (B.C.).

The plant *Abrus precatorius* (*Leguminosae*) originates from South-East Asia but is now found also in other tropical and subtropical regions. The root contains glycyrrhizin and is known as Indian liquorice. The seeds are red with a black spot and are highly glossy and decorative. They are often used as eyes in voodoo dolls produced in tropical countries, and in rosaries (pater-noster-beans), necklaces, and jewelry. In spite of this fairly wide use, intoxication in humans is infrequent, probably because the seeds when not chewed will pass the intestine without being dispersed. The uniformity in size, the hardness, the durability, and the pleasing color of the beans, which are called gumchi in Hindustani (gunja in Sanskrit), are probably the reasons for their use in South-East Asia as weights (rati) for weighing gold and jewels (1 carat = 2 seeds). The exceedingly old Ganda system of weights, which has been widely used, was actually based on multiples of 4 abrus seeds. It appears that several familiar weights can be traced back to this system: the varaha or ducat = 4^2 seeds; the shekel = 4^3 seeds; the old ounce = 4^4 seeds; the ser (India) = 4^5 seeds [1–3]. In India and Sri Lanka the seeds have been used for criminal purposes to intoxicate cattle and even people by introducing under the skin of the victims spikes (sui) prepared from dried paste from abrus seeds [4].

Abrus seeds have from ancient times been used in Asiatic medicine against a wide variety of diseases. In Arabic countries, where the seeds are known as ain-ud-deek (coc's eye), they have been used as an aphrodisiac. Extracts from the seeds have been used traditionally in South America against chronic eye diseases and were introduced in 1882 by deWecker into European medicine. For some years the Jequiriti-infus was used against chronic eye diseases and particularly against trachoma. However, in many cases a strong inflammation (Jequiriti-ophthalmia) and even complete necrosis of the eye was seen [5]. This reaction was first related to alleged

* From ἁβρος (Gr.) graceful, delicate; and *precare* (Lat.) to pray. *Jequirity* is a Brazilian (Indian) name of the plant.

† The etymology is not clear. Possibly from *kikar* (Hebr.) round; via Greek *κικινοσ*. In the classical literature the seeds are also referred to as *κροτων* (see [12]).

Jequiriti-bacilli, but it was later realized that the inflammation was of toxic origin. The protein nature of the abrus toxin was first suggested by Warden and Waddel in 1884 [6], and the toxin was characterized as an albumose by Martin in 1887 [7]. This was in fact the first described example of a toxic protein.

The castor bean plant *Ricinus communis* (*Euphorbiaceae*) originates from Asia and Africa but is now growing also in Europe and America. In tropical regions the plant grows to a high tree which is characterized by its rapid growth and sudden wilting (miracle tree). The kikaion described in the Bible [8] is assumed to have been this variant of *Ricinus communis.* The variants growing in Europe are bushes rather than trees. The castor bean was cultivated for its oil already in ancient Egypt, and castor oil has been used for a number of purposes. It is now used in medicine only as a laxative but in earlier times the oil or the whole seeds as well as other parts of the plant were used against many diseases. The high toxicity of castor beans has been known since ancient times, and at least 700 cases of intoxication in humans have been described [9]. The fact that the toxic properties of castor beans are due to a protein was first realized by Dixson in 1887 [10]. An extensive bibliography on castor bean lectins has been published by Balint [9].

The first extensive study of the castor bean toxin was carried out by Stillmark in Kobert's Laboratory in Dorpat (Tartu, Estonia) in the late nineteenth century [11, 12]. Stillmark obtained evidence that the toxin is a protein, and he suggested the name ricin. By studying the effect of the toxic fraction of castor beans on blood he observed agglutination of the erythrocytes and precipitation of serum proteins, and assumed that these effects could account for the toxic action of ricin on animals. Later Hellin [13], who was also working with Kobert in Dorpat, discovered that an extract containing the abrus toxin also agglutinated erythrocytes and precipitated serum proteins.

The assumption that cell agglutination is the reason for the toxic effect of abrin and ricin was soon challenged by the observation that the agglutinating properties of abrus and ricinus extracts could be separated from the toxic properties. Thus, when normal serum was added to the extracts a precipitate was formed. The toxic property remained in the supernatant, whereas the agglutinating properties disappeared [14, 15]. Furthermore, the agglutinating ability of ricinus extracts was more easily destroyed by treatment with pepsin-HCl than were the toxic properties [16].

The high resistance of abrin and ricin to various proteolytic enzymes had been described already by Stillmark [12] and Hellin [13]. This, together with the finding that purified and diluted but still highly toxic extracts gave no protein reactions by the methods of that time, resulted in a long-lasting discussion as to whether the toxins were in fact proteins [17–19].

Extensive immunological studies on abrin and ricin were carried out by Paul Ehrlich [20]. Since the toxins are about 100 times less toxic when given by mouth than after parenteral injection, he was able to induce a certain immunity by feeding rabbits with small amounts of seeds. Later the immunization was continued by subcutaneous injections. A similar immunization with abrus seeds had in fact been carried out in India for hundreds of years [4]. It had been recognized that if calves were fed small amounts of abrus seeds they were later resistant to abrus intoxication.

The work of Ehrlich on abrin and ricin resulted in the discovery of some of the fundamental principles of immunology. Thus he was able to show that immunization resulted in the formation of serum proteins capable of specifically precipitating and neutralizing the toxins, and that there was a quantitative relationship between the amount of antiserum and the amount of toxin it could neutralize. Also he noted that anti-ricin did not protect against abrin and *vice versa,* demonstrating that the phenomenon of immuno-specificity is associated with the antiserum. He could also show that, during pregnancy, immunity to the toxins is transferred from the mother to the offspring by the blood, and that after birth it may be transferred through the milk.

4.2 TOXIC EFFECTS

4.2.1 Effects in animals and man

Although abrin and ricin are highly toxic even when given by mouth, the toxicity is much stronger after parenteral administration. This difference appears to be due to low absorption from the intestine rather than to enzymatic degradation in the digestive tract. Nevertheless, intoxication of animals by contamination of the feed with castor beans is a problem in veterinary medicine [21–23].

Different animal species exhibit different sensitivity to the toxins. For instance, calculated on a weight basis, the guinea pig is more sensitive to ricin than the mouse [20], and the horse appears to be the most sensitive animal (reviewed in ref. [9]). A certain latent period is always observed during which there is no pathological reaction even after parenteral administration of the toxins. This lag time decreases with increasing amounts of toxin administered. After a lag of some hours the symptoms of intoxication develop, and the animals die after 10 hours or more. Even when very high amounts of toxin (several mg) are injected into mice, they do not die earlier than after about 10 hours (Fig. 4.1). After injection of high doses of toxin the most common findings are hemorrhages in the intestine, mesenterium, and omentum. A diffuse nephritis may also be seen [24]. Before death the body temperature decreases and the animals are shivering in a characteristic way. After injection of smaller doses smaller hemorrhages are seen in the ileum, and a red-spotted edema of the Peyers patches is observed. Multiple necroses appear in the liver and kidney with cytoplasmic vacuolation, pyknosis of the nuclei, and karyorrhexis [25, 26]. In the myocardium the myofibrils undergo degeneration. After subcutaneous injection of sublethal doses, necrosis and induration occur around the site of injection [22]. In the case of abrin, a characteristic alopecia is often seen around the injection site [20].

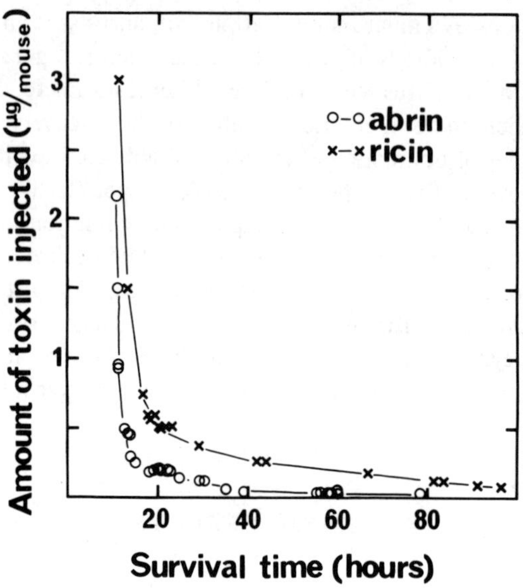

Fig. 4.1 Survival time of mice after injection of various amounts of abrin or ricin [27].

After intravenous injection of radioactively labelled abrin and ricin in mice the highest concentration is found in the liver and the spleen [27]. The toxins are degraded with a half-life of 5 hours (ricin) and 17 hours (abrin).

The body temperature appears to be important for the sensitivity of animals to abrin and ricin. Thus, guinea pigs kept at 39° and fishes kept at 29° died earlier after injection of abrin than control animals kept at lower temperatures [28]. In experiments on frogs Moriyama [29] found that the lethal dose of ricin (scored after 3 days) was 63 ng at 37° and 500 ng at 17°, whereas at 4° the frogs survived for more than two weeks even after injection of 400 μg of ricin. In contrast, the toxic effect of various snake venoms showed very small variations with temperature. Interestingly, ricin is also toxic to lower plants like *Tradescantia* and *Valisnera* [30].

As pointed out above, many cases of intoxication, particularly with castor beans, have been described in humans. One or a few beans of abrus or ricinus seeds may, when chewed and swallowed, be lethal. The most common symptoms are nausea, vomiting, pain in the stomach, colic, diarrhoea, hemorrhage from the anus, anuria, cramps, dilatation of the pupillae, fever, thirst, burning in the throat, headache, and shock symptoms. Death occurs in exhaustion or cramp. The most common autopsy findings are multiple ulcera in the stomach and the small intestine and numerous hemorrhages around these ulcera [31–34]. For forensic and other purposes the presence of abrin and ricin in biological material can be detected by its toxic effect when injected into mice, and by the ability of the specific antitoxins to protect the

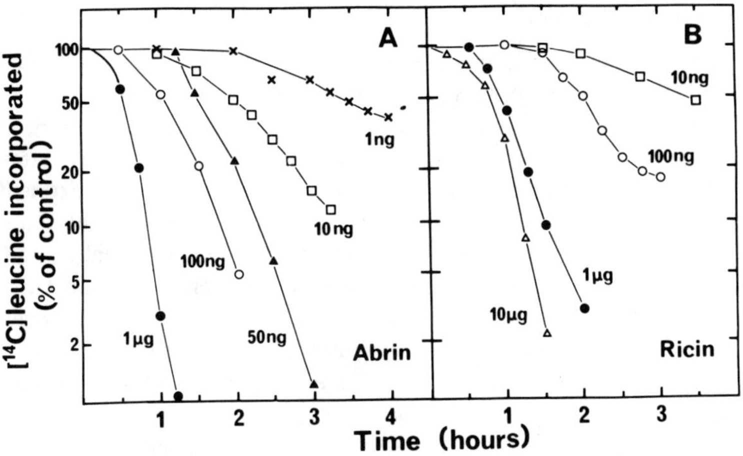

Fig. 4.2 Inhibition of protein synthesis in HeLa cells by (A) abrin and (B) ricin. Samples of HeLa cells were incubated with different toxin concentrations; after the indicated periods of time [^{14}C]leucine was added, and the amount of radioactivity incorporated during 20 minutes was measured [43].

mice against the injected material [21, 35, 36].

4.2.2 Protein synthesis inhibition in cultured cells

The earliest demonstrable effect of abrin and ricin on cells in culture is inhibition of protein synthesis as first shown by Lin *et al.,* [37, 38]. Somewhat later the DNA synthesis decreases, and still later also the RNA synthesis is inhibited. These results were confirmed in several laboratories with various animal and human cell lines [39–42]. Earlier studies had shown that the incorporation of amino acids by microsome preparations is reduced after ricin treatment [26] whereas there is no early effect of abrin and ricin on energy metabolism and oxidative phosphorylation. Thus, the data indicate that the toxic effect is primarily due to inhibition of protein synthesis [40].

After addition of abrin and ricin to the cell culture medium a certain lag time proceeds before protein synthesis inhibition is apparent in the cells (Fig. 4.2). The lag decreases with increasing toxin concentration, but even when high concentrations of toxin are added the lag time cannot be reduced below a minimum of about 20–30 minutes [41, 43].

Intoxicated cells undergo early surface changes. The surface becomes irregular, but the cells continue to exclude trypan blue for several hours after protein synthesis has ceased indicating that their surface membrane remains intact for a long period of time [42, 64].

4.2.3 Anti-tumor properties

A tumor inhibiting effect of ricin was reported by Mosinger in 1951 [44]. In 1969 Lin *et al.*, [45, 46] demonstrated a protective effect of abrin and ricin against Ehrlich ascites tumor cells in mice. Thus, abrin and ricin were reported to be able to prevent the development of ascites tumor even when administered 5 days after inoculation of the tumor cells. The toxins have also been tried in the treatment of human cancer [47], particularly uterine cancer. Local application has been used as well as intratumorous, intra- or retro-peritoneal, or intraarterial administration. Although remissions were observed they appear to have been of short duration. Few side-effects were reported. Also other authors [48, 49] have found anti-cancer properties of ricin or its subunits on Ehrlich ascites tumor and Yoshida sarcoma, but the effect was much less than that reported by Lin *et al.*, [46]. A certain protective effect against experimental leukemia was also reported [50].

Experiments carried out in this laboratory (Φ. Fodstad, unpublished) showed a certain protective effect of abrin against Ehrlich ascites cells, but the effect was less than that observed by Lin *et al.*, [45, 46]. Ricin was less effective than abrin. In all cases the best effect was seen when the amounts of toxin injected were high (about half the LD_{50} dose). Fodstad also showed that abrin had an inhibiting effect on the growth of B16-melanoma in mice, and that the treated animals survived for a 30 per cent longer period than the control animals. On the other hand, abrin did not prolong the life of mice carrying subcutaneously implanted Lewis lung carcinoma although the growth of the tumor was strongly inhibited. The reason for this discrepancy is that abrin does not inhibit the growth of pulmonary metastases of the subcutaneous tumor.

4.3 PURIFICATION OF THE LECTINS

Already the preparations of Stillmark and Hellin contained high concentrations of toxin due to the fact that abrus and ricinus seeds are very rich in these lectins. Germinating castor beans lose most of their ricin content within a few days [51]. Various purification schemes were worked out by extracting seeds with salt solutions and fractionating the proteins by the use of salt precipitation, dialysis against distilled water, alcohol precipitation, and other methods [19, 52, 53]. Crystalline ricin was first obtained by Kunitz and McDonald [54] and by Kabat *et al.*, [55]. By the use of ion-exchange chromatography Takahashi, Ishiguro *et al.*, [56, 57] purified a ricin fraction (ricin-D*) which was more toxic than any ricin preparations earlier

* The name ricin is here used for the toxic castor bean protein with molecular weight of about 65 000; it is assumed to be identical with ricin D [57] and RCA_{II} [60]. The castor bean agglutinin with molecular weight of about 120 000 has been called ricinus agglutinin; it is probably identical with RCA_{I} [60]. Abrin is the name used for the toxic abrus bean protein with molecular weight of about *(continued on page 137)*

described. They also showed clearly that toxin and agglutinin are two different proteins [57]. The first convincing separation of toxin and agglutinin from abrus seeds was done by Khan *et al.*, [58] using paper chromatography.

The purification of abrus and ricinus lectins was greatly improved with the introduction of affinity chromatography on Sepharose 4B [59–62]. Sepharose 4B, which contains β-galactosyl residues, is able to bind abrus and ricinus toxins and agglutinins, whereas the contaminating proteins pass through the column. The bound lectins can later be eluted with galactose. Since both the toxins and the agglutinins bind to the column, additional separation methods are required to separate the two lectins. This can be accomplished by gel filtration, sucrose gradient centrifugation, or ion exchange chromatography [59–63]. Nicolson *et al.*, [64] reported that ricin can be eluted from a Sepharose 4B column with N-acetylgalactosamine, whereas galactose is required for the elution of ricinus agglutinin. In our hands ion exchange chromatography gives better separation than the other methods, and larger amounts of protein can be applied. By the methods described in Figs. 4.3 and 4.4 both abrin and ricin can be obtained as pure proteins. The yield of abrin was ~ 30 mg from 40 g of seeds and that or ricin ~ 120 mg from 100 g of seeds. The LD_{50} in mice of our preparations was measured to be 13 ng of abrin and 65 ng of ricin after intravenous injection (Table 4.1). We could also show that the pure agglutinins are virtually non-toxic in mice [63], whereas purified ricinus agglutinin had some toxic effect on cells in culture [65].

In many abrin preparations some material is retarded on the Sepharose 4B column and is eluted from the column as a peak in the absence of galactose (Fig. 4.3B). This fraction is much less toxic than abrin which sticks to the column. In the Ouchterlony immunodiffusion test the two fractions show reactions of partial identity. Probably, the less toxic fraction corresponds to the abrin A described by Wei *et al.*, [66], whereas the toxin which binds to the column is probably identical with abrin C. Ricin isolated by Lugnier and Dirheimer [67] did not bind to a Sepharose 4B column, whereas ricinus agglutinin did bind. The reason for this is not clear.

The agglutinins in concentrations above 5–10 ng ml^{-1} agglutinate red blood cells, whereas abrin and ricin agglutinate cells only at 100–1000 times higher concentrations. Our data indicate that the low agglutinating ability of the pure toxins is due to dimer formation of the toxin molecules rather than to contamination by the agglutinins [63].

When abrin and ricin are reduced with 2-mercaptoethanol the interchain disulfide bond is split. In simple buffer a considerable part of the chains precipitates upon reduction. We were able to prevent this by adding galactose or lactose to the buffer

(Continued from page 136)

65 000; it is probably identical with abrin C [66]. The abrus seed agglutinin with molecular weight of about 130 000 has been named 'abrus agglutinin'; it is probably identical with the abrus lectin described by Wei *et al.*, [76]. The term *lectin* refers in this review to the monovalent abrin and ricin as well as to the divalent abrus and ricinus agglutinin.

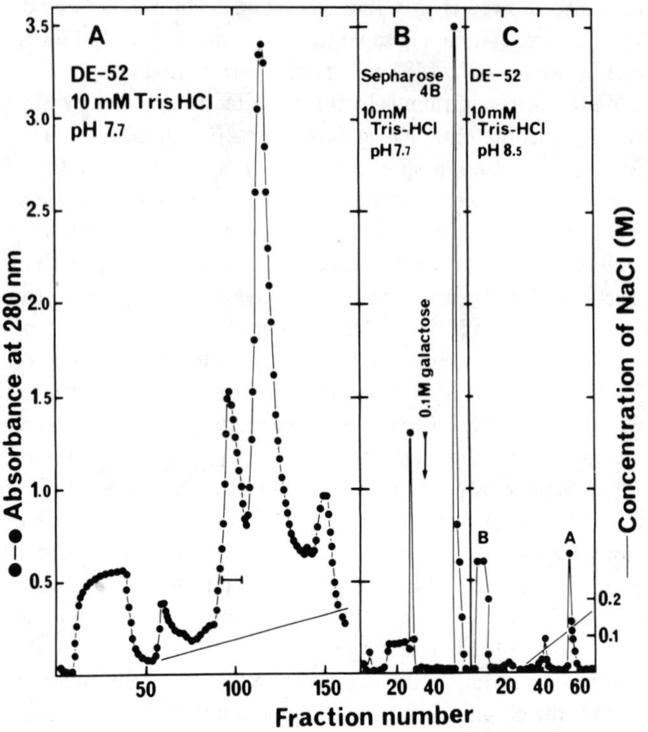

Fig. 4.3 Isolation of abrin and its constituent peptide chains [61]. 40 g of decorticated abrus seeds were homogenized in 5 per cent acetic acid, the extract was dialyzed against water, and the precipitated material was removed. The supernatant was adjusted to 10 mM Tris-HCl (pH 7.7) and chromatographed on a 2.6 x 40 cm DEAE-cellulose (DE-52) column equilibrated with the same buffer and eluted with a 1600 ml NaCl gradient (from 30 to 180 mM) in the same buffer (A). The most toxic fraction (indicated with a bar) was applied to a Sepharose 4B column (B) and, after washing, the bound material was eluted with 0.1 M-galactose in the same buffer.

To separate the two constituent peptide chains, abrin eluted from the Sepharose 4B column with galactose was dilued with an equal amount of H_2O and the pH was adjusted to 8.5. Galactose was added to a final concentration of 0.5 M, and then 5 per cent 2-mercaptoethanol was added. After incubation for 2 hours at room temperature and then a 4° overnight, the material was applied to a 1.6 x 20 cm (DE-52) column equilibrated with 10 mM Tris-HCl (pH 8.5). The B-chain passed through the column, whereas the A-chain was bound and was eluted with a 500 ml NaCl gradient (from 0 to 0.2 M) in the same buffer.

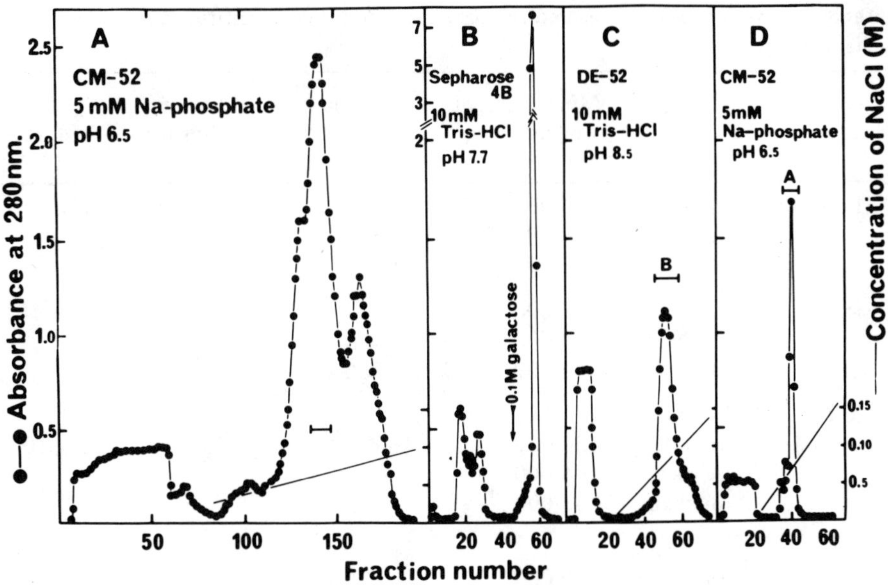

Fig. 4.4 Isolation of ricin and its constituent peptide chains [62]. Decorticated castor beans (100 g) were homogenized in 0.14 M-NaCl and the pH was adjusted to 4. The solubilized material was dialyzed against distilled water and the precipitate was removed. The soluble material was made up to contain 5 mM-Na phosphate (pH 6.5) and applied to a 2.6 x 40 cm carboxymethylcellulose column (CM-52) equilibrated with the same buffer (A). The bound material was eluted with a 1600 ml NaCl gradient (from 10 to 100 mM) in the same buffer. The most toxic fraction (indicated with a bar) was adjusted to pH 7.7 and applied to a Sepharose 4B column (B). After washing, the bound material was eluted with 0.1.M-galactose.

To separate the constituent peptide chains, the material eluted with galactose was diluted with the same volume of H_2O, and the pH was adjusted to 8.5. Galactose was added to a final concentration of 0.5 M and then 2-mercaptoethanol to 5 per cent, and the mixture was incubated for 2 hours at room temperature and then overnight at 4°. It was then applied to a DE-52 column (C) equilibrated with 10 mM Tris-HCl (pH 8.5). The A-chain passed through the column, whereas the B-chain was bound and was eluted with a NaCl gradient (from 0 to 0.15 M). To purify the A-chain further, it was dialyzed against 5 mM-Na phosphate (pH 6.5) and applied to a 1.6 x 20 cm CM-52 column (D) equilibrated with the same buffer. The pure A-chain was eluted from the column with a linear NaCl gradient (from 0 to 0.15 M) in the same buffer.

before the 2-mercaptoethanol [61, 62]. After reduction the two chains can be separated on a DE-52 column where abrin A-chain and ricin B-chain bind to the column and can be eluted with a salt gradient (Fig. 4.3C and 4.4C). Ricin A-chain can be further purified by chromatography on a CM-52 column (Fig. 4.4D). The

Table 4.1 Properties of abrus and ricinus lectins and their constituent peptide chains [63, 72]

Lectin		Molecular weight	pI	Toxicity to mice (No. of LD_{50} μg)	Binding sites for lactose	K_a (M^{-1}) for lactose at 20°
Abrin	Intact	65 000	6.1	77	1	8×10^3
	A-chain	30 000	4.6	<0.1	0	
	B-chain	35 000	7.2	<0.1	1	
Ricin	Intact	65 000	7.1	15	1	1.5×10^4
	A-chain	32 000	7.5	<0.1	0	
	B-chain	34 000	4.8	<0.1	1	
Abrus agglutinin	Intact	134 000	5.2	<0.1	2	8×10^3
	A′-chain	36 000				
	B′-chain	33 000				
Ricinus agglutinin	Intact	120 000	7.8	<0.1	2	1.5×10^4
	A′-chain	34 000				
	B′-chain	31 000				

LD_{50} was scored after 3 days

isolated chains are stable also in the absence of galactose. The constituent chains of ricinus agglutinin can be separated in a similar way [65], whereas the chains of abrus agglutinin are not separated by this procedure.

4.4 STRUCTURE AND PHYSICAL PROPERTIES OF TOXINS AND AGGLUTININS

4.4.1 Structure

Abrin and ricin have molecular weights of 60 000–65 000 [42, 61, 62, 64, 66, 68] and both consist of two polypeptide chains joined by an S–S bond [69–72]. The A-chain or 'effectomer' is slightly shorter than the B-chain or 'haptomer' (Fig. 4.5 and Table 4.1). The amino acid composition of the four peptide chains as found in our laboratory is given in Table 4.2. The amino acid composition of ricin and its constituent peptide chains* has also been determined by other authors [73, 74], and the results are in general agreement with our data. Statistical analysis of the

* The Ile-chain [73] appears to be identical with ricin A-chain, and the Ala-chain with ricin B-chain. In the following we therefore refer to the chains as the A- and B-chains.

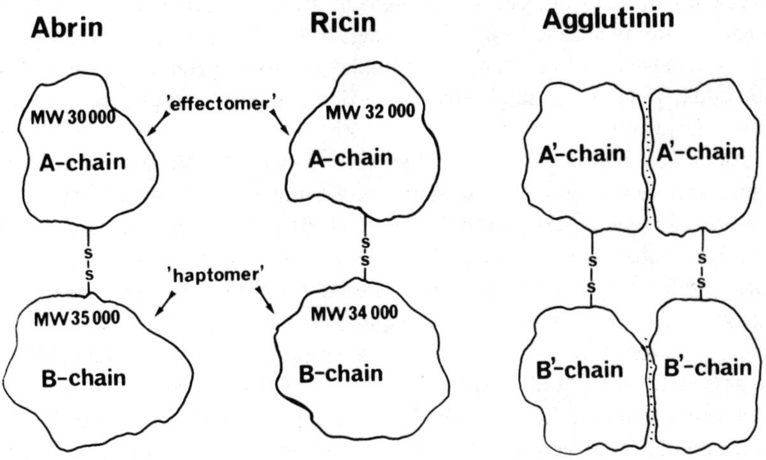

Fig. 4.5 Schematic structure of abrin, ricin, and the agglutinins.

Table 4.2 Amino acid composition of the constituent peptide chains of arbin and ricin [72].

Amino acid	Abrin		Ricin	
	A-chain	B-chain	A-chain	B-chain
Aspartic acid	31.0 (± 0.3)	27.9 (± 0.1)	27.3 (± 0.3)	43.6 (± 0.2)
Threonine	19.8 (± 0.1)	21.8 (± 0.6)	18.1 (± 0.2)	23.4 (± 0.2)
Serine	23.4 (± 0.4)	27.8 (± 0.1)	20.0 (± 0.9)	21.6 (± 0.3)
Glutamic acid	33.3 (± 0.1)	31.4 (± 0.1)	31.2 (± 0.6)	23.9 (± 0.3)
Proline	14.6 (± 0.1)	15.7 (± 0.8)	16.9 (± 0.7)	16.0 (± 0.2)
Glycine	16.4 (± 0.5)	18.3 (± 0.4)	18.5 (± 0.6)	23.0 (± 0.3)
Alanine	17.9 (± 1.0)	22.6 (± 0.2)	25.4 (± 0.2)	17.2 (± 0.1)
Half-cystine	2.0	2.2	1.2	6.0
Tryptophan	N.D.	N.D.	N.D.	N.D.
Valine	15.0 (± 0.6)	18.5 (± 0.3)	13.7 (± 0.4)	16.9 (± 0.6)
Methionine	2.6 (± 0.4)	4.1 (± 0.2)	2.2 (± 0.5)	2.9 (± 0.3)
Isoleucine	13.8 (± 0.8)	12.6 (± 0.3)	19.4 (± 0.1)	16.3 (± 0.1)
Leucine	22.4 (± 3.1)	27.2 (± 0.1)	23.1 (± 0.2)	26.9 (± 0.3)
Tyrosine	10.5 (± 0.5)	16.0 (± 0.4)	14.9 (± 0.9)	12.5 (± 0.5)
Phenylalanine	12.1 (± 2.0)	11.8 (± 0.1)	14.3 (± 0.1)	4.8 (± 0.3)
Histidine	5.8 (± 0.8)	3.9 (± 0.1)	3.6 (± 0.6)	2.3 (± 0.1)
Lysine	4.6 (± 0.4)	7.8 (± 0.0)	1.9 (± 0.7)	7.9 (± 0.3)
Arginine	18.4 (± 1.4)	16.6 (± 0.2)	20.4 (± 0.9)	14.8 (± 0.5)

amino acid composition indicated that the two A-chains are closely related, whereas the two B-chains are more different [72]. Surprisingly, also the A- and B-chains appear to be related. This is consistent with the finding of Gürtler and Horstmann [74] that tryptic digest maps of ricin A- and B-chains are similar, suggesting extensive homologies in amino acid sequence. The same authors found that tryptic digest maps of the two ricinus agglutinin chains were similar to those of the ricin chains. Also, Nicolson *et al.*, [64] found similar patterns when comparing tryptic maps of ricin and ricinus agglutinin.

The amino acid composition of isolated abrin A- and B-chain has not been determined in other laboratories, but the amino acid composition of whole abrin as measured by Lin *et al.*, [75] and by Wei *et al.*, [66] is in general agreement with that obtained by summation of our data for the A- and the B-chain. The differences found may be due to the fact that the seeds used in different laboratories may originate from different parts of the world, and represent different variants of the plant. We have demonstrated that abrus seeds obtained from four different sources contain abrin with slightly different properties (unpublished data).

It was found in several laboratories [72, 75–77] that the intact toxins and agglutinins contain no reactive sulfydryl groups either in the absence or in the presence of urea. In the free A- and B-chains, close to one reactive SH group was found both in the presence and in the absence of urea [72, 78, 79]. Since no free SH group could be detected in the intact toxins, the number of half-cystine residues present in each chain in excess of 1 must be present as intrachain disulfides. It is therefore likely that the half-cystine determinations in Table 4.2 are too low and that the correct values may be 3 residues of half-cystine in each abrin chain, 1 residue in ricin A-chain, and 7 residues in ricin B-chain. This is in general accordance with the findings of Gürtler and Horstmann [74]. However, other authors have found values which differ considerably from these results [64, 66, 75, 80, 81].

Funatsu *et al.*, [78, 82] and Gürtler and Horstmann [74] found the amino-terminal and the carboxy-terminal amino acids of ricin A-chain to be isoleucine and serine, and those of ricin B-chain to be alanine and phenylalanine, respectively. The amino-terminal ends of the A- and the B-chain of ricin have recently been sequenced by Li *et al.*, [83]. The helical content of ricin A- and B-chain was estimated to be 0.3 per cent and 25 per cent, respectively [84].

The published data show considerable differences as to the amount and type of carbohydrate present in abrin and ricin [72, 80, 85–87]. We found (Table 4.3) that in both toxins most of the carbohydrates are present in the B-chains. However, ricin A-chain contains some mannose. The amount of mannose and glucosamine (probably N-acetylglucosamine) present in the B-chains was almost the same in the two toxins. The small amounts of glucose found in intact abrin and ricin, but not in either of their isolated peptide chains, may represent artifacts.

Abrus and ricinus agglutinin also contained mainly mannose and glucosamine, although some glucose was found also in these lectins. Furthermore, our data show that ricinus agglutinin contains about twice as much carbohydrate as ricin, whereas abrus agglutinin contains 5–6 timess as much carbohydrate as abrin [72].

It thus appears that most of the carbohydrates in the toxins are bound to be B-chain, and that abrin and ricin B-chains contain about the same amount and kind

Table 4.3 Carbohydrate content in abrus and ricinus lectins [72]

Protein	Number of sugar residues per molecule of protein or polypeptide		
	Mannose	Glucose	Glucosamine
Abrin	10.0	1.6	1.6
Abrin A-chain	0	0	0
Abrin B-chain	9.7	0	1.9
Ricin	15.6	2.3	2.4
Ricin A-chain	4.3	0	0.4
Ricin B-chain	10.7	0	1.33
Abrus agglutinin	55.8	2.4	6.8
Ricinus agglutinin	25.4	4.4	4.9

of carbohydrates. Also, the two agglutinins contain mannose and glucosamine, although in different amounts compared with the toxins. It should be noted that mannose and glucosamine have been found in a variety of other lectins as well [88].

Funatsu *et al.*, [78] reported that ricin contains 3 oligosaccharide chains, one consisting of $(Glc\ Nac)_2\ (Man)_4$ in the A-chain and two consisting of $(Glc\ Nac)_2\ (Man)_6$ and $(Glc\ Nac)_2\ (Man)_7$ in the B-chain. In every case the oligosaccharide chains were found to be attached to asparagine residues. Precipitation reactions with concanavalin A indicate that mannose is the terminal sugar [89].

The net charge of the lectins was measured by isoelectrofocusing. Our data (Table 4.1), which are in agreement with data from other laboratories [71, 76], show that the pI of abrin (6.1) is only moderately lower that that of ricin (7.1). There is, however, a striking difference in the charge of the constituent peptide chains of the toxins. Thus, the A-chain of abrin and the B-chain of ricin are rather acidic, whereas the B-chain of abrin and the A-chain of ricin are almost neutral. It is thus clear that both toxins consist of an acidic and a neutral polypeptide chain. Since it was shown by Olsnes *et al.*, [90] that highly toxic hybrid molecules can be made from the isolated abrin A-chain and ricin B-chain (which are both acidic) as well as from ricin A-chain and abrin B-chain (both neutral), and since the toxicity of these hybrids is close to that of native abrin and ricin, it is clear that the charge of the toxins is not of much importance for their toxic effect.

Crystallization and X-ray investigation of the lectins have been carried out by several authors. Wei *et al.*, [91] found that abrin C (which is probably identical with the abrin isolated in our laboratory) crystallizes as a monoclinic unit cell of probable symmetry $P2_1$ and parameters a = 113Å, b = 72Å, c = 71Å, and β = 103°. The asymmetric unit contains two protein molecules of molecular weight 63 800 and a solvent content of approximately 45 per cent by volume. The less toxic abrin A was also examined [66, 92] and found to crystallize in an orthorhombic unit cell of symmetry $P2_1 2_1 2_1$ and dimensions a = 75 Å, b = 270 Å, and c = 70 Å with 2 molecules of molecular weight 65 000 in the asymmetric unit. Furthermore, the same authors [76] found that abrus agglutinin crystallizes in a tetragonal unit cell of symmetry $P4_1 2_1 2$ and dimensions a = 140 and c = 210 Å. The asymmetric unit was found to contain two protein molecules of molecular weight 126 000 and to have a solvent content of approximately 41 per cent by volume. Crystallization of abrin was also carried out by McPherson and Rich [87]. According to these authors X-ray diffraction analysis of crystals of abrin suggests that the molecule has four subunits arranged in a square-planar array related by 222 point-group symmetry. Electron microscopic examination of the crystals and X-ray diffraction data indicated that the molecule is roughly square (90 x 90 Å) and approximately 45 Å thick. The subunits are located at the corners of the square and are roughly ellipsoidal with dimensions 30 x 30 x 45 Å. It should be noticed, however, that the abrin used in this work was found to have a molecular wight of about 260 000, which is about 4 times the value found by other authors.

Crystallization of ricin has been carried out by several authors [75, 77, 93–95], but only Wei [92] has so far reported X-ray examinations. He found that ricin crystals belong to space group $C2$ with a = 177, b = 57, c = 92 Å, and β = 105° 40. The asymmetric unit was found to contain one molecule of molecular weight 50 000. The crystal water content was 68 per cent by weight and the density 1.142 g/ cm^3.

4.4.2 Immunochemical properties

As already shown by Ehrlich [20], abrin and ricin are immunologically different. Also, the isolated A- and B-chains of either toxin do not cross-react immunologically [96]. In quantifying the amount of antibodies against abrus and ricinus lectins, it is necessary to add lactose to the sera to prevent coprecipitation of various glycoproteins containing carbohydrate chains which otherwise may react with the lectins [96].

Anti-abrin and anti-ricin sera effectively protect animals against intoxication. We have shown that, in experiments with 2 ml samples of cells in culture, 25 μl of antiserum will completely prevent the effect of 10 μg of ricin while in the absence of antiserum the cells are intoxicated by than 1–5 ng [97]. By increasing the amount of toxin a sharp end-point is found at the equivalence point, and by increasing the toxin concentration slightly more there is a strong toxic effect on the cells [97].

Ehrlich [20 originally started the immunization by feeding the animals with abrus or ricinus seeds. Later, toxoids formed by formaldehyde treatment of abrin and ricin have been used. In contrast to the situation with many other toxins, it is difficult to remove the toxic effect of abrin and ricin with formaldehyde, and treatment for long periods with high concentrations is necessary to give toxoids with sufficiently low toxicity. In our immunization experiments higher titers of antibodies were usually formed against the A-chain than against the B-chain. Also, immunization with the isolated A- and B-chains results in the formation of higher titers are obtained with the A-chain than with the B-chain. Formaldehyde treatment of the chains increases the yield of antibodies.

Not all antibodies formed by immunizing with isolated chains will recognize the A- and the B-chains present in the intact toxins. Apparently, some antigenic determinants on the free chains are not present or exposed in the intact toxins. After immunization with formaldehyde-treated whole toxins, antibodies directed against the A-chain and the B-chain were equally efficient in preventing the toxic effect on animals and on cells in culture [90]. We could also demonstrate that anti-B-chain antibodies efficiently inhibit the binding of the B-chain to cell surfaces, and anti-A-chain antibodies prevent the enzymatic activity of the A-chains on ribosomes in a cell-free system. In fact these properties can be used in the quantitative determination of anti-A- and anti-B-chain antibodies [97].

Already Jacoby [14, 15] had found that immunization with ricin induced antibodies cross-reacting with ricinus agglutinin. Later studies have confirmed that abrin and ricin are indeed related to abrus and ricinus agglutinin [65, 96]. Abrin B-chain appears to be related to, but clearly different from, the binding chain in abrus agglutinin. A partial identity reaction between ricin and ricinus agglutinin is found in Ouchterlony immunodiffusion tests with antisera produced by immunization with either lectin [64, 96, 97]. In analogous experiments with the abrus lectins, antisera produced by immunizing with the agglutinin gave partial identity reactions with the agglutinin and the toxin, whereas anti-abrin did not form precipitates with abrus agglutinin. Clearly, the two abrus lectins differ more than the ricinus lectins.

Anti-ricinus agglutinin sera studied in our laboratory cross-react with ricin, but give very little precipitation with the isolated A- and B-chains. However, in Ouchterlony plates with the A- and B-chains placed in adjacent holes, a star-shaped precipitate is formed (Fig. 4.6), presumably due to interaction of the antibody with intact ricin formed where the two chains meet [97]. Quantitative studies have shown that in this serum most of the antibodies reacting with intact ricin did not recognize and bind to either of the isolated ricin chains. Part of the antibodies do recognize the ricin A-chain in a ricin A/abrin B hybrid [97]. Our data indicate that the antibodies recognize structures in ricin A-chain, ricin B-chain, or both, which are only present or exposed in intact ricin and not in the free chains. The explanation may be that the chains probably undergo extensive conformational changes after their separation. This is in accordance with the finding that only the free A-chain has an inhibitory effect on cell-free protein synthesis [43, 70].

Castor beans contain a strong allergen and many scientists working with ricin suffer from allergic reactions. It is not clear whether this allergen is related to the toxin.

4.4.3 Resistance to physical and chemical treatments

The intact lectins may be stored in the frozen state also in the absence of galactose without loss of activity, and even repeated freezing and thawing of the solution have little influence on their toxicity. In the presence of 0.1 M-galactose, the toxins could

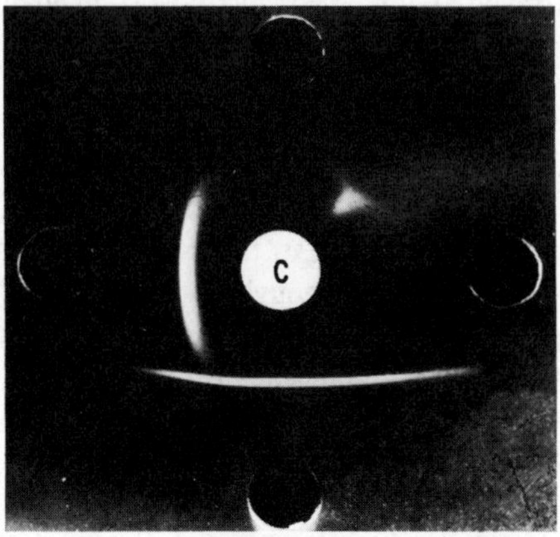

Fig. 4.6 Immunodiffusion of purified anti-ricinus agglutinin with different ricinus lectin fractions. The central well (c) contained anti-ricinus agglutinin serum which had been passed successively through columns containing covalently bound ricin A-chain and ricin B-chain. The peripheral wells contained ricinus agglutinin, ricin, ricin A-chain, and ricin B-chain as indicated [97].

be stored in the refrigerator for several months without loss of activity. The toxins were inactivated by boiling, and abrin is also largely inactivated by incubation at 60° for 1 hour.

The isolated chains are much less stable than the intact toxins. Thus, we found that abrin A- and B-chain and ricin B-chain can only be stored at –20° for long periods if 10 per cent of glycerol is present during the freezing. So far we have not been able to store ricin A-chain in the frozen state under any conditions. Similarly, the isolated chains are less resistant to moderate heating than the intact toxins [72].

The intact lectins are stable over a wide pH range. Thus, they could stand treatment with 0.1 N-acetic acid for 24 hours at room temperature without loss of biological activity, and even after treatment with 0.1 N-HCl only about half of the activity was lost. However, the isolated chains of the toxins are much less stable, at least at acid pH. Thus, treatment with 0.1 N-acetic acid removed most of the biological activity of the isolated chains [72].

Treatment with sodium dodecyl sulfate, which inactivates most proteins, has no effect on the biological activity of the A-chains of either toxin [72]. The B-chains, on the other hand, are sensitive to this treatment. No inhibition of the activity of abrus and ricinus lectins could be observed upon treatment with EDTA, indicating that abrus and ricinus lectins do not require loosely bound divalent cations for biological activity.

Introduction, by reductive alkylation, of about 10 methyl groups into abrin and ricin only slightly impaired their toxic effect (Sandvig and Olsnes, unpublished). Funatsu *et al.*, [8] reported that oxidation of the tryptophan residues of ricin with N-bromo-succinimide did not affect the toxicity whereas iodination (4 residues/molecule) decreased the toxicity to 6 per cent. Hence they concluded that tyrosine may play an important role in the toxic action of ricin. However, using the milder lacto-peroxidase method for iodination, we have been able to label ricin with two iodine residues per molecule without loss of toxicity [98]. It is an old observation that treatment with oxidizing agents inactivates ricin [99, 100] (for review, see [9]). The fact that the B-chain of ricin is very susceptible to chloramine-T treatment may indicate that the oxidation involved in the iodination procedure may be the reason for the loss in activity observed by Funatsu *et al.*, [81].

Lin *et al.*, [85] showed that treatment of abrin with periodate, which oxidizes the sugar moiety of the toxin, results in complete loss of toxic activity. We have confirmed their results with high concentrations of periodate. However, at lower periodate concentrations (10 mM) the toxins still retained much of their toxicity although they were no longer recognized by concanavalin A (unpublished data).

It is an old observation [9, 101] that abrus and ricinus lectins are extremely resistant to treatment with proteolytic enzymes. Among the proteolytic enzymes tested in our laboratory [72] only pronase in high concentrations had some effect. On the other hand, the isolated peptide chains of abrin and ricin were sensitive to proteolytic enzymes, the A-chains being somewhat more sensitive than the B-chains [72]. Funatsu *et al.*, [81] found that ricin is not digested by pepsin or trypsin although it is partially hydrolyzed by nagarase. In this partial hydrolysis only the N-terminal moiety of the A-chain was hydrolyzed releasing about 30 amino acid residues, resulting in a new N-terminal glutamic acid residue. The B-chain remained intact. The partial hydrolysis of the A-chain did not affect the toxicity of ricin, indicating that the N-terminal moiety of the A-chain is not essential for the toxic action of ricin.

4.5 MECHANISM OF ACTION OF ABRIN AND RICIN

4.5.1 Different biological activities of the two constituent polypeptide chains

When we treated abrin and ricin with 2-mercaptoethanol to reduce the interchain disulfide bond, the toxins lost most of their ability to inhibit protein synthesis in intact cells and to intoxicate animals [70]. If the 2-mercaptoethanol was then removed by dialysis, the interchain disulfide bond was re-formed and the toxicity was fully restored [40, 90, 102]. We also found that abrin and ricin inhibited protein synthesis not only in intact cells but also in cell-free systems, and, interestingly, that the reduced toxins were much more potent inhibitors in cell-free systems than the intact toxins [70]. When the A- and the B-chain were tested separately in the cell-free

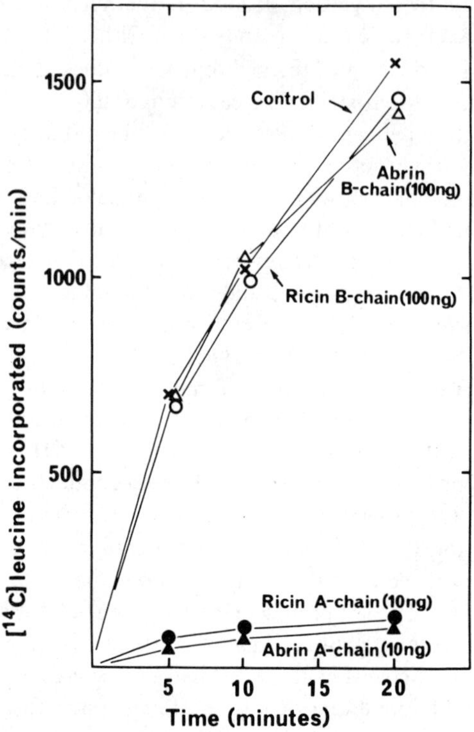

Fig. 4.7 Effect of toxin A- and B-chains on protein synthesis in a cell-free system from rabbit reticulocytes. A cell-free system (0.5 ml) was prepared from a rabbit reticulocyte lysate, and the indicated amounts of abrin and ricin A- and B-chain were added (control: no addition). [^{14}C]leucine was added to all samples, aliquots were taken after the indicated periods of time, and the amount of acid-precipitable radioactivity was measured [61, 62].

protein synthesizing system (Fig. 4.7), only the toxin A-chain inhibited protein synthesis whereas the B-chain had virtually no effect [61, 62].

Since only the intact toxin has a toxic effect on animals and intact cells, it is clear that the B-chain must have some important function. Our first indication that the B-chain binds to cell surfaces, whereas the A-chain does not have this ability, was obtained in indirect hemagglutination experiments [61, 62]. Later we could show that Sepharose particles possessing covalently bound B-chains on their surface bind erythrocytes, whereas particles containing A-chains did not bind erythrocytes (Fig. 4.8) [40]. Furthermore, experiments with radioactive B-chain have shown that the ability of the toxin to bind to cell surface receptors is associated with the B-chain [98]. We also found that when the purified A- and B-chains were applied

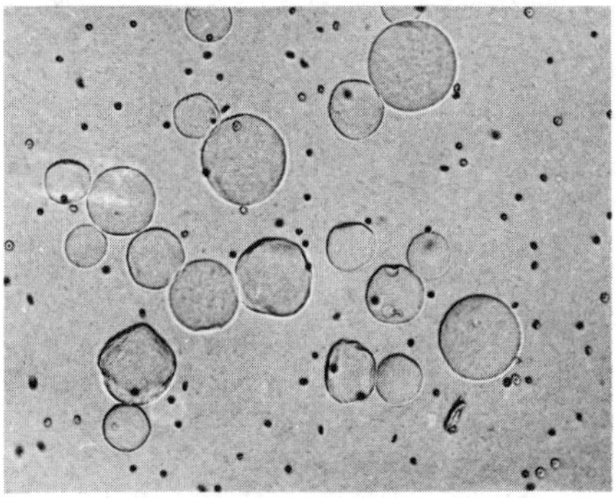

Abrin A-chain

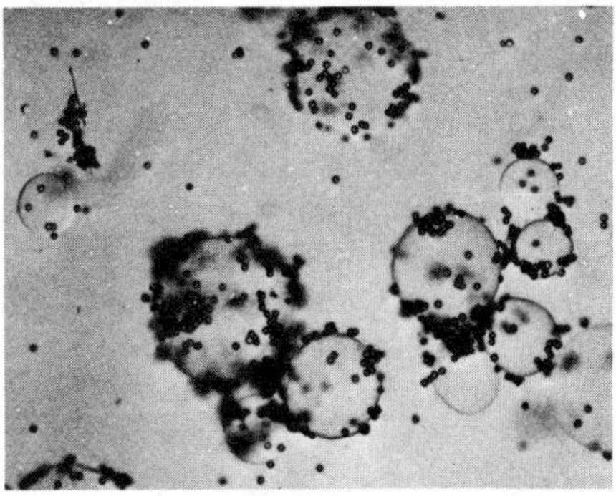

Abrin B-chain

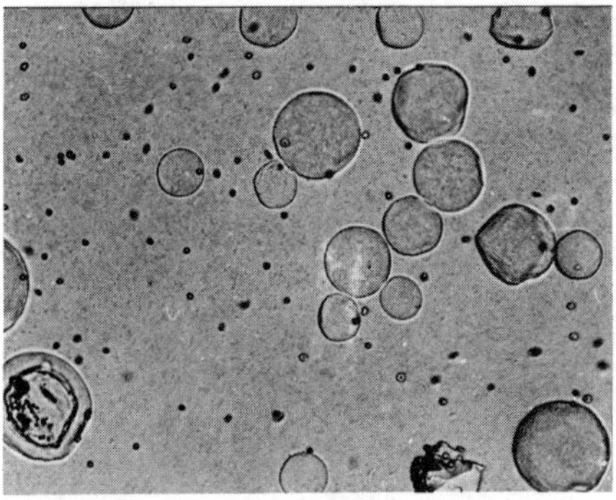

Ricin A-chain

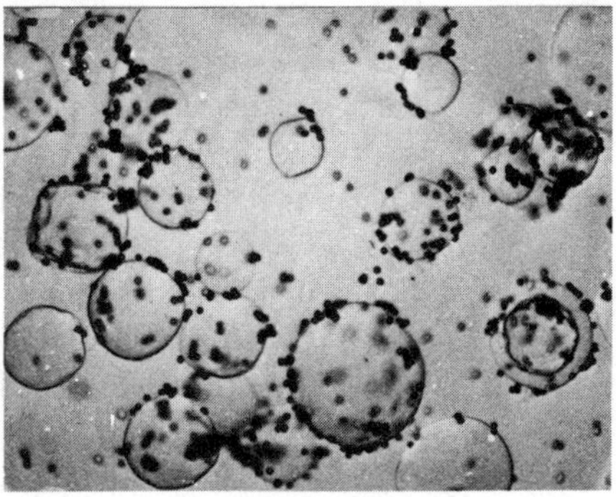

Ricin B-chain

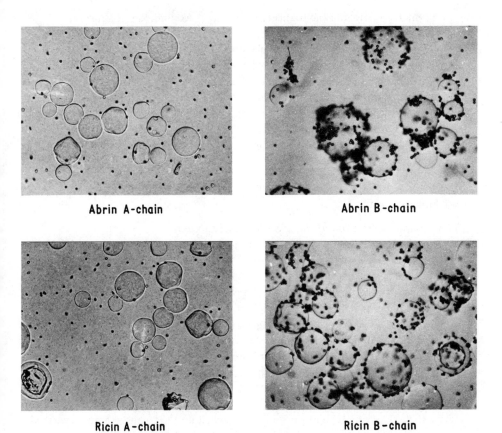

Fig. 4.8 Binding of erythrocytes to Sepharose particles having covalently bound A- and B-chains. Abrin and ricin A- and B-chains were linked covalently to cyanogen bromide activated Sepharose particles, and, after washing, the particles were mixed with human erythrocytes and observed under the microscope [40].

to Sepharose 4B columns only the B-chain was retained and could later be eluted with galactose, whereas the A-chain passed through [40, 62]. In contrast to the A-chain, the B-chain has the same biological activity whether free or present in the intact toxin [98].

The two chains of ricinus agglutinin could be separated in a similar way to that used for the ricin chains [65], and again it was found that the heavy chain has binding properties whereas the shorter chain inhibits cell-free protein synthesis although less efficiently than the A-chain of ricin. We have not yet succeeded in separating the chains of abrus agglutinin without the use of denaturing agents.

When the isolated A- and the B-chain of the toxins are mixed and dialyzed to remove 2-mercaptoethanol the toxins are re-formed. As pointed out above, the reconstituted toxins have the same toxicity as native abrin and ricin. Furthermore, Olsnes *et al.*, [90] showed that, when mixtures of abrin A-chain/ricin B-chain and *vice versa* are dialyzed, the two chains combine-to form hybrid molecules [90]. The hybrids are as toxic to living animals and cells in culture as are native abrin and ricin. Although anti-abrin protects animals and cells in culture only against abrin, and anti-ricin only against ricin, both antisera protect animals and cells against the toxic effect of the hybrid molecules [90]. Furthermore, antibodies directed against either of the two peptide chains in the hybrid molecules are equally effective in protecting against the toxic effect [90].

4.5.2 Function of the A-chain

(a) *Enzymatic inactivation of the 60S ribosomal subunits*

The toxin A-chains inhibit protein synthesis by inactivating the ribosomes [70, 103–105]. This was most convincingly demonstrated in experiments where we treated ribosomes and supernatant separately with the toxin A-chains [104, 106]. After the treatment specific antitoxins were added to prevent further activity of the A-chains, and cell-free systems were reconstituted from toxin-treated ribosomes and untreated supernatant factors and *vice versa*. Only when the ribosomes had been pretreated with toxins was protein synthesis in the cell-free system inhibited [104, 106]. Separate treatment of the 40S and 60S ribosomal subunits with the toxin A-chains (Fig. 4.9) revealed that abrin and ricin specifically inactivated the 60S subunit whereas they had no demonstrable effect on the 40S subunit [106], in accordance with the finding of Sperti *et al.*, [107].

Only the eukaryotic type of ribosome is attacked by the toxins, whereas bacterial ribosomes [104, 108] and mitochondrial ribosomes [108] are resistant. Although ribosomes from certain plants have been found to be sensitive [104], cell-free protein synthesizing systems have been prepared from castor bean seedlings [109, 110]. The protein synthesis in these systems was low. It should be noted that in cell-free systems some protein synthesis is observed even in the presence of high concentrations of toxin A-chains. It is not clear whether the low protein synthesis observed in the cell-free systems from castor beans represents only this residual protein synthesis or if the castor bean ribosomes are indeed resistant to ricin which appears to have been present in high amounts in this system.

Not only protein synthesis on endogenous messenger RNA is inhibited by the toxins, but also poly-U stimulated polymerization of phenylalanine and translation of encephalomyocarditis virus RNA [42, 111–113].

Our early experiments showed that one toxin molecule inactivates a large number of ribosomes [69, 70, 104, 111], indicating that the toxins act by a catalytic mechanism. Later we found [113] that the highly purified toxin A-chains inactivate salt-washed ribosomes in simple buffer solutions at a rate of about 1500 ribosomes per minute per toxin A-chain. The Q_{10} was about 1.8 and the K_m about $1–2 \times 10^{-7}$ M.

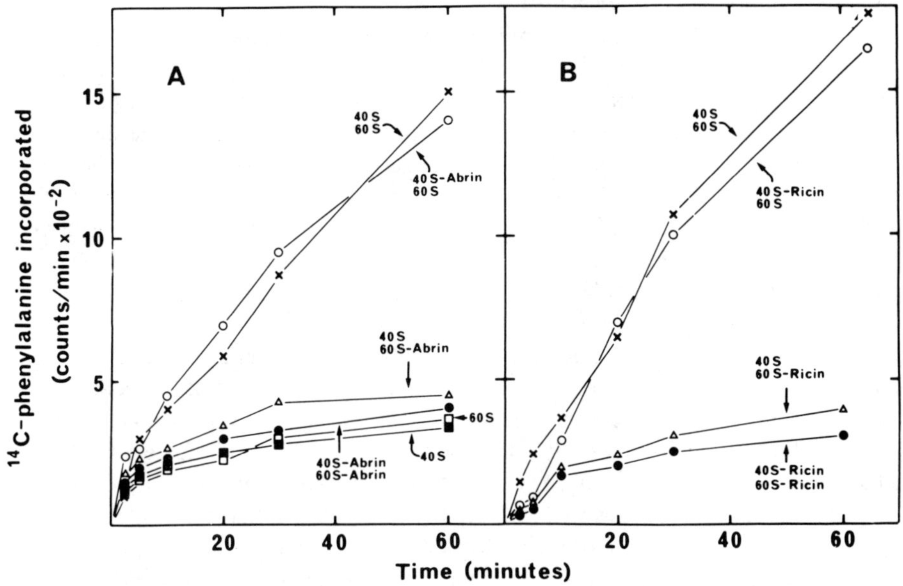

Fig. 4.9 Effect of A-chains from (A) abrin and (B) ricin on isolated ribosomal subunits from rabbit reticulocytes. 40S and 60S ribosomal subunits were incubated separately with and without abrin and ricin A-chains, and then neutralizing amounts of the specific antibodies were added. Parallel samples of subunits were incubated without A-chain. Subsequently toxin treated and untreated subunits were recombined as indicated, and the poly(U) stimulated incorporation of [^{14}C] phenylalanine into acid-precipitable material was measured in aliquots removed at the times indicated [106].

Furthermore, the inactivation could be stopped at any time by adding the specific anti-A-chain antibodies [113]. This clearly demonstrates that the A-chains are enzymes acting directly on the 60S ribosomal subunits. Other authors [105, 114] have also concluded that the toxins act catalytically.

Also, in intact cells the ribosomal 60S subunit appears to be the site of action of the toxins. Thus, Onozaki *et al.*, [115] showed that, when ribosomes were isolated from toxin-treated cells and their ribosomal subunits were separated and tested together with the complementary subunits from untreated cells, the 60S subunits from the toxin-treated cells were inactive whereas the 40S subunits remained active. Also, the supernatant factors remained active in the toxin-treated cells [115]. In contrast to the findings in cell culture, the protein synthesis in animals intoxicated with abrin and ricin decreases only slowly, and in most organs it reaches about half the control value in moribund animals. It is not known whether this moderate reduction in protein synthesis can account for the death of the animals, or if a special tissue is more sensitive to the toxins than the tissues so far examined.

(b) *Steps inhibited in protein synthesis*

To study which of the different steps in protein synthesis is (are) inhibited by the toxins, we added abrin and ricin to cell-free protein synthesizing systems and then analyzed the polysome profile [69, 111]. The polysome structure was preserved in the presence of the toxin A-chains, indicating that the toxins inhibit some step involved in peptide chain elongation. This was further supported by experiments showing that the toxins prevent the disappearance of polysomes in the presence of aurin tricarboxylic acid which selectively inhibits reinitiation of new peptide chains [69, 111]. Further studies have shown that the toxins do not inhibit the binding of puromycin to growing peptide chains, indicating that the toxins do not inhibit the formation of the peptide bond [105, 106, 114].

Carrasco *et al.*, [114] reported that ricin inhibits the enzymatic binding of aminoacyl-tRNA to ribosomes, whereas Montanaro *et al.*, [105] and Nolan *et al.*, [116] could not demonstrate this effect. The toxin-inactivated ribosomes exhibit reduced ability to hydrolyze GTP in the presence of the elongation factors EF-1 [106, 117] and EF-2 [105, 106, 117]. Furthermore, experiments with [^{14}C adenosyl]-diphosphoribose-EF-2 as well as [^{3}H]-EF-2 have shown that the binding of EF-2 in the presence of GTP or the analogue GDPCP is virtually abolished after toxin treatment of the ribosomes [116, 118]. Binding studies with unlabelled EF-2 and labelled GTP, GDP, or the analog GMPCP, gave the same results [114]. On the other hand, it was observed in this laboratory that prebound EF-2 protected the ribosomes against inactivation by toxin A-chains [112]. This indicates that the toxins act on the ribosomes at a site identical with or close to the binding site for EF-2, and changes this site in such a way that the affinity for EF-2 is reduced. High concentrations of EF-2 can partly overcome the protein synthesis inhibition by the toxin A-chains [113], showing that the affinity is only reduced and not completely abolished. An 8S complex which can be released from 60S ribosomal subunits with EDTA, and which consists of 5S RNA and a single ribosomal protein, has been shown to contain some GTPase activity [119]. Our recent results demonstrate that this GTPase is also inhibited by the toxin A-chains, indicating that this is the site of action of the A-chains [106].

Grollman *et al.*, [120] showed that, in HeLa cells incubated with ricin, protein synthesis stops after about 1 hour, and concomitantly the polysome structure disappears and monosomes appear instead. Anisomycin, an inhibitor of peptide bond formation, prevents the ricin-induced breakdown of the polysomes, indicating that protein synthesis is required for this effect to occur.

Lin *et al.*, [121] first showed that in rats treated with abrin and ricin the liver polysomes are broken down in the course of the intoxication, and that also in toxin-treated Ehrlich ascites cells the polysome profile disappears. From studies on rats intoxicated with abrin and ricin, they concluded that the polysomes were broken down by RNAase activated during the intoxication. On the other hand, Grollman *et al.*, [120] did not find evidence for increased ribonuclease activity in HeLa cells, and Urdahl and Olsnes (unpublished data) could not demonstrate increased RNase activity in liver extracts from intoxicated rats, although the polysome profile disappeared in a similar way as described by Lin *et al.*, [121]. Onozaki *et al.*, [115] found that the disaggregation of the polysomes in mouse myeloma cells intoxicated with ricin was too small to account for the inhibition of protein synthesis, and they did not find increased RNase activity in the cytoplasmic fractions.

(c) *Liberation of the A-chain*
In our early experiments [70] we observed that when abrin and ricin were treated with 2-mercaptoethanol their ability to inhibit cell-free protein synthesis was strongly increased. This increase could not be due to requirement of free SH groups since the inhibitory effect of reduced toxins or the isolated A-chains was only moderately lowered after treatment with N-ethylmaleimide which binds irreversibly to free SH groups [72]. Our further studies indicated that in a cell-free system the A-chain inhibits protein synthesis only when it is present in the free state, whereas binding to the B-chain to form intact toxin prevents the inhibitory effect on protein synthesis [43]. When intact toxins were added to cell-free systems, the rate of protein synthesis inhibition was first low and then increased rapidly with time [43, 69, 111]. The kinetics were consistent with the view that the A-chain must be liberated before protein synthesis inhibition occurs. If reducing agents are not present in the system, even high concentrations of intact toxin will not inactivate ribosomes. In rabbit reticulocyte lysate, and probably also in the cytoplasm of other cells, enough reducing agents are present to split the interchain S–S bond in abrin and ricin rapidly. Probably glutathione accounts for most of this reducing activity.

4.5.3 Function of the B-chain

(a) *Binding to simple sugars*
The experiments described above indicated that abrus and ricinus lectins bind to galactose residues in Sepharose 4B particles [59–62]. Agglutination inhibition experiments with simple sugars carried out with ricinus agglutinin in several laboratories indicated that the agglutinin binds to β-D-galactopyranoside residues [60, 63, 122,]. Recent equilibrium experiments carried out in our own and other laboratories have shown that the toxins and agglutinins indeed bind to free galactose or lactose (Table 4.1) with association constants of the order $K_a = 10^4\ M^{-1}$ at 20° [63, 123]. Each toxin molecule has one binding site whereas the agglutinins have two. It was also found [89] that the binding of lactose to ricinus agglutinin is strongly temperature-dependent, the K_a value decreasing with increasing temperature.

Shimazaki *et al.*, [124] studied the effect of sugars on circular dichroism of ricin and ricinus agglutinin. Lactose (and to a lesser extent galactose) was found to induce alterations in the spectrum indicating conformational changes. More alterations were found in ricinus agglutinin than in ricin.

The carbohydrate chains of the lectins do not appear to be involved in the binding to sugars. Thus, complexes consisting of ricinus agglutinin with concanavalin A

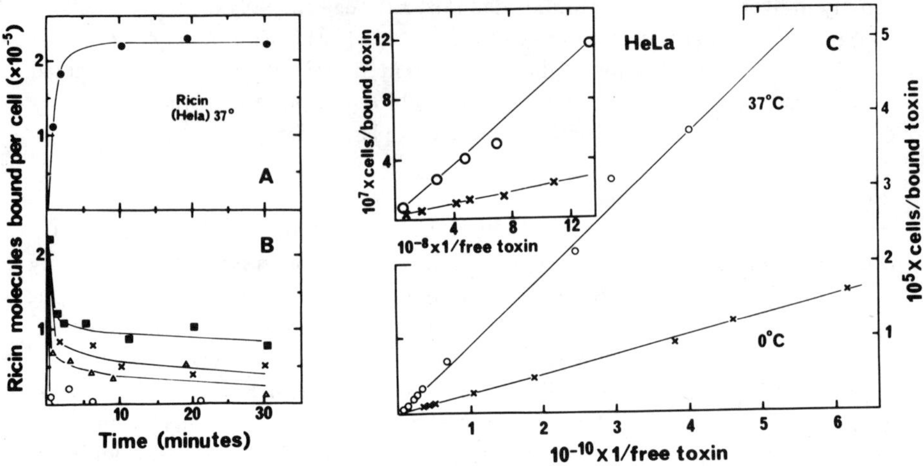

Fig. 4.10 Binding of ricin to HeLa cells.

(A) ^{125}I-ricin was added to HeLa cells, and aliquots were removed after the indicated periods of time. The cells were sedimented, and the amount of bound toxin was measured.

(B) After incubating ^{125}I-labelled ricin with HeLa cells, the suspension was diluted 100-fold with buffer (■), buffer containing unlabelled ricin (x), 10 mM-lactose (△), or 100 mM-lactose (○). After the indicated periods of time, samples were transferred to cellulose acetate filters, and the amount of radioactivity retained by the cells on the filter was measured.

(C) Equilibrium binding of ^{125}I-ricin to HeLa cells. The cells were incubated with different concentrations of ricin for 1 hour at 0° (x) or at 37° (○). Then the cells were sedimented by centrifugation, and the amount of bound radioactivity was measured. The data were plotted according to Steck and Wallach according to the equation:

$$\frac{[\text{cells}]}{[\text{bound toxin}]} = \frac{1}{K \times n} \cdot \frac{1}{[\text{free toxin}]} + \frac{1}{n}$$

where toxin and cells are expressed in M, n is the number of toxin binding sites per cell, and K is the affinity constant between toxin and cell surface receptors. The inserted diagram demonstrates on a different scale the data obtained with high toxin concentrations. The slopes of the curves in the inset are identical with those of the curves in the main diagram.

bound to the carbohydrate chains of the agglutinin will bind lactose to the same extent and with the same strength as free ricinus agglutinin [125]. Furthermore, periodate oxidation of abrus and ricinus lectins does not strongly impair their binding properties (our unpublished data).

Table 4.4 Interaction of abrus and ricinus lectins with HeLa cells [98]

Lectin	Temp. (°C)	k_{+1} ($\times 10^5 M^{-1} s^{-1}$) (association)	k_{-1} ($\times 10^{-4} s^{-1}$) (dissociation)	K_a ($\times 10^7 M^{-1}$)	Binding sites per cell ($\times 10^7$)
Abrin	0			30	3
	37	3.0	39	12	3
Ricin	0	1.3	12	30	3
	37	1.3	140	2.6	3
Ricin B-chain	0			18	2.2
Abrus agglutinin	0			20	3
Ricinus agglutinin	0	0.88		21	1.5

K_a was measured by equilibrium binding as in Fig. 4.10 (C)

(b) *Binding to cell surfaces*

After addition of ^{125}I-labelled abrin and ricin to HeLa cells, toxin is bound very rapidly (Fig. 4.10) and an equilibrium is established between bound toxin and toxin in solution. If the cell suspension is diluted with buffer, part of the bound toxin dissociates off. This dissociation is faster if unlabelled toxin is present in the buffer. If lactose or galactose is present, the toxin is released from the cells almost immediately, showing that the binding is a reversible process [98, 126].

The number of binding sites on cells, and the affinity of the toxins to these sites, have been measured by equilibrium experiments in several laboratories [98, 127–130]. Our data (Table 4.4), which are based not only on experiments with ^{125}I-labelled toxins but also on experiments with unlabelled abrin and ricin, show that HeLa cells possess about 3×10^7 binding sites per cell for both toxins. Erythrocytes have a lower number of sites in accordance with the smaller size of these cells. The association constants were found to be about $10^8 M^{-1}$ at 0° [98]. Thus, the binding to cell surface receptors is much stronger than to free lactose. The association constants between abrus and ricinus lectins and cells are temperature dependent, being highest at low temperature, particularly in the case of ricin, and decreasing with increasing temperature. The pH optimum is between 7 and 8 [98].

The straight lines obtained with HeLa cells in Steck-Wallach plots imply [131] that all the binding sites display a uniform affinity towards abrin and ricin, both at 0° and 37° (Fig. 4.10). The same is found in erythrocytes at 0°, whereas at room temperature there appear to be two groups of receptors with different affinity. The number of binding sites found in our laboratory is in accordance with that found for ricin by other authors [127–129], whereas the values for K_a differ between different laboratories, possibly owing to partial denaturation during labelling procedures involving the use of chloramine-T [98, 72].

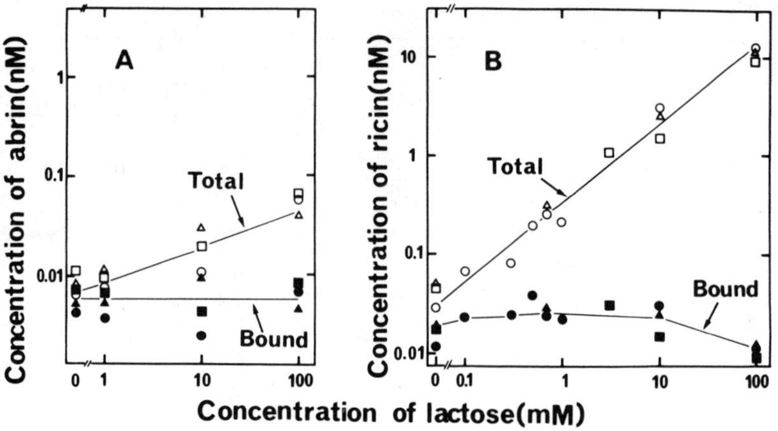

Fig. 4.11 The relationship between lactose concentration and the concentration of toxin necessary to reduce protein synthesis in HeLa cells to half the control value after 3 hours. HeLa cells were incubated in the presence of different concentrations of lactose for 3 hours with increasing concentrations of toxin; then [^{14}C]leucine was added, and the radioactivity incorporated into acid-precipitable material during 1 hour was measured. From the data obtained the total concentration of (A) abrin and (B) ricin required for 50 per cent reduction was plotted against the lactose concentration (open symbols). The amounts of toxin actually bound to the cells at various lactose concentrations (filled symbols) were measured or calculated [43].

Experiments with isolated chains have shown that the binding of the toxins to cell surface receptors involves exclusively the B-chain. In fact the isolated ricin B-chain binds to cell surface receptors with an association constant close to that of intact ricin [98].

(c) *The importance of the binding*

Already Ehrlich [20] assumed that abrin and ricin must be bound before they can intoxicate cells. Several lines of evidence have recently demonstrated that only toxin molecules bound to the cell surface can inhibit protein synthesis and induce cell intoxication. Thus, when lactose is added to the culture medium and as a consequence the equilibrium is changed in favor of free toxin, the toxic effect of abrin and ricin is strongly reduced [39, 40, 43, 96]. In fact, we found (Fig. 4.11) that there is a linear relationship between the concentration of lactose in the medium and the amount of toxin required to induce a certain toxic effect [43]. When the amount of toxin actually bound to the cell surface under the various conditions was measured or calculated, we found that the amount of toxin bound to the cell surface required to induce a certain toxic effect (reduction of protein synthesis to half the control value after 3 hours) was always the same even when the amount of toxin in solution varied by a factor of 300 [43]. Furthermore, competition experiments with the non-toxic ricin B-chain, which binds to cells with the same strength as intact ricin, showed that the toxin-induced inhibition of protein synthesis is reduced to the same extent as the amount of toxin bound to the cells. Recently additional evidence for the importance of the binding has come from studies on resistant cells. Thus, a variant Chinese hamster ovary cell [132] resistant to ricin was found to have only 1–2 per cent of the ricin binding sites of the sensitive cells. Also, some ricin resistant variants from baby hamster kidney cells isolated by Meager *et al.*, [133] had decreased binding capacity for the toxin.

(d) *Nature and specificity of the receptors*

The membrane structure to which abrus and ricinus lectins bind is not known. Most likely they bind to the carbohydrate moiety of membrane glycoproteins. The strong inhibitory effect of erythrocyte membrane glycoproteins and their carbohydrate moieties [130] indicates that similar carbohydrate chains are involved in the binding of the toxins to intact cells. In the erythrocyte membrane two main types of glycoproteins are present, both of which serve as binding sites for lectins. The major sialoglycoprotein (glycophorin), which has a molecular weight of 50 000 and consists of 60 per cent carbohydrate and 40 per cent protein, contains binding sites for lectins from *Lens culinaris, Phaseolus vulgaris, Agaricus bisporus,* and wheat germ [130]. This glycoprotein also carries the blood group A and B characters, and is the receptor for influenza virus. It does not normally bind *Ricinus communis* lectins, but after treatment with neuraminidase a penultimate galactose residue is exposed and binding of ricinus lectins may now occur [130]. This is consistent with the finding that neuraminidase treatment of various cells increases the number of binding sites for ricinus lectins by a factor of 1.5–3 [122, 134, 135]. The normal binding site for abrus and ricinus lectins (and also for concanavalin A) on erythrocytes is the 'band III glycoprotein' [136] which has a molecular weight of about 100 000 and contains only 6–8 per cent carbohydrates and very little sialic acid [137, 138] and increases their sensitivity to abrin and ricin accordingly (K. Sandvig, unpublished). This glycoprotein is believed to be involved in the anion transport and glucose transport across the erythrocyte membrane. The binding sites for ricinus lectins correspond to the intramembranous particles (membrane intercalated particles) [139], and since glycophorin is also present in these particles the data indicate that the intramembrane particles are composed of several proteins [138]. Abrus and ricinus agglutinins agglutinate all the human blood groups to about the same extent [58].

Although it is commonly believed that the cell surface binding sites for lectins consist of glycoproteins, it is possible that other substances such as gangliosides may also act as binding sites. Thus, Surolia *et al.*, [141] showed that ricinus agglutinin binds to liposomes containing ganglioside GM_1 with an association constant of the order 10^6 M^{-1} at 4°.

Not only cell surfaces but also nuclei contain binding sites for ricin [142], and ricin-ferritin complexes were found to bind to the outer surface of granules of polymorphonuclear leucocytes, indicating the presence of carbohydrates on the surface facing the cytoplasm [143].

(e) *Receptor content in resistant cells*
Work in the laboratory of Kornfeld has shown [132] that a ricin-resistant line of Chinese hamster ovary cells which contains only 1–2 per cent as many ricin binding sites as the sensitive parent strain also had a reduced number of binding sites for abrin, as well as for *Phaseolus vulgaris* lectin, lentil lectin, and soy bean lectin. On the other hand, the number of binding sites for mushroom lectin was unchanged and there was a 60 per cent increase in the number of binding sites for concanavalin A. Carbohydrate analysis of the membranes of the resistant cells showed reduced amounts of galactose, N-acetylglucosamine, and sialic acid, and increased amounts of mannose. The growth properties of the resistant cell line were somewhat different from those of the parent cell line, and the resistant line gave a lower plating efficiency. Gottlieb *et al.,* [144] also found that the resistant cells had lost the ability to transfer N-acetylglucosamine from UDP-N-acetylglucosamine to certain (but not all) glycopeptide and glycoprotein complexes containing terminal non-reducing α-linked mannose residues. Since normally galactose and sialic acid are often added to N-acetylglucosamine residues of mammalian glycoproteins, the reduced N-acetylglucosaminyl transferase activity may explain the reduced amounts of galactose and sialic acid in the membrane glycoproteins of the resistant mutant. Furthermore, the membrane glycoproteins of the resistant strain exhibited a marked shift towards glycoprotein species of lower molecular weight.

Hyman *et al.,* [140] isolated ricin resistant strains of murine lymphoma cells with 50–70 per cent of the binding sites on the parental cells. In the laboratory of Hughes, 22 clones of baby hamster kidney cells resistant to ricin have been isolated [133]. Some of these clones contained fewer surface binding sites for ricin than the parent cells, whereas other clones possessed the normal number of surface binding sites. From L-cells 6 clones have been isolated with a close to normal number of ricin binding sites, but strongly reduced sensitivity to the toxin [145]. Apparently, in these cases the resistance is due to a defect later in the uptake mechanism than toxin binding. In our laboratory J. Jacobsen isolated three variant strains of HeLa cells which are resistant to abrin and ricin, although they bind the toxins with the same strength and to the same extent as the parent cells, and the ribosomes appear to be normally sensitive to the toxins. Interestingly these cell lines are also resistant to diphtheria toxin which appears to enter the cell in a similar way to abrin and ricin although it binds to different receptors.

4.5.4 The problem of internalization

Since the toxin A-chains act by inactivating the ribosomes, it is clear that at least the A-chain must penetrate into the cytoplasm. A similar situation applies to diphtheria toxin [146] and cholera toxin [147]. The lag time from addition of and ricin to cells in culture until cellular protein synthesis starts to decrease (Fig. 4.2) is believed mainly to reflect the time required for the uptake mechanism since no lag time is seen in a cell-free system. As shown by Refsnes *et al.*, [41] the lag time is strongly temperature dependent, being about 10 times longer at 20° than at 37°. Furthermore, we found in experiments with antibodies specifically directed against the A- and the B-chains of the toxins that the lag time could be divided into three parts. During the first part anti-A-chain and anti-B-chain antibodies were able to protect the cells against the toxins. Then followed a period when only anti-A-chain antibodies were effective. During the major part of the lag time, antibodies did not have any protective effect. The data are consistent with a 3-step mechanism where the toxins first bind by their B-chain, then the A-chain penetrates the cell membrane and later appears in the cytoplasm where ribosome inactivation may take place.

Attempts have been made to estimate the rate of uptake of toxins by the cell by measuring the irreversible binding. Nicolson *et al.*, [42] studied the binding of ricin to 3T3 mouse fibroblasts. After incubation in the cold, 90–95 per cent of the ricin could be removed by adding lactose to the medium, whereas after incubation at 37° for 60 minutes the major part could not be removed. The fraction which was irreversibly bound was about the same in normal 3T3 cells and SV-40 transformed 3T3 cells which are at least 10 times more sensitive to the toxin. In our laboratory Sandvig *et al.*, [98] found that in normal HeLa cells only about 10 per cent of the cell-bound abrin and ricin became irreversibly bound per hour, and this fraction was similar in variant cell lines ricin-resistant containing the normal number of binding sites. Thus, there is no clear relationship between the amount of toxin irreversibly bound and the toxic effect.

After being bound to the cell surface, the toxins are engulfed into pinocytotic vesicles, and it has been suggested that this is the significant process involved in the penetration of the toxins into the cytoplasm. Oliver *et al.*, [148] showed that binding of ricinus lectins to polymorphonuclear leukocytes induces concentration and internalization of the binding sites. Colchicine and vinblastine prevented the selective removal of lectin binding sites from the surface, although this treatment did not inhibit phagocytosis. The internalization of the binding sites appeared to be specific, since ricinus lectins promoted the internalization of their own binding sites but had no effect upon the concanavalin A binding sites and *vice versa.* One group of receptors was internalized after treatment with either lectin, indicating that these receptors may bind both lectins. Nicolson [149, 150] showed that ferritin-labelled ricin was first bound to the cell surface and then taken up into pinocytotic vesicles. Similarly, Gonates *et al.*, [151] demonstrated the uptake of surface-bound ricin–horse-radish-peroxidase complexes into the Golgi apparatus of sympathetic neurones, but they could not demonstrate release into the cytoplasm. Nicolson [149], on the other hand, claimed that the ferritin-labelled ricin is indeed released from pinocytotic vesicles into the cytoplasm owing to occasional rupture of the vesicles. Furthermore, the time course of this process was such that the toxin appeared in the cytoplasm first when inhibition of protein synthesis became apparent.

Indirect evidence for the role of the pinocytotic uptake mechanism was obtained in experiments with reticulocytes which have little pinocytotic activity. Although the toxins readily bind to these cells, protein synthesis was not inhibited for at least

4 hours under conditions where it virtually stopped after 30–60 minutes in HeLa and Ehrlich ascites cells [41].

To obtain further information on the role of pinocytosis in the uptake mechanism, attempts were made to bind the toxins to the cell surface by a mechanism other than binding by the B-chain to carbohydrate receptors. Refsnes and Munthe-Kaas [152] took advantage of the fact that macrophages, in contrast to HeLa cells, contain receptor sites on their surface with specificity for the Fc part of immunoglobulins [153]. Ricin was reacted with antibodies specifically directed against the B-chain part of the molecule to prevent binding to the specific carbohydrate receptors. As expected, such complexes proved to have no toxic effect on HeLa cells. The complexes were, however, toxic to macrophages. This toxic effect could not be prevented by addition of lactose to the medium. It could, however, be prevented by addition of albumin–anti-albumin complexes which compete for the Fc-receptor sites. Complexes consisting of [ricin anti-ricin A-chain antibodies] were not toxic, in accordance with the fact that anti-A-chain antibodies prevent the inhibitory effect of the A-chain on cell-free protein synthesis [152]. Altogether, the data demonstrate that the uptake of the toxins can be induced by an alternative binding mechanism and result in intoxication of the cell. The efficiency of the alternative mechanism was, however, only 0.1–1 per cent of that seen with the native toxin.

It is possible that in the case of the native toxins there exists a specific and very efficient uptake mechanism which does not involve endocytosis, and that uptake of bound molecules by pinocytosis plays only a minor role in the expression of the toxic effect of abrin and ricin.

4.6 BIOLOGICAL ACTIVITIES OF THE AGGLUTININS

4.6.1 Mitogenic effect

Pure abrus agglutinin has no toxic effect on animals or cells and does not inhibit cell-free protein synthesis. When tested on human lymphocytes, abrus agglutinin induced DNA synthesis in a similar way to *Phaseolus vulgaris* lectin (phytohemagglutinin, PHA). In fact, Closs *et al.*, [154] showed that the required amount of abrus agglutinin is lower than that of *Phaseolus vulgaris* lectin (Fig. 4.12). The stimulation of DNA synthesis can easily be prevented by adding lactose to the medium, and again there is a linear relationship between the amount of lectin added and the concentration of lactose necessary to prevent the stimulation of DNA synthesis from taking place. When lactose is added to the lymphocyte culture later than 4 hours after the addition of abrus agglutinin, the mitogenic effect cannot be completely prevented. In accordance with the findings made with other mitogens, the maximal DNA stimulation is observed after 3–4 days. Abrus agglutinin appears to be specific for thymus-derived lymphocytes. Kaufman and McPherson [155] also found the mitogenic effect of abrus agglutinin. In their experiments a lectin

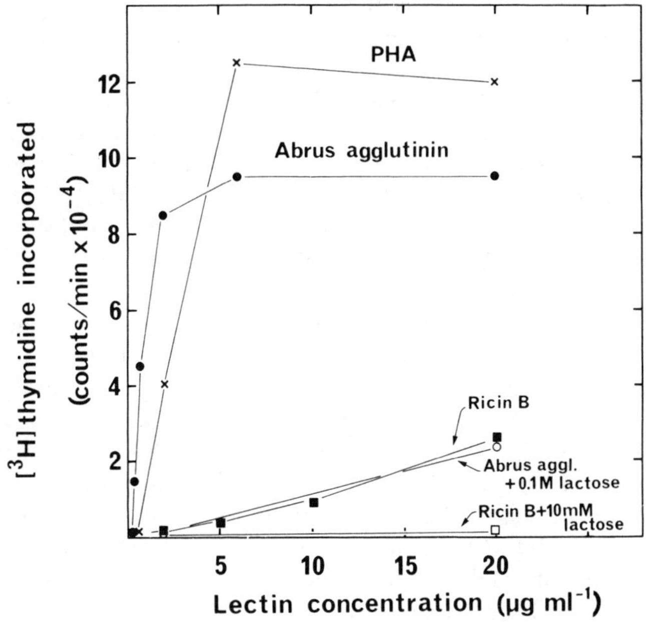

Fig. 4.12 Effect of lectins on thymidine incorporation by human lymphocytes. Different amounts of lectins were added to cultures of human lymphocytes, and the cultures were incubated for 2 days. [^{3}H] Thymidine was added to the cultures 16 hours before harvesting, and the incorporated radioactivity was measured in each sample. In some cases the indicated concentration of lactose was added to the cultures at the beginning of the experiment [154].

preparation aged for several months at 4° was used since the freshly prepared lectin contained too much toxin. Apparently the toxin is destroyed more rapidly than the agglutinin upon storage at 4°. They found a sharp concentration optimum at about 30 μg ml^{-1} whereas higher concentrations were toxic. Our preparations, even in much higher concentration, were completely non-toxic.

The pure ricinus agglutinin is virtually non-toxic to animals but has a certain toxic effect on cells in culture [65] and cannot be used in mitogenesis studies. However, the isolated ricin B-chain is non-toxic and stimulates DNA synthesis in lymphocytes when added in high concentrations (Fig. 4.12). The stimulation was prevented in the presence of low concentrations of lactose.

4.6.2 Toxicity of ricinus agglutinin

Ricinus agglutinin inhibits protein synthesis in HeLa cells, although it has only about 1/30 of the activity of ricin. Studies on cell-free protein synthesizing systems showed that ricinus agglutinin also acts by inactivating the ribosomes [65, 113]. Recently, Saltvedt [65] separated the pair of light chains (the A′-chains) from the pair of heavier chains (B′-chains) by ion exchange chromatography, and tested them separately in cell-free protein synthesizing systems. He found that also in the case of ricinus agglutinin the light chain (the A′-chain) inhibits cell-free protein synthesis. The molecular activity of the agglutinin A′-chain (the number of ribosomes inactivated per minute by one A′-chain molecule) is only 1/15 of that of ricin A-chains, whereas the K_m is similar [113]. Furthermore, only the B′-chain of the agglutinin is able to bind to cell surfaces. It is thus clear that the structure–activity relationship is similar in ricinus agglutinin and ricin.

When tested on cells in culture the toxicity of ricinus agglutinin is found to be 1/30 of that of ricin. However, in mice the difference is much greater (1/2000). The reason for this difference is not clear. It may be related to the fact that ricinus agglutinin forms precipitates with serum proteins and may thus be removed from circulation before it reaches cells of vital importance. Binding of agglutinin to erythrocytes, which at physiological temperature is much stronger than in the case of ricin, may also contribute to the low toxicity of ricinus agglutinin in animals.

Waldschmidt-Leitz and Keller [156, 157] found proteolytic activity in ricinus agglutinin, and they claimed that glycopeptides are split off after agglutination of the cells. This has not been confirmed in other laboratories.

4.7 PRACTICAL APPLICATIONS OF ABRUS AND RICINUS LECTINS

4.7.1 Studies of cell surfaces

Abrus and ricinus agglutinins effectively agglutinate all animal cells so far tested, and as little as 10 ng ml^{-1} will agglutinate 10^7 human erythrocytes. This agglutination is easily prevented by lactose. Lactose also prevents the agglutination of rabbit erythrocytes by ricinus lectins, whereas in experiments with abrus agglutinin lactose has no preventive effect [63]. In contrast to the situation with many other agglutinins, the cells agglutinated by ricinus lectins may be redispersed by adding the competing sugar to the buffer.

The finding [158–160] that malignantly transformed cells are more easily agglutinated by wheat germ agglutinin than the non-transformed parent cells has stimulated much work on the agglutinating ability of different cells by a variety of lectins. SV40-transformed mouse fibroblasts were not only more easily agglutinated than the parent strain (3T3 cells) [60] but they were also 10 times more sensitive to ricin [42] although both cell types bound similar amounts of toxin. Interestingly, in the transformed cell the relative mobility of the ricin receptors was increased, compared with that of the normal 3T3 cells, and it is possible that the endocytosis is also increased [42]. In contrast to this, Miyake *et al.*, [161] found that two non-malignant cells derived from cultured mouse mammary tumor cells, FM3A/D, were more easily agglutinated by ricinus agglutinin than the parent malignant cells, although the density of surface binding sites was about the same in the two strains. It thus appears that the relationship between malignancy of cells and their agglutinability is not clear.

When cells were grown to confluence, the number of binding sites for ricinus agglutinin increased by a factor of 2.5 compared with the value found on sparse cells, whereas in transformed cells (SV 3T3 and 3T12) the number of binding sites did not change under similar conditions [162]. Nomoto *et al.*, [163] showed that binding of ricin to Yoshida ascites sarcoma cells reduced the level of cAMP, probably owing to a change in the K_m of the adenylate cyclase activity. The agglutination process of rat ascites tumor cells by ricinus agglutinin appears to require energy, but not protein synthesis [164]. Ji and Nicolson [165] showed that, when ricinus agglutinin aggregated surface binding sites on resealed human erythrocyte ghosts, the inner surface membrane protein, spectrin, was also aggregated, indicating the existence of transmembrane connections between ricinus lectin binding sites and proteins on the inner surface membrane. In contrast, rabbit sperm plasma membrane binding sites for ricinus agglutinin did not aggreagte upon incubation with the lectin, indicating transmembrane restrictions to the movement of the surface binding sites [166].

Gahmberg and Hakomori [167] found that in hamster fibroblasts labelled with galactose oxidase [^{3}H]borohydride two major galactose proteins were labelled on the surface. The labeling of a 200 000 molecular weight protein was strongly influenced by ricinus lectins whereas the other labelled glycoprotein (molecular weight 130 000) was much less affected. The high molecular weight component was detected only in normal cells and not in several transformed derivatives.

Ricinus agglutinin binds to the spikes of influenza virus. Interestingly, in baby hamster kidney cells infected with influenza virus, the number of binding sites for ricinus agglutinin increased 6-fold over a period of 4–6 hours owing to a process requiring protein synthesis [168].

4.7.2 Ricinus agglutinin in protein purification

It had been shown already by Stillmark [11], Hellin [13], and Kobert [22] that extracts from abrus and ricinus seeds form precipitates with normal serum. Bird [169] and Pardoe *et al.*, [122] observed that *Ricinus communis* extracts form precipitates with salivary proteins, ovary cyst blood group substances, and *Pneumococcus* polysaccharide XIV, provided that terminal non-reducing galactose residues are present. Pardoe *et al.*, [122] demonstrated that immunoglobulins are precipitated by ricin. This was further studied by Harboe *et al.*, [170] who showed that all of 50 monoclonal IgM proteins were precipitated by ricinus agglutinin. The reactive sites were found to be on the Fcμ fragment. Lactose prevented the precipitation.

Since concanavalin A has recently been introduced into protein purification

Table 4.5 Affinity chromatography of human plasma proteins on columns containing insolubilized lectins [170]; + denotes that (at least part of) the protein was bound to the column and subsequently eluted with lactose to appear in peak 2.

Protein	Ricinus agglutinin (Ref. [170]	Concanavalin A (Ref. [171]
Prealbumin	–	–
Albumin	–	–
α_1-Lipoprotein	–	–
α_1-Antitrypsin	–	+
α_1-Acid glycoprotein	–	–
α_1-Antichymotrypsin	–	+
α_2 HS glycoprotein	–	+
Gc-globulin	–	–
Coeruloplasmin	–	+
Haptoglobin	+	+
α_2-Macroglobulin	+	+
β_2-Lipoprotein	–	+
β_{A1}-Globulin	–	+
Transferrin	–	+
IgA	+	+
IgM	+	+
IgG	+	+

owing to its ability to bind glycoproteins containing mannose residues [171–173], it occurred to us that ricinus agglutinin, being specific for terminal galactose residues, may also be valuable in glycoprotein purification. Abrus agglutinin, as well as abrin and ricin, can be used for this purpose, but in our hands ricinus agglutinin has given the best results owing to its strong binding capacity which can easily be prevented by lactose.

When human serum was passed through a column containing covalently bound ricinus agglutinin, most of the proteins passed through the column, whereas a fraction was bound and could be eluted with lactose [170]. The different fractions were then analyzed by fused rocket immunoelectrophoresis in agar plates containing specific antisera against single serum proteins. It was found (Table 4.5) that IgM is completely retained, whereas haptoglobin, IgG, and IgA were partly retained by the column. Most serum proteins passed through. Interestingly, Saltvedt *et al.*, [174] demonstrated that only one of the IgG subclasses (IgG_3) was bound to the column, whereas the other subclasses passed through. In the case of IgA also only some of the molecules are retained by the column, and this is also related to subclasses (E. Saltvedt, unpublished). Ricinus agglutinin thus appears to be a valuable tool in the separation and characterization of immunoglobulins and presumably also in the purification of a variety of other glycoproteins containing terminal non-reducing galactose residues.

Recently Surolia *et al.*, [175] have shown that only small amounts of native fetuin and ceruloplasmin are retained by a ricinus column, whereas removal of the terminal sialic acid resulted in complete binding of the proteins to the column. Ashwell (personal communication) has suggested the use of Sepharose columns containing covalently bound ricinus lectins as a clinical test for the presence of desialated serum-glycoproteins. The desialated forms of these proteins are normally bound and degraded in the liver [176–178], and they never reach high serum levels in healthy persons but may increase in liver deficiency.

4.7.3 Possible applications in the therapy of cancer

As mentioned above, data from several laboratories indicate that abrin and ricin have cancerostatic properties on certain experimental tumors. The reason for the higher sensitivity of cancer cells is not clear. Possibly it is related to altered membrane properties and increased uptake in malignant cells. Interestingly, abrin was found to suppress strongly the growth of a human melanoma and a human fibrosarcoma growing in thymus-less ('nude') mice (Φ. Fodstad, unpublished). It is therefore possible that in the future the cancerostatic properties could be used in the therapy of certain human tumors. The fact that the toxins kill not only dividing cells but also cells in the resting stage could be of advantage in the treatment of slowly growing tumors having low growth fractions. However, on the basis of present knowledge the toxins as such do not appear to possess advantages over the anti-cancer agents commonly used.

One possibility which has attracted our attention is that the toxins would be far more powerful anti-cancer agents if the 'haptomer', the B-chain, could be replaced by tumor-specific antibodies which could serve as homing agents. Experiments with adriamycin, daunomycin, and other cytostatic agents bound to anti-cancer antibodies have been reported from different laboratories, but so far the improvement of the cancerostatic effect has not been very convincing [179, 180]. One reason for this could be that in the case of most anti-cancer agents it is possible that the number of molecules that can be bound per cell by this specific mechanism is insufficient. The advantage of abrin and ricin in such experiments is their extreme toxicity which implies that one or a few molecules may be sufficient to kill a cell. Similar considerations have been made with diphtheria toxin [181] which has a toxicity similar to that of abrin and ricin. However, the complexes of ricin and anti-ricin B-chain antibodies described above proved to have very low toxicity compared with that of the free toxins. It is possible, however, that the uptake of complexes varies, and experiments are in progress in our laboratory to test the cytotoxic effects of different kinds of complexes of toxin and antibodies.

4.8 IMPLICATIONS AND PROBLEMS

A structure–activity relationship similar to that found in abrin and ricin had previously been discovered in diphtheria toxin [146]. Diphtheria toxin is a single peptide chain which is easily cleaved by proteolytic enzymes into two fragments, the A- and the B-fragment. The A-fragment is an enzyme which inactivates elongation factor 2 (EF-2) thus inhibiting protein synthesis, whereas the B-fragment is involved in the binding of the toxin to surface receptors. These receptors have not been identified, but they appear to be present in a relatively small number per cell. Also cholera toxin consists of a binding part which appears to bind to ganglioside GM_1 on the cell surface [182], and an enzymatically active part which in some way stimulates the adenylate cyclase from the inside of the cell membrane [147]. In none of these cases is the mechanism of the entrance of the toxin into the cell understood. It is possible that different uptake mechanisms may exist. Thus, low concentrations of NH_4^+ ions protect cells against diphtheria toxin [183], although they apparently do not inhibit the binding to the cell surface, or the pinocytotic uptake of the toxin, or the activity of the diphtheria A-fragment in a cell-free system [184]. In contrast, NH_4^+ ions do not protect cells against abrin and ricin (unpublished data). It therefore appears that NH_4^+ ions specifically inhibit the penetration of diphtheria toxin through the cell membrane. It is not known whether there exists a similar structure–activity relationship ('haptomer–effectomer') in the case of physiologically important molecules such as certain protein hormones. The hormones consisting of two polypeptide chains, like certain tropic hormones, are particularly interesting in this respect [90, 185]. Studies to test this possibility are hampered by the fact that the entrance into the cytoplasm of one or a few molecules per cell cannot be followed by any of the physical methods currently available. The reason why toxins like abrin and ricin can be traced is their extremely high biological activity and the dramatic consequences of their entrance into the cell. In the case of regulatory proteins like hormones, the effects on the cell are much less, and the mechanism of action on the molecular level is not known. Although certain protein hormones are believed to initiate their action on the cell surface, it is possible that at least some protein hormones penetrate into the cytoplasm. It is an interesting possibility that the toxins have adapted themselves to an uptake mechanism which was originally developed for the uptake of physiologically important proteins. The toxins may therefore prove to be valuable tools in the study of transfer of information through the cell membrane.

Another problem which arises is the role of the toxins in the plant seeds. Apparently the toxins are produced during the maturation of the seeds and disappear after germination [51]. However, as with most lectins their physiological role is not understood. The possibility that the toxins should protect the seeds from being eaten by animals is rather unlikely, at least in the case of abrus seeds. Thus, the seeds are surrounded by a shell which is not completely digested, and the seeds therefore pass the intestine without being dispersed and very little toxin is dissolved. In fact, abrus seeds do not germinate before they have been eaten by birds and have passed their intestinal tract with the result that the surrounding shell is partly digested. In the laboratory it is possible to start the germination by treating the seeds with dilute sulfuric or hydrochloric acid.

These is also the intriguing question of why abrin and ricin are so similar although they are produced in plants which are not taxonomically related (*Leguminosae* and *Euphorbiaceae*). All attempts to detect similar toxins in seeds from a variety of related plants have so far failed. On the other hand, the toxins have many features in common with non-toxic lectins present in related plants. Possibly the A-chain of the toxins has originated as a consequence of a modification in the gene for a peptide chain of a non-toxic lectin.

REFERENCES

1. *A Dictionary of the Economic Products of the Malay Peninsula* (1966) (Ed. by the Ministry of Agriculture and Co-operatives), Kuala Lumpur, Vol. 1, pp. 4–10.
2 Tschirch, A.(1925), *Handbuch der Pharmakognosie*, Tauchnitz, Leipzig, Vol. 2, pp. 749–753.
3. Madaus, G. (1938), *Lehrbuch der biologischen Heilmittel,* Thieme, Leipzig, Vol. 2, pp. 365–369.
4. Stirling, R.F. (1924), *Veterinary J.*, **80**, 473–484.
5. Lewin, L. (1907), in *Realenyclopädie der Gesamten Heilkunde* (Ed. A. Eulenburg), Urban and Schwarzenberg, Berlin, Vol. 1, pp. 107–110.
6. Warden, C.J.H. and Waddel, L.A. (1884), *The Non-Bacillary Nature of Abrus-Poison,* Calcutta.
7. Martin, S. (1887), *Proc. Royal Soc.,* **42**, 331–333.
8. Jonas, **4**, 6.
9. Balint, G.A. (1974), *Toxicology,* **2**, 77–102.
10. Dixon,T. (1887), *Austral. Med. Gazette,* **6**, 137–155.
11. Stillmark, H. (1888), *Über Ricin, ein giftiges Ferment aus den Samen von Ricinus communis L. und anderen Euphorbiaceen,* Inaug.-Dissert., Dorpat, Estonia.
12. Stillmark, H. (1889), *Arb. pharmak. Inst. Dorpat,* **3**, 59–151.
13. Hellin, H. (1891), *Der giftige Eiweisskörper Abrin und seine Wirkung auf das Blut,* Thesis, University of Dorpat, Estonia.
14. Jacoby, M. (1902), *Beitr. Chem. Physiol. Pathol.,* **1**, 51–77.
15. Jacoby, M. (1902), *Beitr. Chem. Physiol. Pathol.,* **2**, 535–544.
16. Jacoby, M. (1901), *Arch. Exp. Pathol. Pharmak.,* **46**, 28–40.
17. Müller, F. (1899), *Arch. Exp. Pathol. Pharmak.,* **42**, 302–322.
18. Hausmann, W. (1902), *Beitr. Chem. Physiol. Pathol.,* **2**, 134–142.
19. Osborne, T.B., Mendel, L.B. and Harris, I.F. (1905), *Amer. J. Physiol.,* **14**, 259–286.
20. Ehrlich, P. (1891, 1892), *The Collected Papers of Paul Ehrlich,* Pergamon Press, London (1957), Vol. 2, pp. 21–44.
21. Clarke, E.G.C. (1953), *J. Pharm. Pharmacol.,* **5**, 458–459.
22. Kobert, R. (1906), *Lehrbuch der Intoxikationen,* Enke, Stuttgart, pp. 695–707.

23. Eisler, M. (1933), *Handbuch der Pflanzenanalyse,* Springer, Wien, Vol. 4, pp. 987–994.
24. Flexner, S. (1897), *J. Exp. Med.* **2**, 197–216.
25. Waller, G.R., Ebner, K.E., Scroggs, R.A., Das Gupta, B.R. and Corcoran, J.B. (1966), *Proc. Soc. Exptl. Biol. Med.,* **121**, 685–691.
26. Dirheimer, G., Haas, F. and Metais, P. (1968), in *Hepatonephrites toxiques,* Mason et Cie, Paris, pp. 45–50.
27. Fodstad, Ø., Olsnes, S. and Pihl, A. Submitted.
28. Lesné, E. and Dreyfus, L. (1908), *Compt. Rend. Soc. Biol.,* **64**, 432–433.
29. Moriyama, H. (1934), *Jap. J. Exptl. Med.,* **12**, 591–599.
30. Jacoby, M. (1924), *Handbuch der Experimentellen Pharmakologie* (Ed. A. Heffter), Springer, Berlin, Vol. 2/2, pp. 1735–1747.
31. Abdülkadir-Lütfi, (1935), *Deutsche Med. Wochenschrift,* **61**, 416–417.
32. Gunsolus, J.M. (1955), *J. Amer. Med. Ass.,* **157**, 779.
33. Dirheimer, G., Haas, F. and Métais, P. (1966), *Compt. Rend. Soc. Biol.,* **160**, 2458–2461.
34. Guggisberg, M. (1968), *Rev. Med. Suisse Romande,* **88**, 206–208.
35. Clarke, E.G.C. and Humphreys, D.J. (1971), *J. Forensic Sci. Soc.,* **11**, 109–112.
36. Clarke, E.G.C. and Humphreys, D.J. (1973), in *Toxins of Animal and Plant Origin* (Ed. A. deVries and E. Kochva), Gordon and Breach, New York Vol. 3, pp. 1075–1080.
37. Lin, J.-Y., Kao, W.-Y., Tserng, K.-Y., Chen, C.-C. and Tung, T.-C (1970), *Cancer Res.,* **30**, 2431–2433.
38. Lin, J.-Y., Liu, K., Chen, C.-C. and Tung, T.-C. (1971), *Cancer Res.,* **31**, 921–924.
39. Onozaki, K., Tomita, M., Sakurai, Y. and Ukita, T. (1972), *Biochem. Biophys. Res. Comm.,* **48**, 783–788.
40. Olsnes, S., Refsnes, K. and Pihl, A. (1974), *Nature,* **249**, 627–631.
41. Refsnes, K., Olsnes, S. and Pihl, A. (1974), *J. Biol. Chem.,* **249**, 3557–3562.
42. Nicolson, G.L., Lacorbiere, M. and Hunter, T.R. (1975), *Cancer Res.,* **35**, 144–155.
43. Olsnes, S., Sandvig, K., Refsnes, K. and Pihl, A. (1976), *J. Biol. Chem.,* in press.
44. Mosinger, M. (1951), *Cont. rend. soc. biol.,* **145**, 412–415.
45. Lin, J.-Y., Lin, L.-T., Chen, C.-C., Tserng, K.Y. and Tung, T.-C. (1970), *J. Formosan Med. Ass.,* **69**, 53–57.
46. Lin, J.-Y., Tserng, K.-Y., Chen, C.-C., Lin, L.-T. and Tung, T.-C. (1970), *Nature*, **227**, 292–293.
47. Tung, T.-C., Hsu, C.-T. and Lin, J.-Y. (1971), *J. Formosan Med. Ass.,* **70**, 569–578.
48. Reddy, V.V.S. and Sirsi, M. (1969), *Cancer Res.,* **29**, 1447–1451.
49. Gürtler, L. and Steinhoff, D. (1972), *Hoppe-Seyler's Z. Physiol. Chem.,* **353**, 1521.
50. Reddy, V.V.S., Choudhry, J.N., Vadlamudi, S., Waravdekar, V.S. and Goldin, A. (1973), *Fed. Proc.,* **32**, Abstr. 2936.
51. Funatsu, M., Hara, K., Ishiguro, M., Funatsu, G., Otsuka, H. and Ide, M. (1973), *Proc. Jap. Acad.,* **49**, 835–839.
52. Mourgue, M., Baret, R., Reynaud, J. and Bellini, J. (1958), *Bull. Soc. Chim. Biol.,* **40**, 1453–1463.

53. Le Breton, E. and Moulé, Y. (1949), *Bull. Soc. Chim. Biol.*, **31**, 94–97.
54. Kunitz, M. and McDonald, M.R. (1948), *J. Gen. Physiol.*, **32**, 25–31.
55. Kabat, E.A., Heidelberger, M. and Bezer, A.E. (1947), *J. Biol. Chem.*, **168**, 629–639.
56. Takahashi, T., Funatsu, G. and Funatsu, M. (1962), *J. Biochem.* (Tokyo), **52**, 50–53.
57. Ishiguro, M., Takahashi, T., Funatsu, G., Hayashi, K. and Funatsu, M. (1964), *J. Biochem.* (Tokyo), **55**, 587–592.
58. Khan, A.H., Gul, B. and Rahman, M.A. (1966), *J. Immunol.*, **96**, 554–557.
59. Tomita, M., Kurokawa, T., Onozaki, K., Ichiki, N., Osawa, T. and Ukita, T. (1972), *Experientia*, **28**, 84–85.
60. Nicolson, G.L. and Blaustein, J. (1972), *Biochim. Biophys. Acta*, **266**, 543–547.
61. Olsnes, S. and Pihl, A. (1973), *Eur. J. Biochem.*, **35**, 179–185.
62. Olsnes, S. and Pihl, A. (1973), *Biochemistry*, **12**, 3121–3126.
63. Olsnes, S., Saltvedt, E. and Pihl, A. (1974), *J. Biol. Chem.*, **249**, 803–810.
64. Nicolson, G.L., Blaustein, J. and Etzler, M. (1974), *Biochemistry*, **13**, 196–204.
65. Saltvedt, E. (1975), Submitted.
66. Wei, C.H., Hartman, F.C., Pfuderer, P. and Yank, W.-K. (1974), *J. Biol. Chem.*, **249**, 3061–3067.
67. Lugnier, A. and Dirheimer, G. (1973), *FEBS Letters*, **35**, 117–120.
68. Ishiguro, M., Takahashi, T., Hayashi, K. and Funatsu, M. (1964), *J. Biochem* (Tokyo), **56**, 325–327.
69. Olsnes, S. and Pihl, A. (1972), *Nature*, **238**, 459–461.
70. Olsnes, S. and Pihl, A. (1972), *FEBS Letters*, **28**, 48–50.
71. Funatsu, M. (1972), in *Proteins, Structure and Function* (Eds. Funatsu, M., Hiromi, K., Imahori, K., Murachi, T. and Narita, K.), Wiley, New York, pp. 103–139.
72. Olsnes, S., Refsnes, K., Christensen, T.B. and Pihl, A. (1975), *Biochim. Biophys. Acta*, **405**, 1–10.
73. Funatsu, M., Hara, K., Ishiguro, M., Funatsu, G. and Kubo, K. (1973), *Proc. Jap. Acad.*, **49**, 771–776.
74. Gürtler, L.G. and Hortsmann, H.J. (1973), *Biochim. Biophys. Acta*, **295**, 582–594.
75. Lin, J.-Y., Shaw, Y.-S. and Tung, T.-C. (1971), *Toxicon.*, **9**, 97–101.
76. Wei, C.H., Koh, C., Pfuderer, P. and Einstein, J.R. (1975), *J. Biol. Chem.* **250**, 4790–4795.
77. Funatsu, M., Funatsu, G., Ischiguro, S., Nanno, S. and Hara, K. (1971), *Proc. Jap. Acad.*, **47**, 713–717.
78. Funatsu, M., Funatsu, G., Ishiguro, M., Nanno, S. and Hara, K. (1971), *Proc. Jap. Acad.*, **47**, 718–723.
79. Funatsu, G. and Funatsu, M. (1970), *Jap. J. Med. Sci. Biol.*, **23**, 342–344.
80. Waldschmidt-Leitz, E. and Keller, L. (1970), *Hoppe-Seyler's Z. Physiol. Chem.* **351**, 990–994.
81. Funatsu, M., Funatsu, G., Ishiguro, M. and Nanno, S. (1970), *Jap. J. Med. Sci. Biol.*, **23**, 264–267.

82. Ishiguro, M., Funatsu, G. and Funatsu, M. (1971), *Agr. Biol. Chem.*, **35**, 729–733.
83. Li, S.S., Wei, C.H., Lin, J.-Y. and Tung, T.-C. (1975), *Biochem. Biophys. Res. Comm.*, **65**, 1191–1195.
84. Funatsu, M., Funatsu, G., Ishiguro, M. and Hara, K. (1973), *Jap. J. Med. Sci. Biol.*, **26**, 30–32.
85. Lin, J.-Y., Cheng, Y.-C., Liu, K. and Tung, T.-C. (1971), *Toxicon*, **9**, 353–360.
86. Nanno, S., Ishiguro, M., Funatsu, G. and Funatsu, M. (1971), *Jap. J. Med. Sci. Biol.*, **24**, 39–42.
87. McPherson, A. and Rich, A. (1973), *FEBS Letters*, **35**, 257–261.
88. Lis, H. and Sharon, N. (1973), *Ann. Rev. Biochem.*, **42**, 541–574.
89. Podder, S.K., Surolia, A. and Bachhawat, B.K. (1974), *Eur. J. Biochem.*, **44**, 151–160.
90. Olsnes, S., Pappenheimer, A.M., Jr. and Meren, R. (1974), *J. Immunol.*, **113**, 842–847.
91. Wei, C.H. and Einstein, R.J. (1974), *J. Biol. Chem.*, **249**, 2985–2986.
92. Wei, C.H. (1973), *J. Biol. Chem.*, **248**, 3745–3747.
93. Ishiguro, M., Funatsu, G. and Funatsu, M. (1971), *Agr. Biol. Chem.*, **35**, 724–728.
94. Lin, J.-Y., Tserng, K.-Y. and Tung, T.-C. (1970), *J. Formosan Med. Ass.*, **69**, 48–52.
95. Lin, J.-Y., Lei, L.-L. and Tung, T.-C. (1969), *J. Formosan Med. Ass.*, **68**, 518–521.
96. Pappenheimer, A.M., Jr., Olsnes, S. and Harper, A.A. (1974), *J. Immunol.*, **113**, 835–841.
97. Olsnes, S. and Saltvedt, E. (1975), *J. Immunol.*, **114**, 1743–1748.
98. Sandvig, K., Olsnes, S. and Pihl, A. (1976), *J. Biol. Chem.* in press.
99. Delga, J. (1954), *Compt. Rend. Soc. Biol.*, **148**, 302–304.
100. Boquet, P. (1939), *Compt. Rend. Soc. Biol.*, **132**, 429–431.
101. Nakamura, K. (1940), *J. Biochem.* (Tokyo), **31**, 311–322.
102. Funatsu, M., Hara, K., Ishiguro, M. and Funatsu, G. (1973), *Proc. Jap. Acad.*, **49**, 754–758.
103. Olsnes, S. (1972), *Naturwissenschaften*, **59**, 497–502.
104. Olsnes, S., Heiberg, R. and Pihl, A. (1973), *Mol. Biol. Rep.*, **1**, 15–20.
105. Montanaro, L., Sperti, S. and Stirpe, F. (1973), *Biochem. J.*, **136**, 677–683.
106. Benson, S., Olsnes, S., Pihl, A., Skorve, J. and Abraham, K.A. (1975), *Eur. J. Biochem.*, **59**, 573–580.
107. Sperti, S., Montanaro, L., Mattioli, A. and Stirpe, F. (1973), *Biochem. J.*, **136**, 813–815.
108. Greco, M., Montanaro, L., Novello, F., Saccone, C., Sperti, S. and Stirpe, F. (1974), *Biochem. J.*, **142**, 695–697.
109. Parisi, B. and Ciferri, O. (1966), *Biochemistry*, **5**, 1638–1645.
110. Parisi, B., Milanesi, G., van Etten, J.L., Perani, A. and Ciferri, O. (1967), *J. Mol. Biol.*, **28**, 295–309.
111. Olsnes, S. and Pihl, A. (1972), *FEBS Letters*, **20**, 327–329.
112. Fernandez-Puentes, C., Benson, S., Olsnes, S. and Pihl, A. (1976), *Eur. J. Biochem.*, in press.

113. Olsnes, S., Fernandez-Puentes, C., Carrasco, L. and Vazquez, D. (1975), *Eur. J. Biochem.*, **60**, 281–288.
114. Carrasco, L., Fernandez-Puentes, C. and Vazquez, D. (1975), *Eur. J. Biochem.*, **54**, 499–503.
115. Onozaki, K., Hayatsu, H. and Ukita, T. (1975), *Biochim. Biophys. Acta,* **407**, 99–107.
116. Nolan, R.D., Grasmuk, H. and Drews, J. (1976), *Eur. J. Biochem.*, Submitted.
117. Sperti, S., Montanaro, L., Mattioli, A. and Testoni, G. (1975), *Biochem. J.*, **148**, 447–451.
118. Montanaro, L., Sperti, S., Mattioli, A., Testoni, G. and Stirpe, F. (1975), *Biochem. J.*, **146**, 127–131.
119. Grummt, F., Grummt, I. and Erdmann, V. (1974), *Eur. J. Biochem.*, **43**, 343–348.
120. Grollman, A.P., Grunfeld, C., Brewer, C.F. and Marcus, D.M. (1974), *Cancer Chemother. Rep.*, **58**, 491–501.
121. Lin, J.-Y., Pao, C.-C., Ju, S.-T. and Tung, T.-C. (1972), *Cancer Res.*, **32**, 943–947.
122. Pardoe, G.I., Bird, G.W.G. and Uhlenbruck, G. (1969), *Z. Immunitätsforsch.*, **137**, 442–457.
123. Van Wauwe, J.P., Loontiens, F.G. and De Bruyne, C.K. (1973), *Biochim. Biophys. Acta,* **313**, 99–105.
124. Shimazaki, K., Walborg, E.F., Jr., Neri, G. and Jirgensons, B. (1975), *Arch. Biochem. Biophys.*, **169**, 731–736.
125. Surolia, A., Ahmad, A. and Bachhawat, B.K. (1974), *Biochim. Biophys. Acta,* **371**, 491–500.
126. Olsnes, S., Sandvig, K., Refsnes, K., Fodstad, Ø. and Pihl, A. (1976), in *Cell Surface Receptors,* Plenum Publishing Co.
127. Kornfeld, S., Eider, W. and Gregory, W. (1974), in *Control of Proliferation in Animal Cells* (Ed. B. Clarkson and R. Baserga), Cold Spring Harbor Laboratory, Cold Spring Harbor, pp. 435–445.
128. Kawaguchi, T., Matsumoto, I. and Osawa, T. (1974), *Biochemistry,* **13**, 3169–3173.
129. Boldt, D.H., MacDermott, R.P. and Jorolan, E.P. (1975), *J. Immunol.,* **114**, 1532–1536.
130. Lee Adair, W. and Kornfeld, S. (1974), *J. Biol. Chem.*, **249**, 4696–4704.
131. Steck, T.L. and Wallach, D.F.H. (1965), *Biochim. Biophys. Acta,* **97**, 510–522.
132. Gottlieb, C., Skinner, S.A.M. and Kornfeld, S. (1974), *Proc. Nat. Acad. Sci.* **71**, 1078–1082.
133. Meager, A., Ungkitchanukit, A., Nairn, R. and Hughes, R.C. (1975), *Nature,* **257**, 137–139.
134. Nicolson, G.L., Lacorbiere, M. and Eckhart, W. (1975), *Biochemistry,* **14**, 172–179.
135. Nicolson, G.L. (1973), *J. Nat. Cancer Inst.*, **50**, 1443–1451.
136. Fairbanks, G., Steck, T.L. and Wallach, D.F.H. (1971), *Biochemistry,* **10**, 2606–2617.
137. Guidotti, G. (1972), *Ann. Rev. Biochem.*, **41**, 731–752.

138. Da Silva, P.P. and Nicolson, G.L. (1974), *Biochim. Biophys. Acta,* **363**, 311–319.
139. Triche, T.J., Tillack, T.W. and Kornfeld, S. (1975), *Biochim. Biophys. Acta,* **394**, 540–549.
140. Hyman, R., Lacorbiere, M., Stavarek, S. and Nicolson, G. (1974), *J. Nat. Cancer Inst.,* **52**, 963–969.
141. Surolia, A., Bachhawat, B.K. and Podder, S.K. (1975), *Nature,* **257**, 802–804.
142. Kaneko, I., Satoh, H. and Ukita, T. (1972), *Biochem. Biophys. Res. Comm.,* **48**, 1504–1510.
143. Feigenson, M.E., Schnebli, H.P. and Baggiolini, M. (1975), *J. Cell Biol.,* **66**, 183–188.
144. Gottlieb, C., Baenzieger, J. and Kornfeld, S. (1975), *J. Biol. Chem.,* **250**, 3303–3309.
145. Kornfeld, S., Lee Adair, W., Gottlieb, C. and Kornfeld, R. (1964), in *Biology and Chemistry of Eukaryotic Cell Surfaces,* (Ed. E.Y.C. Lee and E.E. Smith) Miami Winter Symposium **7**, pp. 291–316.
146. Pappenheimer, A.M., Jr. and Gill, D.M. (1973), *Science,* **182**, 353–358.
147. Gill, D.M. and King, C.A. (1975), *J. Biol. Chem.,* **250**, 6424–6432.
148. Oliver, J.M., Ukena, T.E. and Berlin, R.D. (1974), *Proc. Nat. Acad. Sci.,* **71**, 394–398.
149. Nicolson, G.L. (1974), *Nature,* **251**, 628–630.
150. Nicolson, G.L. (1975), *Amer. J. Clin. Pathol.,* **63**, 677–684.
151. Gonatas, N.K., Steiber, A.S.U., Kim, D., Graham, I. and Avrameas, S. (1975), *Exptl. Cell Res.,* **94**, 426–431.
152. Refsnes, K. and Munthe-Kaas, A. (1976), *J. Exp. Med.,* in press.
153. Huber, H., Douglas, S.D. and Fudenberg, H.H. (1969), *Immunology,* **17**, 7–21.
154. Closs, O., Saltvedt, E. and Olsnes, S. (1975), *J. Immunol.,* **115**, 1045–1048.
155. Kaufmann, S.J. and McPherson, A. (1975), *Cell,* **4**, 263–268.
156. Waldschmidt-Leitz, E. and Keller, L. (1969), *Hoppe-Seyler's Z. Physiol. Chem.,* **350**, 945–950.
157. Waldschmidt-Leitz, E. and Keller, L. (1972), *Hoppe-Seyler's Z. Physiol. Chem.,* **353**, 227–230.
158. Aub, J.C., Sanford, B.H. and Wang, L. (1965), *Proc. Nat. Acad. Sci.* **54**, 400–402.
159. Burger, M.M. and Goldberg, A.R. (1967), *Proc. Nat. Acad. Sci.,* **57**, 359–366.
160. Burger, M.M. (1969), *Proc. Nat. Acad. Sci.,* **62**, 994–1001.
161. Miyake, S., Mizunoe, F., Hozumi, M. Sugimura, T., Koyama, K., Takeuchi, M. and Sugino, Y. (1973), *Gann,* **64**, 383–389.
162. Nicolson, G.L. and Lacorbiere, M. (1973), *Proc. Nat. Acad. Sci.,* **70**, 1672–1676.
163. Nomoto, A., Satoh, H. and Ukita, T. (1975), *J. Biol. Chem.,* **250**, 2703–2708.
164. Kaneko, I., Satoh, H. and Ukita, T. (1973), *Biochem. Biophys. Res. Comm.,* **50**, 1087–1094.
165. Ji, T.H. and Nicolson, G.L. (1974), *Proc. Nat. Acad. Sci.,* **71**, 2212–2216.
166. Nicolson, G.L. and Yanagimachi, R. (1974), *Science,* **184**, 1294–1296.
167. Gahmberg, C.G. and Hakomori, S. (1975), *J. Biol. Chem.,* **250**, 2447–2451.
168. Penhoet, E., Olsen, C., Carlson, S., Lacorbiere, M. and Nicolson, G.L. (1974), *Biochemistry,* **13**, 3561–3566.

169. Bird, G.W.G. (1959), *Vox Sang.*, **4**, 313–317.
170. Harboe, M., Saltvedt, E., Closs, O. and Olsnes, S. (1975), *Scand. J. Immunol.*, **4**, Suppl. 2, 125–134.
171. Bøg-Hansen, T.C. (1973), *Analyt. Biochem.*, **56**, 480–488.
172. Dufau, M.L., Charreau, E.H. and Catt, K.J. (1973), *J. Biol. Chem.*, **248**, 6973–6982.
173. Bessler, W. and Goldstein, I.J. (1973), *FEBS Letters*, **34**, 58–62.
174. Saltvedt, E., Harboe, M., Følling, I. and Olsnes, S. (1975), *Scand. J. Immunol.*, **4**, 287–294.
175. Surolia, A., Ahmad, A. and Bachhawat, B.K. (1975), *Biochim. Biophys. Acta*, **404**, 83–92.
176. Morell, A.G., Gregoriadis, G., Scheinberg, I.H., Hickman, J. and Ashwell, G. (1971), *J. Biol. Chem.*, **246**, 1461–1467.
177. Pricer, W.E., Jr. and Ashwell, G. (1971), *J. Biol. Chem.*, **246**, 4825–4833.
178. Stockert, R.J., Morell, A.G. and Scheinberg, I.H. (1974), *Science*, **186**, 365–366.
179. Ghose, T. and Nigam, S.P. (1972), *Cancer*, **29**, 1398–1400.
180. Hurwitz, E., Levy, R., Maron, R., Wilchek, M., Arnon, R. and Sela, M. (1975), *Cancer Res.*, **35**, 1175–1181.
181. Moolten, F.L. and Cooperband, S.R. (1970), *Science*, **169**, 68–70.
182. van Heyningen, W.E., Carpenter, C.C.J., Pierce, N.F. and Greenough, W.B. (1971), *J. Infect. Dis.*, **124**, 415–418.
183. Kim, K. and Groman, N.B. (1965), *J. Bacteriol.*, **90**, 1552–1556.
184. Bonventre, P.F., Saelinger, C.B., Ivins, B., Woscinski, C. and Amorini, M. (1975), *Infect. Immun.*, **11**, 675–684.
185. Olsnes, S. (1976), *Bull. Inst. Pasteur*, **74**, 85–99.

5 Tetanus Toxin Structure as a Basis for Elucidating its Immunological and Neuropharmacological Activities

B. BIZZINI

The Specificity and Action of Animal, Bacterial and Plant Toxins
(*Receptors and Recognition,* Series B, Volume 1)
Edited by P. Cuatrecasas
Published in 1976 by Chapman and Hall, 11 Fetter Lane, London EC4P 4EE

5.1 INTRODUCTION

Although tetanus has been known since most remote times, it still is today a world-wide problem. Tetanus is a toxi-infection, since the toxin produced by the causal agent is responsible for all the symptoms that accompany the development of the disease. This explains why many attempts have been made to purify the toxin. The isolation of pure tetanus toxin has permitted the investigation of its physico-chemical characteristics and the study of the structure–function relationships. Tetanus toxin is a protein with a fairly large molecular weight which exhibits a wide spectrum of activities. Its most prominent feature is its very high toxicity. The injection of animals with the toxoid resulting from the formaldehyde detoxification of the toxin molecule elicits the formation of two antibody populations, one of which is precipitating and neutralizing, the other being precipitating but not neutralizing. The fragmentation of the toxin molecule and the specific chemical modification of amino acid residues in the molecule have been used for the purpose of further delineating the regions of antigenic reactivity. The results of such work should enable one to understand how neutralization of the toxicity is achieved on combination of the toxin with the antitoxin.

Endowed with an exquisite specificity for the central nervous system, the toxin appears to affect selectively certain types of inhibitory synapses. Therefore, tetanus toxin could serve as a probe into the basic mechanism of the nature of the central inhibition. In addition, tetanus toxin interferes with the neuromuscular apparatus either by blocking the presynaptic release of acetylcholine or altering the function and metabolism of acetylcholine and cholinesterase in striated muscles.

The main questions still to be answered are how the toxin can ascend from the nerve terminals up to the spinal cord, and how lethality is related to binding once the toxin has bound to its receptors. Solution of these questions will throw light on the way in which the central nervous system performs some of its activities, while the unravelling of the mode of action of tetanus toxin will provide one with a means of antagonising specifically the toxin effects in human disease.

5.2 CHEMISTRY OF TETANUS TOXIN

5.2.1 Purification of the toxin

The first step in studying the structure–function relationships of a protein is its isolation in a pure state. Much work has been devoted to the purification of tetanus toxin. However, attempts that were carried out between 1893 and 1945 were unsuccessful in isolating a homogeneous toxin preparation. Consequently, I shall not review those works.

In 1948, Pillemer and associates reported the isolation of the toxin in a crystalline state. The purification was carried out from the culture filtrate using a series of selective precipitations with methanol at low temperature under controlled conditions of pH and ionic strength. The pale-yellow color of the crystalline toxin could indicate that the preparation still contained some extraneous materials. Although it exhibited flocculating activity equal to 3600 Lf per mg N and toxicity equal to 6.6×10^7 MLD per mg N, Pillemer and his group reported hydrodynamic characteristics for it which subsequent authors, except Largier and Joubert (1956), failed to confirm. In the light of recent data, it is likely that the toxin isolated by Pillemer's method may have undergone partial degradation at the temperatures lower than 0°C which are maintained all through the purification schedule. Alterations of the molecular parameters of the toxin molecule on freezing have been observed by Bizzini and Raynaud (1974a, 1975) [see also Section 5.3.3(a) (1)].

The abnormally high value of the flocculating activity of crystalline toxin would further be in accordance with the observation by Bizzini and Raynaud (1974a, 1975) that degraded forms of toxin usually exhibit higher flocculating potencies than intact toxin. An alternative explanation could be that the toxin has undergone moderate proteolysis by proteases released from *Cl. tetani* into the culture medium, enzymatic breakdown of the molecule taking place at locations remote from the toxic site (or regions). In this connection, it appears rather significant that the toxin isolated by Largier (1956), while showing physical characteristics close to those of Pillemer's, had flocculating activity in the range 4300–4800 Lf per mg N. Largier purified the toxin from the culture filtrate by applying a three-stage procedure. The first two steps involved multi-membrane electrodecantation according to Polson's method, carried out first at the isoelectric point of the toxin and then at a pH value removed from it. Final purification was by ammonium sulfate fractionation. Toxicity of the purified preparation amounted to 10^8 MLD per mg N.

Since then, homogeneous toxin preparations have been isolated in various laboratories. Showing toxicities ranging from 10^8 up to 3×10^8 MLD per mg N and flocculating potencies around 3000 Lf per mg N, all the preparations differ definitely in their hydrodynamic features from Pillemer's and Largier's toxin preparations. Purification schedules have taken advantage of the high degree of resolution of ion-exchange chromatography, gel filtration, or preparative polyacrylamine gel electrophoresis. Attempts to isolate toxin by isoelectrofocalization were unsuccessful because of the toxin precipitation inside the column at the end of the separation run (Bizzini, unpublished).

Isolation of pure toxin was performed either from crude filtrates or from bacterial cell extracts. The use of cell extracts for tetanus toxin purification was propounded by Raynaud (1947). It relies on the fact that the toxin is not being discharged in appreciable quantity into the surrounding medium during the logarithmic growth phase. Consequently, the toxin that has accumulated inside the bacterial cells can be extracted from the washed cells before it has diffused outside, with a neutral hypertonic solution consisting of 1M-sodium chloride and 0.1M-sodium citrate.

High concentration of toxin can be achieved in the extracts, which are, in addition, devoid of many contaminants present in the culture filtrates. However, the amount of toxin that is recovered in the extracts is invariably lower than that which would be obtained by allowing toxin to diffuse into the surrounding medium. To account for the just mentioned discrepancies observed between yields of toxin in the extracts and in the crude filtrates, Lettl and coworkers (1966) have postulated that the toxin had to be synthesized during the logarithmic growth phase as a fairly insoluble 'protoxin'. Such protoxin would then be converted into soluble active toxin in the later stages of the cultivation cycle, chiefly during cell lysis. As toxin diffused into the surrounding medium, additional amounts of active toxin would form inside the bacterial cell as a result of autolytic breakdown of the accumulated protoxic structures. However, the existence of a true tetanus protoxin has still to be demonstrated.

Among the numerous papers which have appeared in recent years on tetanus toxin purification, only a few have described isolation of products meeting criteria of homogeneity. When purification was attempted from crude filtrates, apparently homogeneous toxin preparations were isolated by Dawson and Mauritzen (1967), Bizzini *et al.*, (1969), and Holmes and Ryan (1971). The procedure worked out by Dawson and Mauritzen included methanol precipitation of the toxin followed either by a step of ion-exchange chromatography on DEAE-cellulose, the adsorbed toxin being eluted by a stepwise phosphate gradient, or by a combination of DEAE-cellulose and Sephadex G200 chromatography. The latter method yielded products somewhat more toxic (10^8 MLD per mg N) than did the former (5–8 x 10^7 MLD per mg N). Flocculation values of products derived from either technique were in the range 2500–3000 Lf per mg N. Bizzini *et al.*, (1969) have outlined a procedure which applies equally well to purification of toxin from filtrates or extracts. In the case where crude filtrates are employed, purification has to include an additional step of G200 Sephadex filtration to remove toxin aggregates. Holmes and Ryan (1971) succeeded in isolating a highly toxic product by submitting to preparative disc gel electrophoresis the toxin previously precipitated from the culture filtrate by ammonium sulfate at 40 per cent saturation and chromatographed on Bio-Gel P-100. Flocculating activity was not recorded. Toxicity amounted to 3 x 10^8 MLD per mg N. A procedure suitable for toxin purification from both filtrates and extracts was also developed by Matsuda and Yoneda (1974). Combination of ammonium sulfate fractionation, with subsequent ultracentrifugation at 120 000g to remove insoluble materials, and gel filtration on Sephadex G200 resulted in separation of toxin preparations showing toxicities in the range 2–4 x 10^4 MLD per Lf unit (i.e. assuming a mean flocculating value of 3000 Lf per mg N, 6 x 10^7 to 1.2 x 10^8 MLD per mg N).

The toxin extracted from bacterial cells was successfully purified by Murphy and

Miller (1967) who applied successively a step of ammonium sulfate precipitation, DEAE-cellulose and G100 Sephadex chromatography. The highly purified toxin assayed 3000 Lf per mg N and 1.5 x 10^8 MLD per mg N. Subsequently, Bizzini *et al.*, (1969) recovered pure toxin from cell extracts by a procedure including acid precipitation with *in situ* generated metaphosphoric acid, K/K_2 phosphate buffer fractionation at neutral pH, removal of nucleic acids by toxin solubilization in cold 7.5 per cent methanol at pH 4.0, and DEAE-cellulose chromatography. The chromatography step was carried out with 0.01M-EDTA solution (pH 7.6) as toxin solvent and column eluent. Under these conditions toxin passed unretarded through the column, whereas contaminating substances still present were absorbed on DEAE-cellulose. Specific flocculating potency varied from 3050 to 3190 Lf per mg N, and toxicity from 1.5 x 10^8 to 2.0 x 10^8 MLD per mg N. Recently, Robinson *et al.*, (1975), applying the purification procedure of Murphy and Miller (1967) to cell extracts, reported isolation of highly purified toxin preparations characterized by the uncomparably potent toxicity of 6.5 x 10^8 MLD per mg N.

5.2.2 Physico-chemical characteristics of tetanus toxin

Since they differ in some properties, the toxin purified from cell extracts will be referred to further in the text as intracellular toxin while that from culture filtrates will be called extracellular toxin.

Homogeneity of the toxin preparations isolated by Murphy and Miller, Bizzini *et al.*, Holmes and Ryan, Matsuda and Yoneda, and Robinson *et al.*, was supported by disc electrophoretic analysis which revealed the presence of only one band.

Pure toxin is very soluble in neutral buffered solutions, since concentrations as high as 40 000 Lf per ml (8 per cent in protein) can be attained. Solutions are quite stable and can be stored at 4°C for several years without impairment of the activity.

(a) *Optical characteristics*

Absorbancy ratio 280 nm/260 nm measured at 1.0 OD at 280 nm was found to be 2.1 for pure toxin (Murphy and Miller, 1967; Bizzini *et al.*, 1969; Matsuda and Yoneda, 1974).

Absorbance index values reported by various authors for intracellular toxin showed discrepancies. $E_{280\ nm}^{1\%\ protein}$ = 7.8 was found by Murphy and Miller (1967), while Bizzini *et al.*, (1969) and Craven and Dawson (1973) found 12.56 and 11.8, respectively. For extracellular toxin $E_{280\ nm}^{1\%}$ values of 12.54 and 12.4 were indicated by Bizzini *et al.*, and Craven and Dawson, respectively.

(b) *Molecular parameters*

Pillemer *et al.*, (1948) then Largier and Joubert (1956) reported sedimentation coefficients of 4.5S and 3.9S, respectively, for their purified toxin preparations. In addition, Largier and Joubert calculated a molecular weight of 68 000 using the diffusion constant 5.5 x 10^{-7} cm^2/sec determined for their toxin, while they estimated a value of 78 000 for Pillemer's crystalline toxin. However, subsequent investigators failed to isolate a 4.0S tetanus toxin. A large body of evidence has since then accumulated indicating that the sedimentation coefficient of pure tetanus toxin varied in the range 6.0–7.1S (Raynaud *et al.*, 1960; Murphy and Miller, 1967; Dawson and Mauritzen, 1968). Furthermore, Mangalo *et al.*, (1968) determined in the ultracentrifuge a molecular weight of 148 000 ± 8000. This value was further supported by sedimentation velocity studies on tetanus toxin in the strongly dissociating environment of 6 molar guanidine HCl in which a molecular weight of 145 800 ± 5450 was estimated (Bizzini *et al.*, 1973a). Robinson *et al.*, (1975) have recently determined an average molecular weight of 150 000 ± 10 000. The sole divergent value reported in recent years has been that of Dawson and Nichol (1969) who found 176 000 ± 5000.

The molecular weight of tetanus toxin has also been estimated using gel filtration or disc electrophoresis methods. Molecular weights of 127 000 (Murphy *et al.*, 1968) and 140 000 (Holmes and Ryan, 1971) were estimated by gel filtration in the absence of a denaturing agent, and of 150 000 (Bizzini *et al.*, 1973a) in the presence of 6 molar guanidine HCl.

Diffusion constants which have been determined have varied according to the authors and to the method used for the determination. They range from 4.2 to 4.4×10^{-7} cm^2/sec (Murphy *et al.*, 1968) and 4.9×10^{-7} (Dawson and Nichol, 1969) to 5.4×10^{-7} (Mangalo *et al.*, 1968). A frictional ratio of 1.09 was measured by Mangalo *et al.*, (1968) and 1.5 by Dawson and Nichol (1969). Molecular-sieve chromatography allowed calculation of a Stokes radius of 4.74 (Bizzini *et al.*, 1973a).

5.2.3 Chemical characteristics

(a) *Amino acid composition*

The amino acid content of Pillemer's crystalline toxin was incomplete, since it was established using a microbiological method unsuitable for assaying each amino acid species (Dunn *et al.*, 1949).

Since then, several amino acid compositions of homogeneous preparations of tetanus toxin have been published. While analyses reported by Dawson and Mauritzen (1967) and Bizzini *et al.*, (1970), and Murphy *et al.*, (personal communication) and Robinson *et al.*, (1975) were in good accordance with one another, the one presented by Holmes and Ryan (1971) differed essentially from the others. Holmes and Ryan's toxin albeit highly purified, showed an amino acid composition in decisive contrast with the other known analyses of the toxin. Devoid of tryptophan, it has much lower levels of aspartic acid, isoleucine, leucine, tyrosine, phenylalanine, lysine, and arginine residues, whereas the content of glycine is about three times as large as that found by other investigators. Since both physical and toxicological characteristics of Holmes and Ryan's toxin are in the range of those of toxin preparations of similar homogeneity, the question may be raised whether different toxins might not be synthesized by different strains of *Clostridium tetani.* For comparison, amino acid

Table 5.1 Comparative amino acid composition of tetanus toxin (Number of amino acid residues per mole toxin of 150 000 molecular weight)

Amino acids	Dawson and Mauritzen (1967)	Murphy *et al.*, (1969)	Bizzini *et al.*, (1969)	Holmes and Ryan (1971)	Robinson *et al.*, (1975)
Aspartic acid	192	200	200	126	191
Threonine	71	72	67	62	61
Serine	85	98	94	138	86
Glutamic acid	108	111	107	93	104
Proline	52	55	55	43	48
Glycine	55	63	62	185	59
Alanine	46	53	51	79	48
Half-cystine	9	10	10	14	10
Valine	49	61	60	51	48
Methionine	21	23	23	24	19
Isoleucine	106	126	129	74	96
Leucine	102	111	115	72	103
Tyrosine	62	76	81	32	66
Phenylalanine	48	56	58	28	48
Lysine	98	104	106	63	96
Histidine	13	14	14	9	11
Arginine	33	37	35	18	32
Tryptophan	n.d.	13	12	0	7
Total	1150	1283	1279	1111	1133

compositions recently reported are summarized in Table 5.1. In conclusion, tetanus toxin did not show particular features in its amino acid content with respect to that of non-toxic proteins, which could have provided one with a basis for explaining its very high toxicity.

(b) *Number and reactivity of free sulfhydryl groups and number of disulfide links*
The presence and ready availability of free SH groups in the toxin molecule could help in clarifying the conflicting results which have for long obscured the question of the actual molecular state of tetanus toxin. It appears in the light of the most recent work that tetanus toxin does contain 6 free SH groups (Murphy *et al.*, 1968; Bizzini *et al.*, 1970; Craven and Dawson, 1973), which are not directly available for reaction in the native molecule (Murphy *et al.*, 1968; Bizzini *et al.*, 1970). Furthermore, Murphy and associates showed that only 4 among the 6 existing SH groups could be directly titrated after denaturation in 8 molar urea, the presence of 8-hydroxyquinoline-5-sulfonic acid being required for unmasking all 6 SH groups. In contrast, Craven and Dawson found all 6 SH groups to be available in the presence of 8 molar urea, whereas Bizzini and coworkers brought to reaction only 4 SH groups

in 8 molar urea, but they titrated all 6 SH groups in 6 molar guanidine hydrochloride. Although there is agreement on the number of SH groups, the reactivity of such residues is still debated. In addition, Bizzini and associates have obtained, in the course of systematic studies on the reactivity of native toxin, results which suggest that availability of SH groups is dependent on the nature of the reactant used. So, it was observed that no SH group could be detected using the carboxymethylation reaction, while two of such groups reacted on aminoethylation of the toxin molecule. If the reactivity in the latter case is ascribed to a moderate denaturing effect of ethylenimine itself, this does not account for the ready reaction of para-chloromercuribenzoate with all 6 SH groups of the toxin molecule.

Divergence is also to be found between values reported for the number of disulfide links. Determinations carried out by Murphy *et al.*, (1968) and Bizzini *et al.*, (1970), using different techniques, agree on the presence of two disulfide bridges. In contrast, Craven and Dawson (1973) determined the existence of 1.5 S–S bridge when the whole molecule was subjected to analysis, and of only one S–S link when analyses were performed on each of the two separated sub-chains. Such differences are further reflected in the numbers of half-cystine residues reported in amino acid analyses by different investigators.

(c) *Terminal amino groups*

The situation pertaining to the presence of terminal amino groups and their nature is still more controversial. Using intracellular toxin, Murphy *et al.*, (1968) failed to detect the presence of any terminal amino group, whereas Bizzini *et al.*, (1970) identified leucine as the single terminal residue by both dinitrophenylation and dansylation. For extracellular toxin Holmes and Ryan (1971) found glycine to be the unique terminal group by dansylation. More recently, Craven and Dawson (1973) did not succeed in unmasking any amino group in either extracellular or intracellular toxins. The latter results are all the more puzzling if transformation of intracellular into extracellular toxin is being assumed to occur via the breakdown of one or more peptide bonds [see Section 5.3.3(a) (1)], by which amino acid residue(s) would have to be unmasked. No explanation has as yet been afforded to reconcile these conflicting results. It can only be regretted that either Bizzini *et al.*, (1973a) or Craven and Dawson (1973) did not identify terminal amino residues of the toxin sub-chains for which they reported complete amino acid compositions. Such information would certainly have helped to throw light on this much debated question.

(d) *Polarity and hydrophobicity of tetanus toxin*

From the data of the amino acid composition of their highly purified toxin, Robinson *et al.*, (1975) calculated the polarity and hydrophobicity of tetanus toxin. Using the method described by Capaldi and Vanderkoi (1974), they estimated a polarity of 51 per cent (i.e. per cent polar residues), which suggests that tetanus toxin is a high polarity soluble protein. From amino acid compositions reported by

Bizzini *et al.*, (1973a) and Craven and Dawson (1973), respectively, for the high and low molecular weight chains, Robinson *et al.*, calculated similar polarities as for the whole toxin. Using Goldsack's values (1970) for side-chain free energies of transfer, these authors evaluated tetanus toxin hydrophobicity as 1167 cal. When the transfer free energies reported by Nozaki and Tanford (1971) were used for the calculations, hydrophobicity was found equal to 907 cal.

Hence, tetanus toxin appears to be quite polar relative to other proteins, whereas its hydrophobicity is comparable to that of various classes of soluble proteins as reported in the literature.

5.2.4 Structure of tetanus toxin

Although much effort has been expended on elucidating the secondary structure of tetanus toxin, there still exist disagreements between the models presented by various investigators. Besides Bizzini *et al.*, (1973a) other workers who investigated intracellular toxin concluded that it consisted of a single polypeptide chain (Murphy *et al.*, 1968; Craven and Dawson, 1973; Matsuda and Yoneda, 1974). Murphy and associates (private communication) decided this from the results of Sephadex G100 filtration of the reduced and carboxymethylated toxin in the presence of 8 molar urea. Craven and Dawson arrived at the same conclusion by studying the behavior on Bio-Gel A-15M in urea of intracellular toxin subjected either to oxidative sulfitolysis in urea or to mercaptoethanol reduction in 6 molar guanidine HCl. They further substantiated the results of the gel filtration experiments by analysing the toxin in SDS-polyacrylamide gel electrophoresis after treatment in urea either with sulfite or mercaptoethanol. Matsuda and Yoneda (1974) based their conclusion on the results of SDS-disc gel electrophoresis analysis of the intracellular toxin reduced with dithiothreithol. Contradicting the above results are those of Bizzini and coworkers who demonstrated that intracellular toxin had a complex structure. They accumulated evidence for a sub-chain structure of intracellular toxin by studying variously modified toxin derivatives by ultracentrifugation, gel filtration, and polyacrylamide gel electrophoresis under variable environmental conditions. These authors reported gel filtration in urea to be ineffective in resolving the toxin into its constituent sub-units even though it has previously been reduced and carboxymethylated in the presence of 6 molar guanidine HCl. This could explain why neither Murphy *et al.*, nor Craven and Dawson succeeded in dissociating the reduced toxin in urea. However, the possibility cannot be excluded that the latter investigators have been working with structurally different toxins. Bizzini and associates suggested that the toxin is a dimer of molecular weight 146 000 in which the monomers are held together by strong non-covalent interactions. Each monomer consists of two polypeptide chains of molecular weights 52 000 and 21 000, respectively. Although this structural model would allow one to account for the variable molecular weights which were determined for the toxin, it still awaits confirmation.

On the contrary, extracellular toxin has been found to be composed of two non-identical polypeptide chains. However, the size of the sub-units varied somewhat according to the authors. Craven and Dawson (1973) resolved extracellular toxin on gel filtration, after it was subjected to oxidative sulfitolysis in urea or to mercapto-ethanol reduction in guanidine HCl, into one heavy chain of molecular weight 95 000 and one light chain of molecular weight 55 000. Matsuda and Yoneda (1974) estimated by SDS-polyacrylamide gel electrophoresis a molecular weight of 160 000 ± 5000 for the toxin, and molecular weights of respectively 107 000 ± 4000 for the heavy chain (Fragment β) and 53 000 ± 3000 for the light chain (Fragment α) generated by dithiothreitol reduction. A method for the isolation of a stable light chain was described by Helting and Zwisler (1975) consisting of mild formaldehyde treatment of the toxin prior to its reduction with dithiothreitol in urea. This procedure led to separation of a light chain preparation of molecular weight in the range 46 000–48 000, which is somewhat lower than was reported by other authors. Isolation of the light chain was also possible when formaldehyde treatment was omitted, but the yield was slightly reduced. Craven and Dawson (1973) presented amino acid analyses of the heavy and light chains which they separated. The chains show compositional differences. They also indicated that each heavy and light chain contains 4 SH groups one of which is involved in disulfide bond formation. Subsequently, Dawson (1975) extended the work on sulfite action on the toxin molecule, thus showing that considerable scission of both intracellular and extracellular toxins could be brought about when sulfite acted in the presence of copper ions. SDS-polyacrylamide gel electrophoresis analysis of the cleavage products led Dawson to suggest that sulfitolysis in the presence of Cu^{+2} ions was likely to involve cleavage of at least three sites in addition to disulfide breakdown. Furthermore, ethylenediamine tetra-acetate and glycine were shown to prevent the cleavage of both intracellular and extracellular toxins by sulfite.

In the light of the results presented by Craven and Dawson (1973) and Matsuda and Yoneda (1974) it would appear that extracellular toxin may differ from intra-cellular toxin by the breakage of one or more peptide bonds. In support of this assumption Matsuda and Yoneda reported to have 'nicked' intracellular toxin by mild trypsinization. The thus nicked toxin was shown to behave as extracellular toxin on SDS-polyacrylamide gel electrophoresis in the presence of a reducing agent. Recently, Matsuda and Yoneda (1975) reported the isolation of the two consistuent sub-chains from both reduced extracellular and 'nicked' intracellular toxins. The original toxin molecule could be reconstituted from the two separated sub-chains. In Fig. 5.1 the various structures which can be derived from the reports in the literature are illustrated diagrammatically.

5.2.5 Tetanus toxin conformation

Apart from Robinson and associates (1974), nobody has investigated tetanus conformation. These authors studied the circular dichroic spectrum of the toxin between 200 and 310 nm to explore its secondary and tertiary structure. The far-

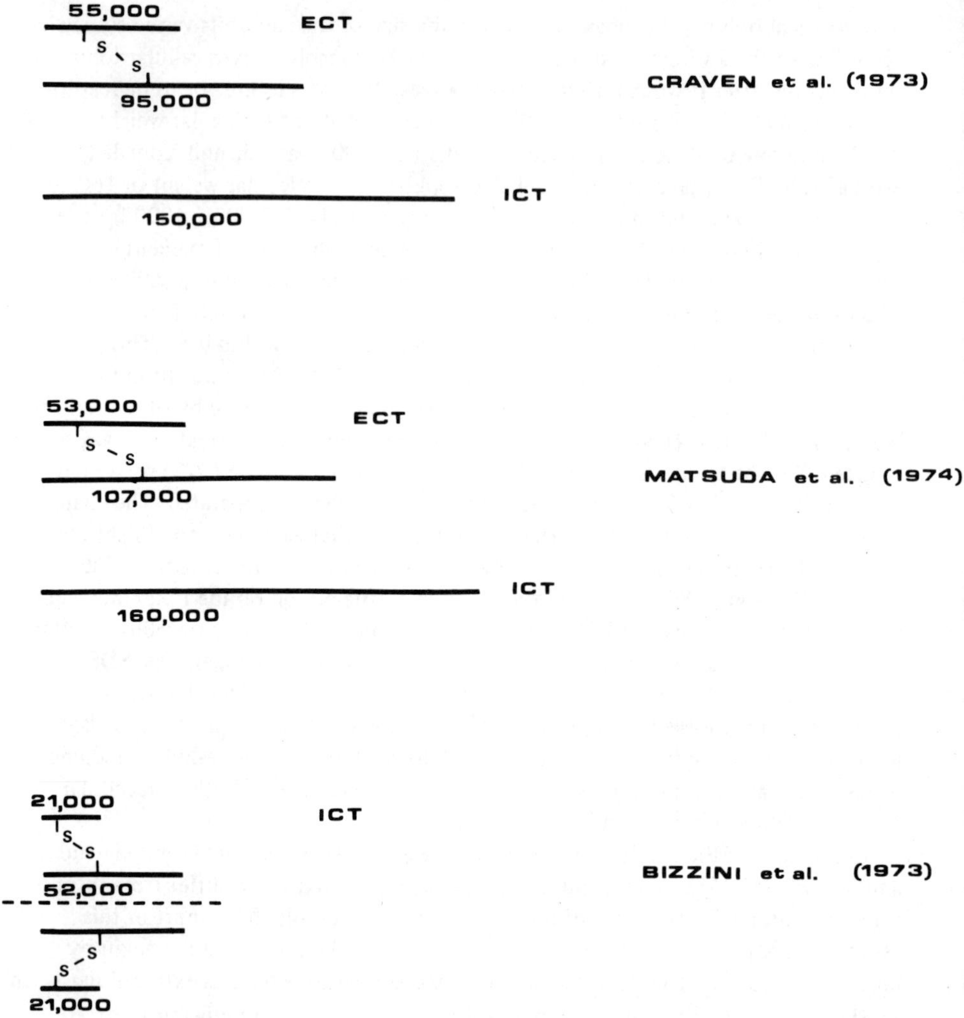

Fig. 5.1 Tentative representation of the structures of extracellular and intracellular tetanus toxins as interpreted from the data of the literature.

ultraviolet circular dichroic spectrum of tetanus toxin has been shown to be characterized by negative extrema at 208 nm and 217 nm with a shoulder at 223 nm. The mathematical treatment of the data gave an estimate of 20 per cent α-helix, 23 ± 3 per cent β-structure, and 57 ± 3 per cent aperiodic structure for the toxin molecule. It thus appears that around 43 per cent of amino acid residues participate in forming a highly ordered secondary structure. Furthermore, careful analysis of the circular dichroic spectrum above 250 nm revealed that many of the aromatic residues of tetanus toxin are constrained in position. The toxin is likely to present a rather rigid conformation adjacent to a highly flexible conformation.

5.3 STRUCTURE – FUNCTION RELATIONSHIPS

Significant advances in characterizing the tetanus toxin physico-chemically have been made recently. Although the primary and tertiary structure of the toxin molecule are as yet unknown, the study of the structure – function relationships has been entered upon using three fruitful approaches. The first has looked for the chemical and conformational changes brought about in the toxin molecule as it was converted into toxoid with formaldehyde. The second approach has investigated the effects of the specific modification of particular amino acid residues on both the immunochemical reactivity and the toxicity of the toxin, and the third has studied the impairment of these activities when the toxin was broken into fragments.

At least three functional groups have been shown to be carried by the toxin molecule (Kryzhanovsky, 1973; Bizzini *et al.*, 1973b). One of these groups is concerned with binding of the toxin to the cell-receptor in the central nervous system. Another group, the toxophore group, the existence of which is still questionable (Habermann, 1973a), would be responsible for the toxicity. The third functional group is involved in binding of the toxin to its antitoxin. However, the last group is complex. It is represented by several antigenic determinant groups. While one was found to be related to the formation of precipitating and neutralizing antibodies by Bizzini *et al.*, (1973b), Nagel and Cohen (1973) identified at least four antigenic determinant groups related to such antibodies. Furthermore, Bizzini and associates demonstrated the presence of an antigenic determinant group responsible for the formation of precipitating but non-neutralizing antibodies.

5.3.1 Chemical and conformational changes induced in the toxin molecule as a result of its conversion into toxoid

It has long been known that proteic toxins lose their toxicity on formaldehyde treatment. Fraenkel-Conrat *et al.*, (1945, 1946, 1947, 1948a,b, 1949) inferred, from the results of systematic studies carried out on model compound, that methylenic bonds may be intramolecular or intermolecular. These authors further demonstrated that methylenic bond formation would result from condensation reactions involving the ϵ-amino group of lysyl residues, on the one hand, and various functional groups present in proteins, such as amide, guanidine, imidazole, phenol, and indole groups, on the other hand. In an attempt to elucidate the mechanism of proteic toxin detoxification by formaldehyde, Blass *et al.*, (1965, 1967, 1969) have investigated the composition of toxoid acid hydrolyzates. The appearance in the hydrolyzates of at least four new acid-resistant compounds, which could not be identified as known amino acids, was established. After isolation and purification, one of these compounds could be chemically characterized and identified as a derivative with the formula shown in Fig. 5.2(a).

The structure of compound (a), which has originated from the linkage of the ϵ-amino group of a lysyl residue to a tyrosyl residue in position 3 of the phenol ring,

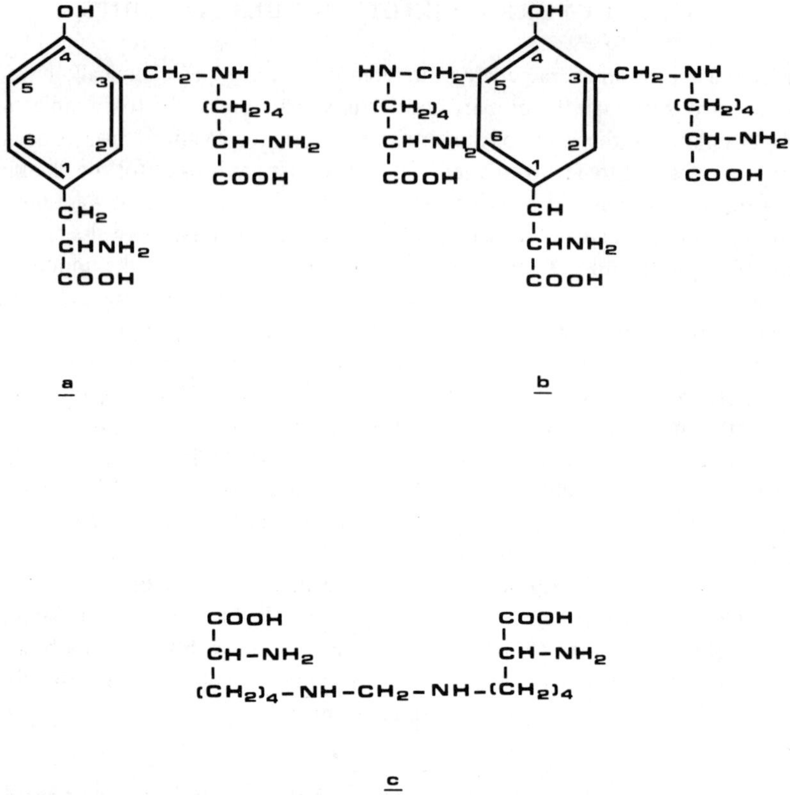

Fig. 5.2 Formulas of new acid-resistant compounds appearing in the acid hydrolyzates of toxoids.

was confirmed by synthesis. This compound can be estimated by automatic amino acid analysis; it is eluted in a position midway between the ammonia and arginine peaks. Blass *et al.*, (1968) further showed the irreversible incorporation of ^{14}C-lysine into the toxin molecule in the presence of formaldehyde. In addition, the presence of exogenous lysine in the toxin detoxification mixture with formaldehyde causes a sharp increase in the amount of one of the new compounds revealed in the toxoid acid hydrolyzates. This derivative was tentatively identified as the compound shown in Fig. 5.2(b). Independently, Dawson and Mautitzen (1969) observed the incorporation of both ^{3}H-lysine and ^{3}H-leucine into tetanus toxin as a result of formaldehyde action. They also stated that the lysine linkage to the toxin molecule was less stable than that of leucine. French and Edsall (1945) have suggested that reaction of formaldehyde with many proteins might lead to the formation of iminomethylol derivatives involving ϵ-amino groups of lysyl residues. However, the formation of iminomethylol derivatives cannot be postulated to account for the lability of linkage of ^{3}H-lysine to the toxin as seen by Dawson and Mauritzen. Indirect evidence was also obtained by Bizzini and Raynaud (1974b) for the formation of the compound shown in Fig. 5.2(c). The amino acid content of tetanus toxoid shows deficits in some amino acid residues relative to the toxin, chiefly in lysyl and tyrosyl residues (Bizzini *et al.*, 1970). The elution of compounds (a)–(c) in separate positions on an automatic amino acid analysis system has permitted the determination of their respective levels. Thus the losses of tyrosyl and lysyl residues from the toxin could fully be accounted for. Furthermore, since the contents of histidyl and arginyl residues are also decreased, it appears probable that these amino acids may be involved in the reaction with formaldehyde.

As was shown by Murphy (1967), conversion of toxin into toxoid results in polymerization. However, isolation by Raynaud *et al.*, (1971) of a 7S 'monomer' tetanus toxoid, which is in the same molecular state as the toxin, clearly indicates that detoxification was not brought about by polymerization. Polymerization which accompanies the detoxification process leads to a decrease of the flocculating activity of the toxoid relative to the toxin. This impairment of the flocculating activity is not observed in the 'monomer' toxoid. It was further demonstrated by Bizzini *et al.*, (1971) that the 'monomer' toxoid is a more potent immunogen after adsorption on to calcium phosphate than toxoids containing polymerized forms. Polymerization during the detoxification process also substantiates the possibility of intermolecular bond formation. This has a direct and important implication in toxoid preparation, because extraneous compounds can be incorporated into the toxoid proper when detoxification is carried out on crude or partly purified toxin samples. As is largely documented, undesirable reactions to toxoid in vaccination are mostly due to impurities still present in the preparations.

Comparison of the circular dichroic spectrum of the tetanus toxoid between 250 and 310 nm with that of the toxin shows that conformational changes have occurred in the toxoid. In fact, the amount of ordered secondary structure of the toxoid is somewhat modified relative to native toxin, since its helical content was found to be equal to 17 ± 1 per cent α-helix, 21 ± 7 per cent β-structure, and 62 ± 8 per cent aperiodic structure. The conformational changes were shown to take place in the vicinity of tryptophyl, tyrosyl, and phenylalanyl residues (Robinson *et al.*, 1975). The most probable modification is the local environment of those tyrosyl residues which enter reaction with formaldehyde to yield methylenic bonds with ϵ-amino groups of closely located lysyl residues.

5.3.2 Effect of the specific modification of particular amino acid residues on the activities of tetanus toxin

Experiments aiming at elucidating the detoxification mechanism of proteic toxins by formaldehyde have pointed to the probable involvement of tyrosyl and lysyl residues in the toxicity of these proteins. In fact, it could not be safely concluded from the accumulated evidence whether detoxification could be ascribed to bolting through methylenic bonds of regions of toxicity expression, or to conformational changes originating from methylenic bond formation, or to modification of tyrosyl and/or lysyl residues located in the toxic site(s), or to cooperation of several of the invoked mechanisms.

The investigation of the effects on the toxin activities of the specific modification of particular amino acid residues has thrown some light on this problem.

(a) *Specific modification of tyrosyl residues*

Modification of tyrosyl residues of tetanus toxin was carried out under apparently non-denaturing conditions of pH and temperature using the highly specific tetranitromethane reaction (Bizzini *et al.*, 1973b). The pattern of reaction has indicated that tyrosyl residues in the molecule were not equally reactive. Effectively about 20 tyrosyl residues out of 81 present were readily available for reaction. However, when the nitration reaction was prolonged for long enough in the presence of a large excess of tetranitromethane, practically all tyrosyl residues (80 out 81) were modified. To account for this result it was postulated that conformational changes had to occur, as nitration was proceeding, which made increasingly more tyrosyl residues available for reaction. It was further assessed that the reaction with tetranitromethane was not restricted to tyrosyl residues but involved tryptophyl residues to an extent varying as a function of the degree of nitration. Polymerization as a result of tetranitromethane action was also observed. However, tryptophyl residue modification and polymerization were shown not to be related to alteration of toxin activity as induced by nitration.

Toxicity was very rapidly decreased as nitration progressed until it attained a residual value when 6 out of 81 tyrosyl residues had been nitrated. At this level of modification conformational changes could not be detected by measuring the reactivity of sulfhydryl groups in the nitro-derivatives relative to that in native toxin. However, the atoxic nitro-derivatives have also lost the ability to fix themselves to the specific cellular receptors for toxin in the central nervous system. Thus, it cannot be decided whether disappearance of the toxicity is not merely consequent upon inability of nitro-derivatives to attach to target cells. Furthermore, low-nitrated toxin derivatives which still expressed some toxicity were found to launch symptoms of poisoning quite different from those elicited by unmodified toxin, since a flaccid rather than a spastic paralysis was observed.

Impairment of the immunological reactivity of the toxin was effective only when substantial nitration had taken place. Qualitatively, changes of immunological reactivity were first detected by formation of a spur of the nitro-derivative (18 to 44 tyrosyl residues modified) with native toxin on double gel diffusion against antitoxin directed to native toxin. However, a second very faint precipitation line was formed with nitro-derivatives in which at least 33 tyrosyl residues were nitrated. When all tyrosyl residues were modified, the thus formed nitro-derivative did not appreciably precipitate with tetanus antitoxin. Quantitative analysis of the reaction of nitro-derivatives with antitoxin yielded results which did not correlate with qualitative data. The amount of antibody nitrogen precipitated from a serum directed to native toxin was not modified with respect to intact toxin until 44 nitro groups had been introduced into the toxin molecule. At this level of nitration the precipitating ability of the nitro-derivative was decreased by about 50 per cent relative to native toxin. At the same time as the precipitating capability was lowered, the ability to give rise to neutralizing antibodies was drastically diminished. The immunogenicity was completely lost in the fully nitrated toxin.

Nitration in the toxoid of those tyrosyl residues which are not involved in a methylenic bond was accompanied by the loss of most of the immunogenicity. This seems to indicate that the toxophore group, if it actually exists, is not directly related to the appearance of neutralizing antibodies.

Bizzini and associates inferred from the above-mentioned results that the toxin molecule was likely to contain 3 kinds of functional groups. Group 1 would be directly associated with the toxicity, being either a toxophore group distinct from or linked to the attachment group. Group 2 would represent antigenic determinants capable of giving rise to the formation of precipitating and neutralizing antibodies, whereas group 3 would elicit the formation of precipitating but non-neutralizing antibodies.

Although toxicity disappearance was not accompanied by chemically detectable conformational changes, one cannot exclude the possibility that such changes have occurred locally. However, it seems highly probable that tyrosyl residues may participate directly in toxicity. This assumption is further substantiated by the observation that transformation of nitro- into amino-derivatives by reduction remains without effect on toxicity. In contrast, the changes noted in the immunological reactivity are likely to be related to conformational changes consequent upon the modification. Such conformational changes are clearly demonstrated in the situation in which formaldehyde treatment of the toxin nitro-derivatives (33 or more tyrosyl residues modified) results in products without immunological relatedness to native tetanus toxoid. A change of charge and/or hydrophobicity of reactive regions could also contribute to alteration of the immunochemical reactivity.

(b) *Specific modification of lysyl residues*

Several groups of investigators have studied the effects of the specific modification of lysyl residues on the toxin activities. These groups have used various methods to carry out the modification. The results obtained have varied according to whether the reaction used did or did not entail changes of the charge or the conformation.

Reactions of the first type included the maleylation and carbamylation reactions; those of the second type included reductive alkylation, amidination, and acetylation.

Reductive alkylation led to somewhat different results according to different authors. Stein and Biel (1973) did not observe any alteration of either the flocculating activity or the toxicity until 72 per cent of lysyl residues had been modified, some loss of toxicity appearing at a level of 73 per cent of modification. Bizzini *et al.*, (1975) did not record any loss of toxicity at a methylation level of 40 per cent, while the toxicity was decreased by 97 per cent when 59 per cent of lysyl residues were involved in methylation. Robinson *et al.*, (1975) reported that the loss of toxicity was dramatic when the extent of methylation was 65 per cent. At the same time the latter investigators determined by circular dichroic spectral analysis a notable loss of the secondary ordered structure in methylated derivatives relative to that of native toxin. However, these authors do not rule out the possibility that some of the structural modifications may be contributed to by denaturation inherent in the methylation process.

Disappearance of the toxicity was achieved by Habermann (1973a) on acetylation of tetanus toxin. However, the extent of modification achieved was not determined. This author further ascertained that only trace amounts of radioactivity accumulated in the spinal cord and the blood serum following injection of ^{125}I-labelled acetylated toxin, whereas acetylated toxin still retained the capacity to react with brain matter *in vitro*. To account for the loss of toxicity, Habermann has suggested that acetylated toxin elimination was accelerated and therefore did not attain its target cells in sufficient amounts. The assay of acetylated toxin by the solid-phase antitoxin method indicated that the immunoreactivity was preserved in acetylated toxin. The latter fact also furthers Kryzhanovsky's view that immunoreactive sites and binding sites to the central nervous system are distinct.

Amidinated toxin and toxoid derivatives have been extensively investigated by Bizzini *et al.*, (1975). They showed that a reduction of toxicity was detectable only in highly amidinated toxin derivatives and was not complete even though the extent of amidination involved as many as 85 per cent of available lysyl residues. These results agree with those obtained on reductive alkylation of tetanus toxin (Bizzini *et al.*, 1975; Robinson *et al.*, 1975). Moreover, amidination did not significantly alter the *in vitro* immunological reactivity as measured by qualitative and quantitative methods. The immunogenicity was also found to be retained in the toxoids prepared from amidinated derivatives or in the amidinated toxoid. At the utmost, a slightly decreased immunogenic potency was exhibited by the toxoid deriving from a toxin derivative in which the extent of amidination was equal to 85 per cent.

Modification of lysyl residues with reactants which affect the charge and/or the conformation of the molecule resulted in more or less pronounced alteration of the toxin activities. Maleylation (Bizzini *et al.*, 1975) was quickly followed by a decrease of toxicity which was completely abolished when the extent of maleylation averaged 50 per cent. However, extents of maleylation which destroyed the toxicity did not impair the immunoreactivity both *in vitro* and *in vivo*. The immunological reactivity was altered only in highly maleylated toxin derivatives with a level of maleylation of 85 per cent. Immunogenicity of the toxoid was also abolished by maleylation.

As was reported by Robinson *et al.*, (1975) carbamylation of 74 per cent of lysyl residues resulted in complete loss of toxicity and immunogenicity of the toxin. Paradoxically, these investigators found that the secondary ordered-structure of carbamylated toxin was less altered than that of the methylated derivative with respect to toxin.

It appears safe to speculate from the above data that lysyl residues are not likely to participate directly in toxicity. In fact, modification of such residues using reactions which do not affect either the charge and/or the conformation of the toxin molecule did not lead in any case to complete disappearance of the toxicity. On the contrary, modification entailing charge and/or conformational changes were followed by early and complete loss of the toxicity. Moreover, the toxic site seems to be much more sensitive than the immunoreactive sites to changes of local environment. So, the statement by Robinson *et al.*, (1975) that the secondary ordered-structure was more altered in the methylated than in the carbamylated derivatives could be accounted for in two ways, first, as was suggested by Robinson *et al.*, (1975), by assuming that the reductive alkylation process brought about some denaturation of the protein, or secondly by postulating that the charge change due to carbamylation when acting on particular lysyl residues could alter in some way the spatial arrangement of discrete regions related to toxicity. Moreover, the loss of immunogenicity of the toxoid consequent to maleylation would further indicate that the expression of this activity is conveyed by a region outside the toxophore group. Although it cannot as yet be ruled out that lysyl residues may play some direct role in toxicity, it appears more probable that such a role is indirect. These residues would interfere with the mechanism of detoxification by formaldehyde by allowing methylenic bonds to establish themselves with amino acid residues critical for toxicity, such as tyrosine or histidine.

(c) *Specific modification of tryptophyl residues*

The modification of tryptophyl residues of tetanus toxin has been carried out independently by Stein and Biel (1973) and by Bizzini *et al.*, (1973b) using different methods. Photo-oxidation of tetanus toxin by ultraviolet irradiation in the presence of methylene blue resulted in complete loss of toxicity. If tyrosine was added to the toxin before photo-oxidation, toxicity was not abolished (Stein and Biel, 1973). Bizzini *et al.*, (1973b) modified tryptophyl residues with Koshland's reagent. They noted that the toxicity was not appreciably impaired until 6 out of 13 tryptophyl residues present in the toxin molecule were modified. In contrast, a dramatic loss of toxicity was recorded when 10 tryptophyl residues had been modified.

However, it is as yet difficult to decide whether the impairment of the toxicity is specifically ascribable to tryptophan modification. In fact, in experiments carried out by Stein and Biel, photo-oxidation could have involved, in addition to tryptophan, tyrosyl and histidyl residues presumably critical for toxicity. On the other hand, results by Bizzini and associates could have originated from steric hindrance effected by the bulky substituent, which the 2-hydroxy-5-nitrobenzyl group represents, or from conformational changes which this might induce.

(d) *Specific modification of histidyl residues*

Implication of histidyl residues in the toxicity of the toxin molecule appears very likely from the data presented by Stein and Biel (1973). Carbethoxylation of tetanus toxin with ethoxyformic anhydride, under conditions of pH which will bring preferentially histidyl residues to reaction, completely abolished the toxicity. However, the lack of data concerning the specificity of the reaction and the extent of the modification does not allow definitive conclusions to be drawn.

5.3.3 Toxic and immunological activities of the sub-chains and fragments of tetanus toxin

Demonstration by Peetoom and van der Veer (1967) and by Cohen *et al.*, (1970) that the tetanus toxin molecule could be degraded by freezing with preservation of some of its activities has provided investigators with an invaluable tool for studying structure–function relationships in toxin. Subsequently, the finding that tetanus toxin had a sub-chain structure has raised the question of the contribution of each polypeptide chain to the toxin activities.

(a) *Structure–function relationships in fragments of tetanus toxin*

(1) *Degradation by freezing.* Freezing of tetanus toxin at –20°C for 4 days resulted in alteration of its antigenic structure (Peetoom and van der Veer, 1967). Degraded toxin exhibited a reduced sedimentation coefficient and gave only a cross-reaction with the unfrozen toxin and the toxoid. At the same time as the flocculation potency was strongly decreased with respect to intact toxin by freezing, toxicity was completely lost. To account for these results the authors postulated that the immunological identity of tetanus toxin and toxoid depended upon their sharing at least two antigenic determinants. One of the antigenic determinants, designated antigen-(1), was destroyed by freezing. Hence, the disappearance of toxicity could go together with loss of part of the antigenic structure in the case where the toxic site was identical with an antigenic determinant. Alternatively, the close location of the toxic site and the antigenic determinant on the toxin molecule could explain why their destruction may occur concomitantly. Peetoom and van der Veer (1967), then Cohen *et al.*, (1970), further reported that antibodies to antigen-(1) and antigen-(2) were both protective, although antigen-(2) was not related to toxicity. To account for the neutralizing activity of both types of antitoxins, Cohen *et al.*, (1970) have speculated that neutralization has to occur by steric inhibition of the toxic site of the molecule. Antigen-(2) was shown to absorb only 25 per cent of the antitoxic activity of rabbit, human, and horse tetanus antitoxins.

Finally, these investigators have established that degradation of the toxin was not caused but merely revealed by freezing, since the presence of degraded toxin could be demonstrated in fresh unfrozen culture filtrates.

Applying to crude toxin the freezing–thawing procedure described by the abovementioned authors, Bizzini and Raynaud (1974a; 1975) found that the toxin molecule was cleaved into two antigenic moieties, cleavage going together with a decrease of 75 per cent in the toxicity but with preservation of the flocculating activity. On immunodiffusion frozen toxin yielded two precipitation lines which fused with the line given by pure toxin. Fractionation of the frozen crude toxin led to isolation and purification of a series of antigenic fragments differing from one another in their toxic and immunological properties. Fragment B-I with molecular weight close to 110 000 daltons still presented toxic and flocculating activities in the range of those of pure toxin. Another fragment with molecular weight of about 100 000 daltons, while showing decreased toxicity, exhibited flocculating activity around 5 000 Lf units per mg N. Fragments A-II and B-II with molecular weights close to 30 000 daltons exhibited only residual toxicity but enhanced flocculating activity in the range of 8 000 to 10 000 Lf units per mg N. While fragment B-I gave a reaction of identity with unfrozen toxin, fragments A-I, A-II, and B-II only cross-reacted. Two further fragments A-III and B-III showed molecular weights lower than 10 000 daltons. Fragment A-III still cross-reacted with intact toxin and displayed residual toxicity. Fragment B-III was devoid of both toxicity and immunoreactivity. Fragment A-II and B-II behaved as defective antigens, precipitating from a monovalent antitoxin about one-fourth of the antibody nitrogen that intact toxin did. These were nevertheless shown to be immunogenic *in vivo*, although to a lesser degree than the whole toxin molecule. Expectedly, the antigenically complete fragment B-I turned out to be definitely more immunogenic than defective fragments A-II and B-II. In experiments aiming at reconstituting a toxoid molecule from the fragments of the toxin, it was further shown that fragment B-III, although not being flocculating, precipitating, or toxic, appeared to contribute to immunogenicity of the reconstituted toxoid, certainly by permitting the fragments to assume the orientation that they have in native toxoid.

Freezing of pure toxin did not result in cleavage of the molecule. Hence, Bizzini and Raynaud (1975) postulated that enzymatic breakdown of one or more peptide bonds was likely to have taken place so as to make freezing successful. This view was further supported by the observation that splitting of pure toxin could be achieved, provided that the toxin was incubated with a fresh culture filtrate prior to freezing. However, the level of enzymatic activity in different culture filtrates varies in a wide range, since it was verified that these were not all suitable for conditioning toxin to freezing. The latter fact would also explain some conflicting results in the literature which indicate that the crude toxin was completely or partly degraded. Moreover, it seems safe to predict that, if an enzymatic attack of the toxin molecule actually occurs, the attack might concern changeable peptide bonds. Therefore, it can be foreseen that more than one split-product will arise.

(2) *Enzymatic breakdown.* Cleavage of the toxin molecule by papain has been reported independently by Bizzini and Raynaud (1974a) and by Helting and Zwisler (1974). Upon papain digestion of pure toxin the former investigators obtained the cleavage of the molecule into two antigenic moieties, each one giving a reaction of identity with each of the antigens released by freezing of crude toxin. Helting and

Zwisler effected complete degradation of the toxin molecule by papain digestion. The resulting antigenic fragment was isolated and purified (Fraction C). The atoxic Fraction C showed a cross-reaction with its parent toxin and exhibited a flocculating activity close to 10 000 Lf units per mg N. Its molecular weight was estimated to be in the range 40 000–45 000 daltons. Fraction C turned out to be a potent immunogen. The latter fragment would represent a portion of the toxin heavy chain.

(b) *Structure–function relationships in polypeptide sub-chains of the toxin*
Helting and Zwisler (1975) have reported the isolation from the extracellular toxin of a stable light chain preparation for which they determined a molecular weight of 48 000 ± 5%. On immunodiffusion this preparation gave a single line of precipitation which cross-reacted with intact toxin. Furthermore, the light chain preparation was shown to be immunologically non-identical with Fraction C released from the toxin molecule by digestion with papain.

Previously, dissociation of the molecule of extracellular but not of intracellular toxin was brought about by sulfite in the presence of denaturants. Two sub-chains, with molecular weights 95 000 and 55 000 daltons respectively, were separated (Craven and Dawson, 1973). Subsequently, Dawson (1975) found that considerable scission of both extracellular and intracellular toxins could be caused by sulfite in the presence of Cu^{+2} ions. Cleavage had to occur by a mechanism other than that of disulfide scission, since it involved three additional sites. The toxins treated with sulfite in the presence of Cu^{+2} ions retained their flocculating activity, although the flocculation time was greatly lengthened. While toxicity was decreased by a factor of 10 000, immunogenicity of the sulfitolysed toxins was hardly changed despite the considerable degree of cleavage. But the toxin structure and activities were preserved from the effects of sulfitolysis by glycine and ethylenediamine tetra-acetate.

5.4 MECHANISM OF ACTION OF TETANUS TOXIN

Tetanus toxin exerts its pathogenic action predominantly on the central nervous system (CNS). Its central action is made manifest in a spastic paralysis with convulsions. It has long been recognized that the toxin interacts with brain matter, and its receptor in nervous cells has been identified as ganglioside. The toxin migrates selectively from the entrance door, where the causing agent produces it, up into the CNS. It has been shown that the spinal cord is the target organ in tetanus intoxication. Tetanus toxin accumulates specifically in the spinal cord, and it has been suggested that it might act at the presynaptic level by interfering with the release of the inhibitory neurotransmitter substance, presumably glycine. In addition to its central action tetanus toxin may exercise obvious peripheral actions.

5.4.1 Tetanus toxin receptor in the nervous cell

As early as 1898, Wassermann and Takaki made known that toxicity of a tetanus toxin solution could be greatly reduced by adsorption of the toxin on to an homogenate of nervous tissue. Marie (1898) further indicated that grey matter displayed a greater toxin-fixing capacity than white matter. Later on, Landsteiner and Botteri (1906) isolated from nervous tissue a complex fraction endowed with enhanced fixing capacity which they designated protagon. This lasted for a half-century until van Heyningen (1959a, b, c) succeeded in isolating and characterizing the toxin-receptor substance in the nervous tissue as a ganglioside.

Gangliosides

Gangliosides are water-soluble sialomucolipids. Structurally they consist of a non-polar ceramide moiety (predominantly sphingosine-stearic acid) linked to a polar moiety composed of a polysaccharide chain with one or more N-acetylneuraminic acid groups. The many gangliosides which have been characterized up to now differ in the polysaccharide moiety composition. In the nervous tissue four types of gangliosides have been found to predominate. Their polysaccharide chain contains two galactose, one N-acetylglucosamine, and one to three N-acetylneuraminic acid residues. These are also the gangliosides endowed with toxin-fixing activity. Fixation of the toxin requires the presence of at least two N-acetylneuraminic acid residues joined to each other by a neuraminidase-sensitive bond, one of the acid residues being attached to the lactose residue (van Heyningen and Miller, 1961). However, it appears that both neuraminidase-sensitive as well as neuraminidase-insensitive gangliosides may participate in toxin binding. In addition, neuraminidase-sensitive gangliosides treated with neuraminidase retained the toxin-fixing capacity in proportion to their residual N-acetylneuraminic acid level (Bernheimer and van Heyningen, 1961). Removal of the N-acetylneuraminic acid residues or methylation of the carboxyl groups of the latter residues resulted in abolition of toxin-fixing capacity (Mellanby and van Heyningen, 1967). Once fixed to the ganglioside, tetanus toxin does not interfere with the rate of release of N-acetylneuraminic acid by neuraminidase. Importance of N-acetylneuraminic acid residues for toxin fixation is reflected in the observation by Kryzhanovsky (1973) that tetanus toxin could be released from its association with protagon by neuraminidase action. Moreover, there was a strict correlation between the relative amounts of both toxin and N-acetylneuraminic acid liberated from the toxin–protagon complex by neuraminidase.

Toxin-fixing capacity of gangliosides was enhanced by complex formation with cerebroside. The enhancing effect was optimal for a well-defined proportion of cerebroside to ganglioside of 3 to 1 (van Heyningen and Mellanby, 1968). Hence, it appeared most likely that glanglioside exists in nervous tissue as insoluble complexes formed by the association of gangliosides with insoluble compounds, such as cerebrosides or possibly phospholipids, cholesterol, and proteins. Such ganglioside complexes are probably part of the receptor for tetanus toxin in nervous tissue.

Toxin binding to ganglioside was not found to be essentially affected either by temperature or pH changes (van Heyningen, 1959b) although fixation was better at pH 7.0 than at pH 5.0. In contrast, it was shown that calcium ions interfered with the binding process. In the latter case, raised temperature and raised calcium concentration acted cooperatively to reduce the amount of bound toxin (Mellanby and Pope, 1975). Toxin binding was also not substantially altered when toxin was combined with antitoxin prior to its addition to protagon (Bondartchuk *et al.*, 1971).

Some degree of inactivation of various bacterial toxins was reported by van Heyningen (1959c), North and Doery (1961), and van Heyningen (1963). However, fixation by ganglioside of tetanus toxin appeared to be specific, since inactivation was shown not to be related to the binding process. It was further demonstrated that tetanus toxin fixation did not result in its inactivation (van Heyningen and Mellanby, 1973).

Kryzhanovsky *et al.*, (1970) showed that toxin bound to protagon could be neutralized by the specific antitoxin through a three-stage process. Addition of a small amount of antitoxin led to ready neutralization of about half of the bound toxin, most probably the toxin which was fixed on the protagon surface. Further addition of a large excess of antitoxin did not effect neutralization until beyond a sharp limit where complete neutralization occurred on addition of only a minute amount of antitoxin. The appearance of a plateau in the course of the neutralization was ascribed to steric hindrance effects, probably because of the burying of the toxin inside the protagon structure. This view was supported by the observation that $F(ab')_2$ and $F(ab')$ antitoxin fragments were much more effective in neutralizing the toxin bound to ganglioside–cerebroside complex, since the plateau seen in the course of neutralization with the whole antitoxin molecule was shortened, and the smaller the molecular weight of the antitoxin fragments the greater was the shortening (Kryzhanovsky *et al.*, 1972).

However, the fixing capacity of gangliosides is not restricted to the toxin. Tetanus toxoid was found to fix to ganglioside, although to a lesser degree than the toxin (Fulthorpe, 1956; van Heyningen, 1959b; Mellanby and van Heyningen, 1965). Binding of serotonin and strychnine to isolated gangliosides was also reported by van Heyningen (1963) and by Woolley and Gommi (1965), whereas fixation failed to occur to ganglioside–cerebroside complex (van Heyningen and Mellanby, 1968). Much interest has focused on gangliosides themselves. They are highly acid substances. The presence of both hydrophobic and hydrophilic groups in gangliosides would account for their propensity to associate in micelles. This particular feature of ganglioside structure would also be consistent with their occurrence in biological membrane structures, to the activity of which they contribute by specifically interfering with ion movements (McIlwain 1960, 1961). This author have further postulated that ganglioside function in biological membranes would not only be to serve as mobile ion-carriers but also to provide hydrophilic pathways across the lipids of the membrane.

Spence (1969) has reported that gangliosides could no longer be extracted from tissues depleted in sodium and potassium ions. He suggested that the monovalent cations acted by associating themselves with the ionic carboxyl groups of N-acetylneuraminic acid residues. Hence, molecules would interfere with gangliosides by ionic interaction involving the negatively charged groups of N-acetylneuraminic acid and the positively charged groups of the interfering molecules.

Clowes *et al.*, (1972) have investigated the interaction of tetanus toxin with artificial model membranes. The membranes consisted of ganglioside included in egg-yolk lecithin and egg-yolk lecithin–cerebroside bilayers. Measurements of electrical and optical properties of such membranes after application of the toxin have indicated that the molecule is likely to have assumed a flat configuration on the surface of the membrane. Moreover, the toxin molecules were distributed as a single layer on the membrane surface. Despite the flattening of the molecule, it must have retained its essential structure, since it still reacted with its specific antitoxin. However, the mode of binding of the toxin to ganglioside is still unclear. It appeared likely that more than mere electrostatic attraction was involved. In fact, binding was observed under conditions of pH at which both toxin and ganglioside were negatively charged. So as to account for binding it was postulated that electrostatic repulsion was overcome by the presence in the toxin binding site of positively charged groups. If one follows the hypothesis of Clowes *et al.*, (1972), the toxin–ganglioside membrane interaction should be limited at the membrane surface without disruption of it. Therefore, if saccharide or N-acetylneuraminic acid residues do play a role in synaptic transmission, it may be speculated that tetanus toxin will interfere at this level passively rather than actively. This view is strengthened by the lack of any detectable lesion of the CNS as a result of tetanus intoxication.

5.4.2 Tetanus toxin pathways to the central nervous system

For many years the question of how the toxin reaches its target cells in the CNS remained unanswered. However, Marie and Morax (1902), Meyer and Ransom (1903), and Friedmann *et al.*, (1939, 1941) had suspected that tetanus toxin migration occurred through a neural pathway. Subsequently, convincing evidence has accumulated in support of the view that tetanus toxin movement from its site of injection or formation to its site of primary action is by neural spread (Wright, 1955, 1956; Kryzhanovsky, 1965, 1966, 1967; Fedinec, 1965, 1967; King, 1970; Habermann, 1970; Gardner, 1972). However, according to Wright and to Fedinec the toxin did not move along axons but through the endoneurium of such axons. Whatever the case may be, Kryzhanovsky traced with the aid of pharmacological methods the spread of the toxin from the injected muscle up to the CNS. The toxin injected in the muscle was taken up by the associated nerve endings and migrated along the corresponding nerve trunk via the ventral roots and the fibers of ventral roots within the spinal cord to be accumulated in the anterior horns of the respective spinal segments. This neural ascent of the toxin was established both in tetanus *descendens* and *ascendens* using various animal species (Kryzhanovsky *et al.*, 1961; Kryzhanovsky,

1965, 1966, 1967). Habermann (1970, 1971, 1972) and Wellhöner *et al.*, (1973a) with the aid of isotopic techniques confirmed and extended the results of the Russian investigators. Habermann (1970) found that the toxin moved along the nerves at a rate of about 5 mm per hour. Moreover, Habermann and Dimpfel (1973) and Stöckel *et al.*, (1975) ascertained that toxin injected intravenously could not cross the blood–brain barrier. The latter authors postulated that the accumulation of the toxin injected intravenously could be accounted for by its uptake from the blood circulation in motor nerve terminals to be transported retrogradely from several parts of the body to the CNS (Stöckel *et al.*, 1975). Such a spread would further be supported by the slight, specific labelling of motor neurones on the contralateral side of injection. In fact, the demonstration has been repeatedly made that the toxin does not move in the spinal cord from the side ipsilateral to the toxin injection to the contralateral side (Sherrington, 1905; Acheson *et al.*, 1942; Brooks *et al.*, 1957; Wilson *et al.*, 1960; Kryzhanovsky and Dyakonova, 1964; Sverdlov and Alekseeva, 1966; Kano and Takano, 1969; Takano and Henatsch, 1973, Paar *et al.*, 1974). Wellhöner *et al.*, (1973a) corroborated this view by providing it with quantitative ground.

Although there is general agreement that tetanus toxin moves up to the spinal cord via the ventral roots, only recently has information been presented on which tissue compartment in the ventral roots is involved in the toxin ascent. In 1974, King and Fedinec convincingly localized the toxin in the nerve in its epineurium, while the endoneural compartment may be free of toxin. Erdmann *et al.*, (1975), Price *et al.*, (1975), and Stöckel *et al.*, (1975) have provided direct evidence for a retrograde intra-axonal transport of tetanus toxin. Such demonstration has been based on autoradiographic studies. The strategy adopted has varied according to the investigators. Price and his colleagues have crushed proximally and distally, at various time intervals, the nerve supplying the muscle injected with toxin before removing the nerve for autoradiographic examination. Thus, they could visualize the intra-axonal but not the extra-axonal labelling of the nerve. They also established that only the toxin injected intramuscularly and not that injected along the course of the nerve was taken up in the intra-axonal compartment. A rate of transport of the toxin in excess of 75 mm per day could be estimated on the basis of the time of appearance of tracer in the nerve cell body. So as to ascertain the retrograde axonal transport of tetanus toxin, Stöckel and associates carried out experiments in which they measured the differential accumulation of label in the nerve cell bodies between injected and non-injected sides, on the one hand, and tested transections of the nerve or blockade of the fast axoplasmic transport by colchicine, on the other hand. These authors inferred from their results that the retrograde transport of tetanus toxin may depend on properties of the neuronal membrane which are common to all neurones, such as the presence of structural gangliosides. In effect, it was verified that the toxin retrograde transport could be interrupted by simultaneous injection of ganglioside. Hence, the toxin transport might depend on the binding of the toxin to gangliosides of the neural membrane. The fact that administration of neuraminidase prior to that of tetanus toxin blocked the transport of the latter provides further support for this view. Like the above investigators, Erdmann and his coworkers arrived at the conclusion that tetanus toxin movement is via a retrograde intra-axonal route. Furthermore, they observed that the bulk of radioactivity in the ventral roots was confined to a few axons in which the label was absent from the endoneurium and from the flat mesothelium covering the ventral roots. In the spinal cord radioactivity was also detected in only a few motor neurones, and more precisely in the half-segments supplying the injected muscle. The perikaryon and dendrites were particularly rich in label, the cell nucleus less so. In the peripheral nerves the bulk of radioactivity appeared to be located in the epineurium. As in Stöckel's experiments, application of colchicine to the nerve abolished the neural ascent of the toxin.

It seemed likely (Price *et al.*, 1975; Schwab and Thoenen, 1975) if the toxin acts presynaptically that a mechanism should enable the toxin to pass across the synapse from motor neurones to inhibitory nerve-endings. Based on autoradiographic and morphometric data convincing evidence has been provided by Schwab and Thoenen (1975) of such a trans-synaptic movement of the toxin.

Localized tetanus is well-known in experimental animals after injection of the toxin either intramuscularly or subcutaneously. Conversely, most often generalized tetanus has developed in man when diagnosis is established. Therefore, the question has been raised as to whether the toxin in human tetanus also reached the CNS through neural spread. This seems to be the case, since Seib *et al.*, (1973) could demonstrate that, in cat generalized tetanus provoked by intravenous injection of the toxin, the toxin ascent occurred along the motor axons up to the CNS. However, toxin can also move along nerves devoid of motor fibers, as was reported by Hensel *et al.*, (1973).

Although the neural pathway of toxin spread is soundly established, the mechanism by which the toxin can move along the nerves is still to be elucidated. In the past, it has been hypothesized (Kryzhanovsky, 1966) that the neural ascent of the toxin may rely on intrinsic anatomical, functional, and biochemical properties of the pathway itself and on the interference of the latter with the toxin. In support of this assumption Stöckel and his colleagues (1975) stated that the transport may depend on the binding of the toxin to gangliosides of the neuronal membrane.

5.4.3 Fate, distribution, and localization of the toxin

The distribution of ^{125}I-labelled toxin and toxoid in various animal species has been extensively studied by Habermann and his associates. Administration of tetanus toxin intravenously into the rat resulted in the disappearance of about half of the injected toxin from the blood stream within thirty minutes. Most of the toxin was taken up by the liver, whereas the residual toxin was distributed among bones, muscles, and skin. At the same time only trace amounts of toxin were detected in the brain and spinal cord. After one day 20 per cent of the label was still present in the blood plasma, a large amount of it being excreted in urine and faeces. Much of the label had appeared in the spinal cord and the caudal part of the brain stem, whereas both cerebrum and cerebellum remained essentially free of radioactivity (Habermann, 1970; Habermann and Wellhöner, 1972; Habermann and Dimpfel, 1973). Tetanus toxin was found to be more avid for grey matter than for white matter (Habermann, 1973b), and a five-fold to six-fold enrichment in label of grey matter with respect to white matter was measured (Habermann *et al.,* 1973). However, the label was not distributed evenly in grey matter, the greater amount of it being located in the ventrolateral part rich in motor neurones.

Mellanby *et al.,* (1965) have demonstrated that the toxin binds preferentially to the crude synaptosomal fraction of brain and spinal cord homogenates. Habermann *et al.,* (1973) substantiated this observation by showing that the label in the spinal cord of rats intoxicated with ^{125}I-labelled toxin was to be found in a fraction consisting of mitochondria, synaptic structures, and other unidentified granula. The bound label was not displaced by cold toxin, and only poorly so by antitoxin.

Resorption of the toxin injected intramuscularly into rats occurred with delay in comparison with intravenously injected toxin. Thus, one-tenth of the toxin was still present at the site of injection 46 hours later. Intramuscular injection was reflected by an invariably low toxin blood level. The spinal cord was seen to accumulate increasing amounts of label as a function of time, a large quantity of it being excreted in urine and faeces (Habermann, 1970). Independently, Wellhöner *et al.,* (1973a) have studied distribution of the toxin in the spinal cord, in the dorsal and ventral roots and in the peripheral nerves after intramuscular or intracutaneous injection. While the toxin persisted for long at the site of injection, it was observed that the label after 24 hours was strictly confined to the spinal cord segments supplying the injected muscle. Ventral roots were highly labelled, whereas dorsal roots had fixed only a low amount of radioactivity. It was further demonstrated that toxin injected subcutaneously provoked localized tetanus, although only the skin had been damaged. Intravenous injection into cats of ^{125}I-labelled toxin resulted in a gradient of radioactivity in the peripheral nerves which increased from the periphery up to the CNS. On ligating the sciatic nerve an accumulation of radioactivity formed at the ligature (Hensel *et al.,* 1973). Gardner and Fedinec (1975) have investigated by immunohistochemical methods the localization in the rat sciatic nerve of intramuscularly injected toxin. They demonstrated by bioassay that even 16–24 hours after injection the nerves still contained biologically active toxin. The toxin was shown to be present in the epineurium, perineurium, endoneurium, and axons of both proximal and distal nerve segments. Furthermore, Fedinec (1975) published the highly significant observation that the toxin localized in the axons and in the endoneurium was antigenic but non-toxic, as ascertained by bioassay, whereas the one fixed to the epineurium and perineurium was both antigenic and toxic. In addition, it was reported that a long period of unilateral stimulation at different sites led to a more intense enrichment in label of the pertinent spinal cord half segments (Wellhöner *et al.,* 1973b).

Further evidence of the specific labelling of the motor neurones was provided by

Dimpfel *et al.*, (1975) who observed that exposure of primary tissue cultures derived from embryonic mouse brain and spinal cord to ^{125}I-labelled tetanus toxin induced preferential labelling of neuronal cells with respect to glial cells and fibroblasts. From their results Dimpfel and his associates inferred that the toxin *in vivo* could reach only those neurones that have direct connection with the periphery. This would explain why the pathology of tetanus will adopt physiological features pertaining to those neurones that are accessible to toxin.

The great majority of recent work has been carried out with ^{125}I-labelled toxin. Therefore, the question has been raised as to whether the labelled toxin does actually substitute for native toxin. In fact, the sub-chain structure of tetanus toxin might justify the suspicion that the label, which is traced in cellular structures, could represent simply a degradation product of the toxin. However, the fact that Habermann (1973b) and Habermann *et al.*, (1973) failed to demonstrate degradation into trichloroacetic-soluble split-products after treatment of ^{125}I-labelled toxin with brain and spinal cord homogenates in the physiological pH range furthers the view that both ^{125}I-labelled toxin and native toxin are similar. Further support was provided by SDS-polyacrylamide gel analysis of the label extracted from the tissues of animals intoxicated with ^{125}I-labelled toxin which indicated it to be indistinguishable from native toxin. Moreover, the tissue-bound toxin had retained its immunoreactivity. However, according to Habermann *et al.*, (1973), until we are able to separate labelled from unlabelled toxin it cannot be ruled out that the ^{125}I-labelled toxin may represent a 'double' of the toxin. In this connection it should be recalled that the bound toxin in the CNS amounts to less than 1 per cent of the injected toxin. This merely means that 1 DL_{50} of toxin for the rat, assuming an even distribution of it between the spinal cord and brain stem, would correspond to as little as 10^{-15} mole.

5.4.4 Mode of action

In localized tetanus of the rat, the toxin has been shown to be present in the CNS before the onset of the first symptoms of intoxication, and to persist there for several weeks after cessation of the symptoms (Habermann, 1972). A similar dissociation between the onset of the symptomatology and the occurrence of label in the CNS was also demonstrated in generalized tetanus. In fact, symptoms became apparent at a time when the radioactivity in the CNS had already begun to decrease, while the toxin was still demonstrable in the CNS after complete disappearance of the symptoms. Friedmann *et al.*, (1939) and Laurence and Webster (1963) had previously argued, from the observation of the similarity of the toxin's behavior in both localized and generalized tetanus, that the latter might well reflect a multiple localized tetanus. This view has been strengthened by a recent report of Habermann and Dimpfel (1973).

It is now well documented that the spastic paralysis with convulsions in tetanus intoxication is the result of the action of the toxin on the central reflex apparatus. This action would exert itself either by blocking presynaptically the release of the inhibitory neurotransmitter(s) [glycine, γ-aminobutyric acid (GABA)] or by interfering with the postsynaptic action of these neurotransmitters (Curtis and de Groat, 1968; Fedinec and Shank, 1971; Curtis *et al.*, 1973). In addition, tetanus toxin affects the neuromuscular apparatus, either by blocking the presynaptic release of acetylcholine (Ambache *et al.*, 1948a, b; Muchnik and Rubinstein, 1967; Mellanby *et al.*, 1968; Diamond and Mellanby, 1970; Mellanby, 1971; Mellanby and Thompson, 1972) producing a flaccid paralysis, or by impairing the metabolism of acetylcholine or choline-acetyltransferase (Leonardi, 1961; Leonardi *et al.*, 1972; Duchen and Tonge, 1973; Leonardi, 1973; Tonge *et al.*, 1975).

Takano and Kano (1973) have also provided evidence that a concomitant action of the toxin on both gamma and alpha motor neurones, while increasing gamma bias, is likely to play a role in spastic paralysis. According to Mellanby (1975) it would therefore be a gross oversimplification to consider the spastic paralysis in tetanus intoxication as the mere result of a blockade of postsynaptic inhibition in the spinal cord.

(a) *Spinal inhibition*

From the observation that both strychnine and tetanus toxin exercise comparable effects on the CNS, it was concluded that the two substances acted in the CNS by similar mechanisms (Sherrington, 1905). In the same way as it was demonstrated for strychnine (Bradley *et al.*, 1953), tetanus toxin was shown to cause disinhibition of the spinal cord motor neurones by depressing the process by which impulses in inhibitory presynaptic fibers generate an inhibitory postsynaptic potential of motor neurones (Brooks *et al.*, 1957). Moreover, Sverdlov and Alekseeva (1965) reported the prolonged presynaptic inhibition of monosynaptic reflex discharges in motor neurones by tetanus toxin. Subsequently, Curtis and de Groat (1968) have indicated that action of tetanus toxin on the CNS differs from that of strychnine in that the former blocks only synaptic inhibition whereas the latter blocks also the inhibitory effect of glycine. So, on the assumption that glycine or a glycine-like amino acid is the actual inhibitory neurotransmitter, tetanus toxin should interfere only with the release of this amino acid from inhibitory synaptic terminals without reducing the sensitivity of the postsynaptic membrane of motor neurones to glycine. Recently, Curtis *et al.*, (1973) found that tetanus toxin, while lowering the synaptic release of glycine, exercised its effects apparently at the presynaptic level on both presynaptic GABA-mediated and postsynaptic inhibition in the cerebellum.

(b) *Excitatory and inhibitory neurotransmitter substances*

Electrophysiological studies have confirmed that tetanus toxin affects synaptic inhibition (Brooks *et al.*, 1957; Curtis and de Groat, 1968; Curtis *et al.*, 1973). Furthermore, Eccles (1964) suggested that inhibition was mediated via interneurones. The iontophoretic studies by Curtis and Watkins (1960) have also strongly furthered the view that particular amino acids may act as synaptic inhibitory or excitatory synaptic neurotransmitters. The excitatory effect of both L-glutamic acid and L-aspartic acid, and the inhibitory effect of glycine and GABA, was demonstrated by application of these amino acids in the proximity of spinal motor neurones. Further support for the role of glycine and GABA as putative neurotransmitters was provided by their occurrence to high concentrations in spinal cord grey matter and roots (Werman, 1965; Aprison and Werman, 1965). However, the distribution pattern of these amino acids has not allowed a definite conclusion to be drawn as to whether either (or both) of the compounds is actually an inhibitory neurotransmitter. Although both glycine and GABA, when applied locally by iontophoresis, did inhibit the firing of motor neurones, the loss of interneurones was accompanied by a decrease in the concentration of glycine but not of GABA (Davidoff *et al.,* 1967a). This would suggest that GABA, which is not released from interneurones, is not likely to represent the major synaptic neurotransmitter involved in spinal inhibition (Davidoff *et al.,* 1967b). Conversely, GABA would be a sound candidate for mediating inhibition in the cerebral cortex, cerebellum, and lateral vestibular nucleus (Aprison and Werman, 1968; Krnjevic, 1970; Fonnum *et al.,* 1972).

The discovery of specific uptake systems in the CNS for these amino acids has strengthened further their role as putative neurotransmitters in the CNS. Such a system has been described for glycine in the spinal cord but not in the cerebral cortex (Neal and Pickles, 1969; Johnston and Iversen, 1971; Logan and Snyder, 1971). This high affinity system for glycine was localized in a spinal cord area which is rich in synaptosomal structures. Recently, the binding of strychnine to synaptic membranes isolated from the spinal cord and the brain stem has been reported by Young and Snyder (1974). Such binding was reduced by the presence of glycine. It appears likely that the binding of strychnine may represent an interaction with the postsynaptic receptor for the neurotransmitter actions of glycine. Moreover, the fact that the affinity of glycine for receptor binding is about 10 000 times lower than that of strychnine is consistent with existing neurophysiological evidence that the affinity of strychnine for glycine receptor exceeds that of glycine.

GABA binding to synaptosomal membranes of rat brain has also been reported (Zukin *et al.,* 1974). In contrast to glycine, GABA showed higher levels of binding in the cerebellum and lower levels in the medulla oblongata-pons and the spinal cord. Furthermore, the ability of amino acids to mimic the synaptic inhibitory properties of GABA was paralleled by their inhibitory potency with regard to the binding of GABA to synaptic membranes.

Investigating the effect of localized tetanus in cats on glycine level in the spinal cord, Semba and Kano (1969) noted a significant decrease in the glycine content of grey matter on the toxin injected side, whereas concentrations of GABA and other amino acids remained essentially unchanged. This again furthers the view that glycine but not GABA is involved, at least in cat, in mediating spinal inhibition. Concurrently, Osborne and Bradford (1973) designed an ingenious method to test the release of amino acids from deposits of synaptosomes on nylon gauze (beds of synaptosomes) on electrical stimulation. They ascertained that the stimulated release of amino acids was not affected by incubating the synaptosome beds in the

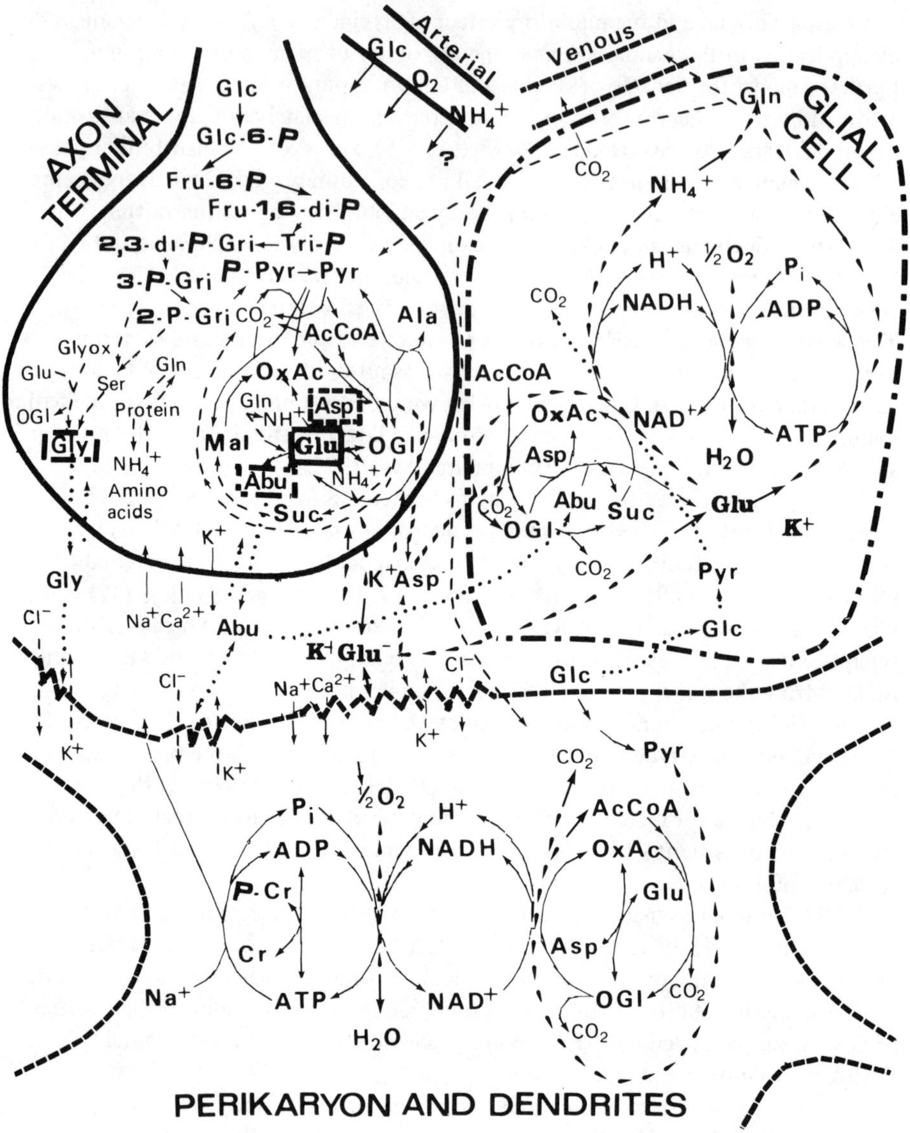

Fig. 5.3 Metabolic pathways induced in the release, action, and uptake of putative amino acid transmitters in the central nervous system. Redrawn after Watkins (1972).

presence of high toxin concentrations. Conversely, stimulated release of glycine and to a lesser degree of GABA was reduced when beds were used which consisted of synaptosomes isolated from the spinal cord of rats previously injected with tetanus toxin. The concomitant reduction in the level of the excitatory amino acid aspartate which occurred was accounted for by assuming that the release of the natural neurotransmitters may involve common mechanisms. Osborne and Bradford did not assess any change in levels of endogenous amino acids, except those of glutamate and GABA, which were increased.

(c) *Metabolic activities in the central nervous system*

The role of glycine and GABA as putative inhibitory neurotransmitters seems now to be well documented. Many questions are still debated with regard to the biosynthetic pathways of formation of these amino acids, the mechanism involved in their release from presynaptic membranes, and the modes of their uptake and action on the postsynaptic membrane. Another question is how these different processes are related to the overall metabolic activity in the CNS. Resolving these questions will, no doubt, greatly help in elucidating the pathogenesis of tetanus. An exhaustive review has been devoted to the metabolic regulation of the putative amino acid neurotransmitters (Watkins, 1972), and the reader is referred to it for further details. Reproduced from the above-quoted review is Fig. 5.3 which depicts the complex relationships existing between the protein metabolism of glial and neuronal cells, the putative amino acid neurotransmitters, either excitatory or inhibitory, and ionic movements at the synaptic membranes.

Inhibitory neurotransmitter substances do appear to exercise their action by hyperpolarizing neuronal perikaryal membranes as a result of an increase in the membrane permeability to both Cl^- and K^+ ions (Fig. 5.4). In fact, electrical stimulation or a high K^+ concentration in the medium entails a selective output of the excitatory amino acids glutamate, aspartate, and GABA, the latter acting as excitatory under certain circumstances. Movements of these amino acids and ions are accompanied by marked metabolic changes which are reflected in enhanced incorporation of the carbon units of ^{14}C-labelled glucose into these amino acids. According to Watkins (1972) many problems would be clarified if changes in amino acids were coupled to energetic metabolic pathways. In fact, uptake of amino acids is an active process requiring energy. In addition, there exists a complex ionic dependence of the uptake process. For instance, the uptake of GABA and glycine appears to be dependent on the presence of Na^+ ions.

An interesting observation has been reported by Kryzhanovsky and associates (1974a) who noted that in both localized and generalized tetanus there was a significant increase in activity of the transport Na, K-ATPase as measured in subcellular fractions isolated from the spinal cord of intoxicated rats. Na,K-ATPase activity was found to be enriched in the crude mitochondrial and microsomal fractions. A change of enzymatic activity was ascertained only on the side of toxin injection in localized tetanus, and not on the side opposite to it. Concurrently, the Ca-ATPase activity was found to be unaltered, whereas the Mg-ATPase activity was increased only in generalized tetanus.

The fact that tetanus toxin did not affect the activity of the various ATPases,

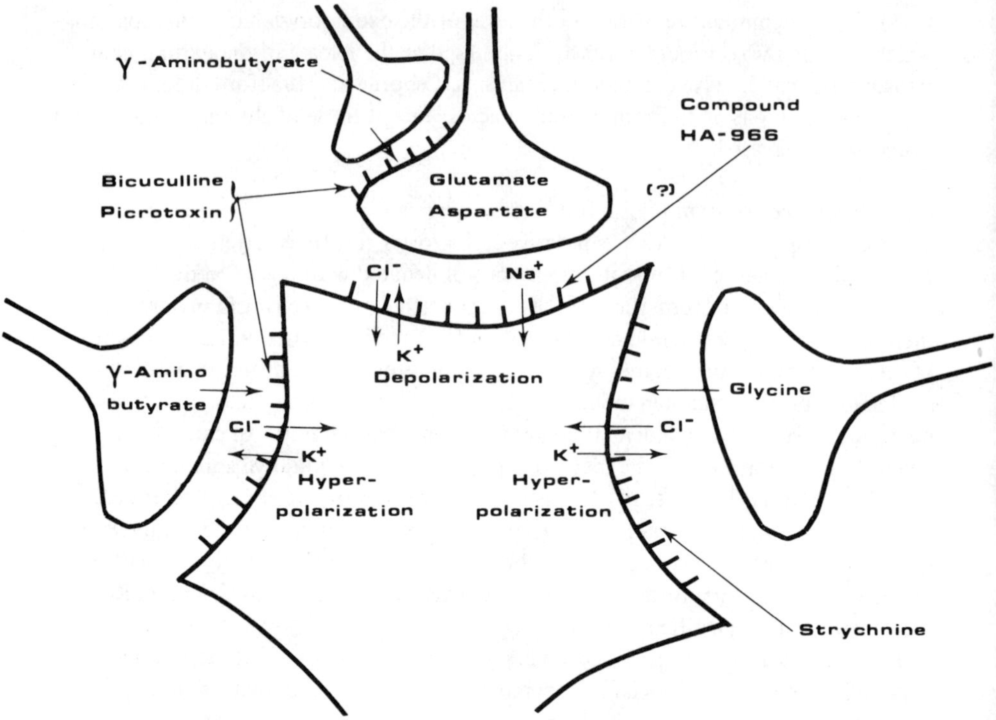

Fig. 5.4 Putative transmitter function of amino acids in the central nervous system. Redrawn after Watkins (1972).

when it was added to subcellular fractions isolated from the spinal cord of healthy animals, leaves open the question as to whether the toxin in tetanus intoxication influences the membrane Na,K-ATPase directly or indirectly.

Additional metabolic changes in the course of tetanus intoxication were also reported by the Russian group of Kryzhanovsky. So, although the significance of the actomyosin-like protein (AMLP) for the function of nervous tissue is still unknown, it appears likely that this protein may play a role both in storage and excretion of synaptic neurotransmitter substances. The actomyosin-like protein was found to be localized in the membranes of synaptic vesicles (Puszkin and Beri, 1972). The sharp increase in the passive transport of Ca^{2+} ions into synaptoplasm during nervous impulse transmission could elicit, at the expense of ATP hydrolysis, the contraction of AMLP which would lead to a secretion of neurotransmitter substances within the synaptic cleft at the contact of the synaptic vesicles with presynaptic membranes.

In addition to its contractile ability, AMLP is endowed with Ca,Mg-ATPase activity and choline acetyltransferase activity, the latter being linked to its myosin-like moiety. It was first demonstrated that tetanus toxin was capable of binding specifically to AMLP. Binding resulted in a decrease in contractile ability associated with the inhibition of Ca,Mg-ATPase activity. It was further shown that these effects of the toxin could be antagonized by high levels of ATP. Hence, Kryzhanovsky *et al.*, (1974b) suggested that the toxin might act by blocking the ATP-hydrolyzing center of the myosin-like moiety. On the other hand, according to these authors the fact that tetanus toxin abolishes the contractile ability of AMLP would make it likely that AMLP is involved in the neurotransmitter secretion process.

Additional changes in the protein composition of brain synaptic structures in rats with localized tetanus were also reported by Kryzhanovsky *et al.*, (1975a). They used polyacrylamide gel electrophoresis to test the protein content of extracts of synaptic structures isolated from the spinal cord of rats in the course of tetanus intoxication. The presence of new proteins was brought to light in Triton X-100 extracts consisting of soluble and membrane-bound proteins, more precisely in the extracts of synaptic membranes, whereas no change was revealed in sodium dodecylsulfate extracts containing structural proteins. Simultaneously it was shown that tetanus toxin affected the *in vitro* protein biosynthesis in synaptosomes by stimulating the incorporation of ^{3}H-labelled lysine into the total proteins of the rat brain cortex. A significant effect was achieved by a toxin concentration as low as 15 μg (1000 MLD, mouse) per ml. Further support for these observations was provided by Kryzhanovsky *et al.*, (1975b) who reported the occurrence of proteins exhibiting new antigenic properties in the synaptosomal fraction isolated from the spinal cord of rats with localized tetanus. Demonstration of the formation of these new antigens was made by adsorbing antisera directed to synaptosomes isolated from the spinal cord of rats suffering from localized tetanus with synaptosomes deriving from the spinal cord of healthy animals. Since such sera still reacted with synaptosome or synaptic membrane fractions of the spinal cord but not of the brain from rats intoxicated with the toxin, it appeared likely that new antigenic substances have appeared as a result of the tetanus intoxication.

Recently, evidence has accumulated that cyclic nucleotides may be involved in the functioning of the CNS. In particular, they may maintain the integrity of the synaptic membranes (for reviews: Greengard and Costa, 1970; Robison *et al.*, 1971; Greengard and McAfee, 1972). In addition, it now seems likely that cyclic-AMP does play a role in the physiology of the synaptic transmission within the CNS (Greengard *et al.*, 1972). In this connection, evidence has been provided for a biochemical action of cyclic AMP in nervous tissue (Miyamoto *et al.*, 1969a,b). It was suggested that such action might be mediated by a cyclic AMP-dependent protein kinase whose function would be to catalyze the phosphorylation of various proteins by ATP. The precise point in regulation of neural function would be the phosphorylation of key proteins. Kuo and Greengard (1969a,b) have postulated a unifying hypothesis that cyclic AMP might exercise its wide variety of actions by regulating the activity of protein kinase. However, the proteins which are involved in reaction with such kinases are as yet unidentified.

Research in this direction has most often remained beyond the scope of experimental means. Within the past few years, Greengard and McAfee (1972) have taken advantage of a preparation of mammalian superior cervical sympathetic ganglion which relative to brain is much more simple in its neural composition and its synaptic organization. Using this they could demonstrate that transmission in the inhibitory pathway was blocked by the same pharmacological agents as prevented an increase in cyclic AMP in response to preganglionic nerve stimulation. Conversely, they observed that blocking the excitatory pathway did not prevent the increase in the content of cyclic AMP.

In recent experiments, King and Fedinec (1975) have revealed that tetanus toxin could affect cyclic nucleotide metabolism. In toxin-induced paralysis of the iris they observed the temporary reversal towards myosis of the toxin action by theophylline, a phosphodiesterase inhibitor, as well as by cyclic nucleotides. In contrast, isoproterenol, (N-isopropylnoradrenaline), an antagonist of cyclic AMP action, produced enhanced dilatation of the pupil. According to these authors, one of the fundamental toxophoric effects of tetanus toxin derives from toxin interference with cyclic nucleotide metabolism.

(d) *Morphologic alterations*

Attempts have been made to correlate such metabolic changes with morphologic alterations of the synaptic structures. Electron miscroscope autoradiographic studies have revealed that tetanus intoxication is accompanied by changes in the motor neurone structures. Kryzhanovsky *et al.,* (1973) have reported that the nucleus, the cytoplasm, and the Golgi apparatus were affected by tetanus toxin. An accumulation of flattened vesicles in the presynaptic terminals of axosomatic synapses in the passive state was often observed, whereas the membrane of presynaptic endings was frequently somewhat thickened. Schwab and Thoenen (1975) have further demonstrated that the synapses containing mainly 'flat' vesicles in contrast to those containing mainly round vesicles were not influenced by tetanus toxin. Round vesicles, at the same time as showing a decrease in density, were converted to some degree into flattened vesicles.

Uchizono (1965) has suggested that synapses containing 'flat' vesicles might represent inhibitory synapses. In fact, Matus and Dennison (1971) found that ^{3}H-labelled glycine incorporation in spinal cord slices occurred in terminals containing 'flat' vesicles but not in those containing only round vesicles.

(e) *Peripheral effects*

In addition to its major central action, tetanus toxin is also likely to exercise peripheral effects. The effect of toxin bound to skeletal muscle in producing ionic flux disturbances was demonstrated by Zacks and Sheff (1965). Subsequently, these authors found, using immunocytochemical methods, that tetanus toxin was localized both in the T-system and in the sarcotubular system of muscle directly involved in the relaxation–contraction cycle (Zacks and Sheff, 1968). They have maintained the view that tetanus toxin interferes in muscles with ionic flux in the sarcotubular system. Apart from an increase in the number of synaptic vesicles, which often filled the whole free space of the axon terminal, Kryzhanovsky *et al.,* (1971) have not observed other gross structural alterations. In support of a neuromuscular action of tetanus toxin, Kaeser and Saner (1969) reported that neuromuscular transmission may be blocked in severe tetanus. This blockade is similar to that following systemic administration of magnesium sulfate. In that instance the muscle becomes unresponsive to indirect, but remains fully responsive to direct, electrical stimulation. Duchen and Tonge (1973) also achieved the inhibition of neuromuscular transmission in slow muscle (*soleus*) by intramuscular injection of the toxin, whereas fast muscle (*extensor digitorum longus*) was not affected. Furthermore, tetanus toxin induced a drop in the wet weight of both slow and fast muscles lasting for 14–28 days. At the same time the content of choline acetyltransferase increased until it attained its maximum by 28 days. Augmentation was more pronounced in fast than in slow muscle. According to these authors, the effect of tetanus toxin at the neuromuscular junction is not likely to be due to a deficiency in the rate of acetylcholine synthesis.

Injury of the nephron was found to be associated with the administration of very high doses of tetanus toxin. Hence, it was postulated that the kidney might well be a major target organ in tetanus intoxication. The acute tubular necrosis that developed could cause electrolyte imbalance which could further contribute to the neuromuscular manifestations (Sottiurai and Fedinec, 1975).

5.5 CONCLUSION

The considerable achievements which have been made within the past few years in our knowledge of the chemistry of tetanus toxin have opened an approach to the understanding of many of its activities. The isolation of the toxin in a pure state has helped in determining unequivocally the fate and the distribution of the toxin in the animal species into which it was injected. Pure toxin also made possible the detailed exploration of its neural spread to the CNS. The use of pure toxin will also enable one to search for the formation of degradation products, once the toxin has fixed to the CNS.

The localization in the toxin molecule of discrete regions related to the expression of a definite type of activity will no doubt contribute to solving fundamental problems pertaining to the immunological and neuropharmacological activities of the toxin. Thus, the recognition that a part of the toxin molecule is concerned with its attachment to the receptor in the target cell assumes great significance. The first aim will be at isolating and characterizing chemically the attachment site. The second aim will be at answering the question at to whether the part of the toxin molecule involved in binding is: (1) still toxic; (2) antigenic; (3) still endowed with the capacity to move along the nerves; (4) capable of interfering with the *in vivo* binding of the toxin.

Should the binding part of the toxin still be toxic, then this could mean that either the isolation procedure did not succeed in excluding the toxophore group or the attachment of this part of the molecule is itself responsible for determining pathogenic disorders. In the case of the toxic site being antigenic, we would have the possibility of raising antibodies specifically directed to the attachment site of the molecule; it is safe to predict that such antibodies would most probably be endowed with high therapeutic efficiency. Furthermore, injection into animals of the complex formed after reaction of the toxin with these antibodies should provide a means of answering the still debatable question (van Heyningen, 1975) of how the toxin binding is related to its toxicity.

Kryzhanovsky's view that tetanus toxin is a polysystemic disease has recently received many confirmations. The widespread activities of tetanus toxin make it not only an invaluable tool for investigating the functioning of the CNS but also a sound candidate for studying the disturbance mechanisms launched in many other systems of the organism. An example among others is illustrated by the observation reported by Mellanby and George (1975) that characteristic experimental epilepsy could be induced in rats by injecting the toxin into the hippocampus of these animals. Thereby, a means is provided for studying the genesis and control of clinical epilepsy.

Moreover, the central effects of tetanus toxin have been ascribed to the formation of 'generators of the pathologically enhanced excitation' (GPEE) (Kryzhanovsky, 1975). Thus, muscle rigidity would originate from GPEE in the spinal efferent output and convulsions from GPEE in the system of spinal afferent input. A main challenge for future research is to explain why such GPEE arise.

No doubt our knowledge of the chemistry of tetanus toxin and that of its interference with its substrate(s) in the organism will grow and therapy of tetanus will be considerably improved on becoming more and more specific.

REFERENCES

Acheson, G.H., Ratnoff, O.D. and Schoenbach, E.B. (1942), *J. Exp. Med.*, **75**, 465–480.

Ambache, N., Morgan, R.S. and Wright, G.P. (1948a), *J. Physiol.* (Lond.), **107**, 45–53.

Ambache, N., Morgan, R.S. and Wright, P.G. (1948b), *Brit. J. Exp. Path.*, **29**, 408–418.

Aprison, M.H. and Werman, R. (1965), *Life Sci.*, **4**, 2075–2083.

Aprison, M.H. and Werman, R. (1968), *Neurosciences Research*, Vol. 1 (S. Ehrenpreis and O.C. Solnitzky, Eds.), Academic Press, New York.

Bernheimer, A.W. and van Heyningen, W.D. (1961), *J. Gen. Microbiol.*, **24**, 121–127.

Bizzini, B., Turpin, A. and Raynaud, M. (1969), *Ann. Inst. Pasteur*, **116**, 686–712.

Bizzini, B., Blass, J., Turpin, A. and Raynaud, M. (1970), *Europ. J. Biochem.*, **17**, 100–105.

Bizzini, B., Turpin, A. and Raynaud, M. (1971), *Ann. Inst. Pasteur*, **120**, 791–799.

Bizzini, B., Turpin, A. and Raynaud, M. (1973a), *Naunyn-Schmiedeberg's Arch. Pharmacol.*, **276**, 271–288.

Bizzini, B., Turpin, A. and Raynaud, M. (1973b), *Europ. J. Biochem.*, **39**, 171–181.

Bizzini, B. and Raynaud, M. (1974a), *Compt. Rend. Acad. Sci.*, **279**, 1809–1812.
Bizzini, B. and Raynaud, M. (1974b), *Biochimie* (Paris), **56**, 297–303.
Bizzini, B. and Raynaud, M. (1975a), *Ann. Inst. Pasteur,* **126**, 159–176.
Bizzini, B., Turpin, A., Carroger, G. and Raynaud, M. (1975b), *Proc. Fourth Internat. Conf. on Tetanus,* Dakar, April 6–12 pp. 145–158. Ed. Fondation Mérieux, Lyon.
Blass, J., Bizzini, B. and Raynaud, M. (1965), *Compt. Rend. Acad. Sci.*, **261**, 1448–1449.
Blass, J., Bizzini, B. and Raynaud, M. (1967), *Bull. Soc. Chim. Fr.*, **10**, 3957–3965.
Blass, J., Bizzini, B. and Raynaud, M. (1968), *Ann. Inst. Pasteur,* **115**, 881–898.
Blass, J., Bizzini, B. and Raynaud, M. (1969), *Ann. Inst. Pasteur,* **116**, 501–521.
Bondartchuk, N.G., Kirilenko, O.A., Kryzhanovsky, G.N. and Rozanov, A.Ya., (1971), *Byull. Eksp. Biol. Med.*, **72**, 71–74.
Bradley, K., Easton, D.M. and Eccles, J.C. (1953), *J. Physiol.*, **122**, 474–488.
Brooks, V.B., Curtis, D.R. and Eccles, J.C. (1957), *J. Physiol.* (Lond.), **135**, 655–672.
Capaldi, R.A. and Vanderkoi, G. (1974), *Proc. Nat. Acad. Sci. USA,* **69**, 930–932.
Clowes, A.W., Cherry, R.J. and Chapman, D. (1972), *J. Mol. Biol.*, **67**, 49–57.
Cohen, H., Nagel, J., van der Veer, M. and Peetoom, F. (1970), *J. Immunol.*, **104**, 1417–1423.
Craven, C.J. and Dawson, D.J. (1973), *Biochim. Biophys. Acta,* **317**, 277–285.
Curtis, D.R. and Watkins, J.C. (1960), *J. Neurochem.*, **6**, 117–141.
Curtis, D.R. and de Groat, W.C. (1968), *Brain Res.*, **10**, 208–212.
Curtis, D.R., Felix, D., Game, C.J.A. and McCulloch, R.M. (1973), *Brain Res.*, **51**, 358–362.
Davidoff, R.A., Shank, R.P., Graham, L.T., Jr. Aprison, M.H. and Werman, R. (1967a), *Nature,* **214**, 680–681.
Davidoff, R.A., Graham, L.T., Jr. Shank, R.P., Werman, R. and Aprison, M.H. (1967b), *J. Neurochem.*, **14**, 1025–1031.
Dawson, D.J. (1975), *FEBS Letters,* **56**, 175–178.
Dawson, D.J. and Mauritzen, C.M. (1967), *Aust. J. Biol. Sci.* **20**, 253–263.
Dawson, D.J. and Mauritzen, C.M. (1968), *Aust. J. Biol. Sci.*, **21**, 559–568.
Dawson, D.J. and Mauritzen, C.M. (1969), *Aust. J. Biol. Sci.*, **22**, 1217–1227.
Dawson, D.J. and Nichol, L.W. (1969), *Aust. J. Biol. Sci.*, **22**, 247–255.
Diamond, J. and Mellanby, J.H. (1970), *J. Physiol.* (Lond.), **210**, 186–187.
Dimpfel, W., Neale, J.H. and Habermann, E. (1975), *Naunyn-Schmiedeberg's Arch. Pharmacol.*, **290**, 329–333.
Duchen, L.W. and Tonge, D.A. (1973), *J. Physiol.* (Lond.), **228**, 157–172.
Dunn, M.S., Camien, M.L. and Pillemer, L. (1949), *Arch. Biochem. Biophys.*, **22**, 374–377.
Eccles, J.C. (1964), *The Physiology of Synapses,* Academic Press, New York.
Erdmann, G., Wiegand, H. and Wellhöner, H.H. (1975), *Naunyn-Schmiedeberg's Arch. Pharmacol.*, **290**, 357–373.
Fedinec, A.A. (1965), *Recent Advances in the Pharmacology of Toxins,* pp. 125–138, Pergamon Press, New York.
Fedinec, A.A. (1967), *Principles on Tetanus* (Ed. L. Eckmann), Huber, Berne.
Fedinec, A.A. (1975), Proc. Fourth Internat. Conf. on Tetanus, April 6–12, Dakar, pp. 123–144. Ed. Fondation Mérieux, Lyon.
Fedinec, A.A. and Shank, R.P. (1971), *J. Neurochem.*, **18**, 2229–2234.

Fonnum, F., Storm-Mathisen, J. and Walberg, F. (1972), *Brain Res.*, **21**, 259–269.

Fraenkel-Conrat, H., Cooper, M. and Olcott, H.S. (1945), *J. Am. Chem. Soc.*, **67**, 950–954.

Fraenkel-Conrat, H. and Olcott, H.S. (1946), *J. Am. Chem. Soc.*, **68**, 34–37.

Fraenkel-Conrat, H., Brandon, B.A. and Olcott, H.S. (1947), *J. Biol. Chem.*, **168**, 99–118.

Fraenkel-Conrat, H. and Olcott, H.S. (1948a), *J. Am. Chem. Soc.*, **70**, 2673–2678.

Fraenkel-Conrat, H. and Olcott, H.S. (1948b), *J. Biol. Chem.*, **174**, 827–835.

Fraenkel-Conrat, H. and Mecham, D.K. (1949), *J. Biol. Chem.*, **177**, 477–486.

French, D. and Edsall, J.T. (1945), *Adv. Protein Chem.*, **2**, 277–335.

Friedmann, U., Zuger, B. and Hollander, A. (1939), *J. Immunol.*, **36**, 473–484, 485–488.

Friedmann, U., Hollander, A., and Tarlov, I.M. (1941), *J. Immunol.*, **40**, 325–364.

Fulthorpe, A.J. (1956), *J. Hyg.* (Lond.), **54**, 315–327.

Gardner, D.P. (1972), Doctoral Dissertation, University of Tennessee Medical Units, Memphis.

Gardner, D.P. and Fedinec, A.A. (1975), Proc. Fourth Internat. Conf. on Tetanus, April 6–12, Dakar, pp. 163–169, Ed. Fondation Mérieux, Lyon.

Goldsack, D.E. (1970), *Biopolymers*, **9**, 247–252.

Greengard, P. and Costa, E. (1970), *Role of Cyclic AMP in Cell Function*, Raven Press, New York.

Greengard, P. and McAfee, D.A. (1972), *Biochem. Soc. Symp.*, **36**, 87–102.

Greengard, P., McAfee, D.A. and Kerabian, J. (1972), *Advances in Cyclic Nucleotide Research*, Vol. 1, pp. 337–355, Raven Press, New York.

Habermann, E. (1970), *Naunyn-Schmiedeberg's Arch. Pharmacol.*, **267**, 1–19.

Habermann, E. (1971), *Naunyn-Schmiedeberg's Arch. Pharmacol.*, **269**, 124–135.

Habermann, E. (1972), *Naunyn-Schmiedeberg's Arch. Pharmacol.*, **272**, 75–88.

Habermann, E. (1973a), *Med. Microbiol. Immunol.*, **159**, 89–100.

Habermann, E. (1973b), *Naunyn-Schmiedeberg's Arch. Pharmacol.*, **276**, 341–359.

Habermann, E. and Wellhöner, H.H. (1972), *Radioactive Tracers in Microbial Immunology*, pp. 67–76, IAEA, Vienna.

Habermann, E. and Dimpfel, W. (1973), *Naunyn-Schmiedeberg's Arch. Pharmacol.*, **276**, 327–340.

Habermann, E., Dimpfel, W. and Räker, K.O. (1973), *Naunyn-Schmiedeberg's Arch. Pharmacol.*, **276**, 361–373.

Helting, T.B. and Zwisler, O. (1974), *Biochem. Biophys. Res. Commun.*, **57**, 1263–1270.

Helting, T.B. and Zwisler, O. (1975), Proc. Fourth Internat. Conf. on Tetanus, April 6–12, Dakar, pp. 639–646, Ed. Fondation Mérieux, Lyon.

Hensel, B., Seib, U.C. and Wellhöner, H.H. (1973), *Naunyn-Schmiedeberg's Arch. Pharmacol.*, **276**, 395–402.

Holmes, M.J. and Ryan, W.L. (1971), *Infect. Immun.*, **3**, 133–140.

Johnston, G.A.R. and Iversen, L.L. (1971), *J. Neurochem.*, **18**, 1951–1961.

Kaeser, H.E. and Saner, A. (1969), *Nature*, **223**, 842.

Kano, M. and Takano, K. (1969), *Jap. J. Physiol.*, **19**, 1–10.

King, L.E. (1970), Doctoral Dissertation, University of Tennessee Medical Units, Memphis.

King, L.E. and Fedinec, A.A. (1974), *Naunyn-Schmiedeberg's Arch. Pharmacol.*, **281**, 391–401.

King, L.E. and Fedinec, A.A. (1975), Abstract, Fourth Internat. Conf. on Tetanus, April 6–13, Dakar.

Krnjevic, K. (1970), *Nature*, **228**, 119–124.

Kryzhanovsky, G.N. (1965), *Recent Advances in the Pharmacology of Toxins*, pp. 105–111, Pergamon Press, New York.

Kryzhanovsky, G.N. (1966), *Tetanus*, State Publ. House 'Medicine', Moscow.

Kryzhanovsky, G.N., (1967), *Principles on Tetanus*, pp. 155–168, Huber, Berne.

Kryzhanovsky, G.N. (1973), *Naunyn-Schmiedeberg's Arch. Pharmacol.*, **276**, 247–270.

Kryshanovsky, G.N. (1975), Proc. Fourth Internat. Conf. on Tetanus, April 6–12, Dakar, pp. 189–199, Ed. Fondation Mérieux, Lyon.

Kryzhanovsky, G.N., Pevnitsky, L.A., Grafova, V.N. and Polgar, A.A. (1961), *Byull. Eksp. Biol. Med.*, **52**, No.3, 42–49; No. 8, 31–37; No. 11, 35–43; No. 12, 30–37.

Kryzhanovsky, G.N. and Dyakonova, M.V. (1964), *Byull. Eksp. Biol. Med.*, **58**, 1021–1024.

Kryzhanovsky, G.N., Alekseev, L.P. and Rozanov, A. Ya. (1970), *Byull. Eksp. Biol. Med.*, **7**, 1036–1038.

Kryzhanovsky, G.N., Pozdnyakov, O.M., Dyakonova, M.V., Polgar, A.A. and Smirnova, V.S. (1971), *Byull. Eksp. Biol. Med.*, **7**, 27–31.

Kryzhanovsky, G.N., Alexeev, L.P., Kulberg, A.Ya. and Tarkhanova, I.A. (1972), *Byull. Eksp. Biol. Med.*, **74**, 66–69.

Kryzhanovsky, G.N., Grafova, V.N., Tumanov, V.P. and Vtyurin, B.V. (1973), *Byull. Eksp. Biol. Med.*, **7**, 100–105.

Kryzhanovsky, G.N., Glebov, R.N., Dmitrieva, N.M., Grafova, V.N. and Sakharova, O.P. (1974a), *Byull. Eksp. Biol. Med.*, **1**, 45–47.

Kryzhanovsky, G.N., Glebov, R.N., Dmitrieva, N.M. and Fedorova, V.I. (1974b), *Byull. Eksp. Biol. Med.*, **4**, 24–27.

Kryzhanovsky, G.N., Kulygina, R.M., Glebov, R.N., Fedorova, V.I. and Sakharova, O.P. (1975a), *Byull. Eksp. Biol. Med.*, **5**, 22–25.

Kryzhanovsky, G.N., Fedorova, V.I., Glebov, R.N., Kulygina, R.M. and Sakharova, O.P. (1975b), *Byull. Eksp. Biol. Med.*, **4**, 19–22.

Kuo, J.F. and Greengard, P. (1969a), *J. Biol. Chem.*, **244**, 3417–3419.

Kuo, J.F. and Greengard, P. (1969b), *Proc. Nat. Acad. Sci. USA*, **64**, 1349–1355.

Landsteiner, K. and Botteri, A. (1906), *Zbl. Bakt. Abt. Orig.*, **42**, 562–568.

Largier, J.F. (1956), *Biochim. Biophys. Acta*, **21**, 433–438.

Largier, J.F. and Joubert, F.J. (1956), *Biochim. Biophys. Acta*, **20**, 407–408.

Laurence, C.R. and Webster, R.A. (1963), *Clin. Pharmacol.*, **4**, 36–72.

Leonardi, G. (1961), *Bull. Sci. Med.*, **133**, 381–384.

Leonardi, G. (1973), *J. Inf. Dis.*, **128**, 235–237.

Leonardi, G., Marcolongo, R., Tiezzi, C. and Di Tommaso, I. (1972), *Ann. Sclavo*, **14**, 758–765.

Lettl, A., Nekvasilova, K., Moravec, K. and Stejskal, A. (1966), *Ann. Inst. Pasteur*, **111**, 622–629.

Logan, W.J. and Snyder, S.H. (1971), *Nature*, **234**, 297–299.

Mangalo, R., Bizzini, B., Turpin, A. and Raynaud, M. (1968), *Biochim. Biophys. Acta.*, **168**, 583–584.
Marie, A. (1898), *Ann. Inst. Pasteur,* **12**, 91–95.
Marie, A. and Morax, V. (1902), *Ann. Inst. Pasteur,* **16**, 818–832.
Matsuda, M. and Yoneda, M. (1974), *Biochem. Biophys. Commun.*, **57**, 1257–1262.
Matsuda, M. and Yoneda, M. (1975), *Infec. and Immun.*, **12**, 1147–1153.
Matus, A.I. and Dennison, M.E. (1971), *Brain Res.*, **32**, 195–197.
McIlwain, H. (1960), *Biochem. J.*, **76**, 16.
McIlwain, H. (1961), *Biochem, J.*, **78**, 24–32.
Mellanby, J.H. (1971), *J. Physiol.* (Lond.), **218**, 68–69P.
Mellanby, J.H. (1975), Abstract, Fourth Internat. Conf. on Tetanus, April 6–12, Dakar.
Mellanby, J. and van Heyningen, W.E. (1965), *Recent Advances in the Pharmacology of Toxins,* pp. 121–123, Pergamon Press, New York.
Mellanby, J., van Heyningen, W.E. and Whittaker, V.P. (1965), *J. Neurochem.*, **12**, 77–79.
Mellanby, J. and van Heyningen, W.E. (1967), *Principles on Tetanus,* pp. 177–187, Huber, Berne.
Mellanby, J., Pope, D. and Ambache, N. (1968), *J. Gen. Microbiol.*, **50**, 479–486.
Mellanby, J.H. and Thompson, P.A. (1972), *J. Physiol.* (Lond.), **224**, 407–421.
Mellanby, J.H. and George, G. (1975), Proc. Fourth Internat. Conf. on Tetanus, April 6–12, Dakar, pp. 201–204, Ed. Fondation Mérieux, Lyon.
Mellanby, J.H. and Pope, D. (1975), Proc. Fourth Internat. Conf. on Tetanus, April 6–12, Dakar, pp. 187–188, Ed. Fondation Mérieux, Lyon.
Meyer, H. and Ransom, F. (1903), *Naunyn-Schmiedeberg's Arch. Pharmacol.*, **49**, 369–416.
Miyamoto, E., Kuo, J.F. and Greengard, P. (1969a), *Science,* **165**, 63–65.
Miyamoto, E., Kuo, J.F. and Greengard, P. (1969b), *J. Biol. Chem.*, **244**, 6395–6402.
Muchnik, S. and Rubinstein, E.H. (1967), *Acta Physiol. Lat. Amer.*, **17**, 166–174.
Murphy, S.G. (1967), *J. Bact.*, **94**, 586–589.
Murphy, S.G. and Miller, K.D. (1967), *J. Bact.*, **94**, 580–585.
Murphy, S.G., Plummer, T.H. and Miller, K.D. (1968), *Fed. Proc.*, **27**, 268.
Nagel, J. and Cohen, H. (1973), *J. Immunol.*, **110**, 1388–1393.
Neal, M.J. and Pickles, H.G. (1969), *Nature* **222**, 679–680.
North, E.A. and Doery, H.M. (1961), *Brit. J. Exp. Path.*, **42**, 23–29.
Nozaki, Y. and Tanford, C. (1971), *J. Biol. Chem.*, **246**, 2211–2217.
Osborne, R.H. and Bradford, H.F. (1973), *Nature New Biol.*, **244**, 157–158.
Paar, G.H., Wiegand, H.J. and Wellhöner, H.H. (1974), *Naunyn-Schmiedeberg's Arch. Pharmacol.*, **281**, 383–390.
Peetoom, F. and van der Veer, M. (1967), *Principles on Tetanus,* pp. 237–244, Huber, Berne.
Pillemer, L., Wittler, R.G., Burrell, J.I. and Grossberg, D.B. (1948), *J. Exp. Med.*, **88**, 205–221.
Price, D.L., Griffin, J., Young, A., Peck, K. and Stocks, A. (1975), *Science,* **188**, 945–947.
Puszkin, S. and Berl, S. (1972), *Biochim. Biophys. Acta,* **256**, 695–709.
Raynaud, M. (1947), *Compt. Rend. Acad. Sci.* **225**, 543–544.

Raynaud, M., Turpin, A. and Bizzini, B. (1960), *Ann. Inst. Pasteur,* **99**, 167–172.
Raynaud, M., Bizzini, B. and Turpin, A. (1971), *Ann. Inst. Pasteur,* **120**, 801–815.
Robinson, J.P., Holladay, L.A., Picklesimer, J.B. and Puett, D. (1974), *Mol. Cell. Biochem.,* **5**, 147–151.
Robinson, J.P., Pickesimer, J.B. and Puett, D. (1975), *J. Biol. Chem.,* **250**, 7435–7442.
Robison, G.A., Nahas, G.G. and Triner, L. (Eds.) (1971), *Cyclic AMP and Cell Function,* Ann. N.Y. Acad. Sci, Vol. 185.
Schwab, M.E. and Thoenen, H. (1976), *Brain Res.,* accepted for publication in 1975.
Seib, U.C., Hensel, B., Wiegand, H. and Wellhöner, H.H. (1973), *Naunyn-Schmiedeberg's Arch. Pharmacol.,* **276**, 403–411.
Semba, T. and Kano, M. (1969), *Science,* **164**, 571–572.
Sherrington, C.S. (1905), *Proc. Roy. Soc.* B, **76**, 269–297.
Sottiurai, V. and Fedinec, A.A. (1975), Proc. Fourth Internat. Conf. on Tetanus, April 6–12, Dakar, pp. 211–218, Ed. Fondation Mérieux, Lyon.
Spence, M.W. (1969), *Can. J. Biochem.,* **47**, 735–742.
Stein, P. and Biel, H. (1973), *Z. Immun-Forsch.,* **145**, 418–431.
Stöckel, K., Schwab, M. and Thoenen, H. (1975), *Brain Res.,* **99**, 1–16.
Sverdlov, Yu. S. and Alekseeva, V.I. (1966), *Fed. Proc.* (Trans. Suppl.), **25**, T931–936.
Takano, K. and Henatsch, H.D. (1973), *Naunyn-Schmiedeberg's Arch. Pharmacol.,* **276**, 421–436.
Takano, K. and Kano, M. (1973), *Naunyn-Schmiedeberg's Arch. Pharmacol.,* **276**, 413–420.
Tonge, D.A., Gradidge, T.J. and Marchbanks, R.M. (1975), *J. Neurochem.,* **25**, 329–331.
Uchizono, K. (1965), *Nature* **207**, 642–643.
van Heyningen, W.E. (1959a), *J. Gen. Microbiol.,* **20**, 291–300.
van Heyningen, W.E. (1959b), *J. Gen. Microbiol.,* **20**, 301–309.
van Heyningen, W.E. (1959c), *J. Gen. Microbiol.,* **20**, 310–320.
van Heyningen, W.E. (1963), *J. Gen. Microbiol.,* **31**, 375–387.
van Heyningen, W.E. (1975), Proc. Fourth Internat. Conf. on Tetanus, April 6–12, Dakar, pp. 183–186, Ed. Fondation Merieux, Lyon.
van Heyningen, W.E. and Miller, P.A. (1961), *J. Gen. Microbiol.,* **24**, 107–119.
van Heyningen, W.E. and Mellanby, J. (1968), *J. Gen. Microbiol.,* **52**, 447–454.
van Heyningen, W.E. and Mellanby, J. (1973), *Naunyn-Schmiedeberg's Arch. Pharmacol.,* **276**, 297–302.
Wassermann, A. and Takaki, T. (1898), *Berl. Klin. Wschr.,* **35**, 5–6.
Watkins, J.C. (1972), *Biochem. Soc. Symp.,* **36**, 33–47.
Wellhöner, H.H., Hensel, B. and Seib, U.C. (1973a), *Naunyn-Schmiedeberg's Arch. Pharmacol.,* **276**, 375–386.
Wellhöner, H.H., Seib, U.C. and Hensel, B. (1973b), *Naunyn-Schmiedeberg's Arch. Pharmacol.,* **276**, 387–394.
Werman, R. (1965), *J. Theor. Biol.,* **9**, 471–477.
Wilson, V.J., Diecke, F.P.J. and Talbot, W.H. (1960), *J. Neurophysiol.,* **23**, 659–666.
Woolley, D.W. and Gommi, B.W. (1965), *Proc. Nat. Acad. Sci. USA,* **53**, 959–968.
Wright, P.G. (1955), *Pharmacol. Rev.,* **7**, 413–465.

Wright, P.G. (1956), *Guy's Hosp. Rep.*, **105**, 57–79.
Young, A.B. and Snyder, S.H. (1974), *Proc. Nat. Acad. Sci. USA*, **70**, 2832–2836.
Zacks, S.I. and Sheff, M.F. (1965), *Acta Neuropathol.*, **5**, 267–273.
Zacks, S.I. and Sheff, M.F. (1968), *Science*, **159**, 643–644.
Zukin, S.R., Young, A.B. and Snyder, S.H. (1974), *Proc. Nat. Acad. Sci. USA*, **71**, 4802–4807.

6 Cell Membranes and Cytolytic Bacterial Toxins

J. E. ALOUF

Acknowledgments
This study was supported in part by grants n^0 7270208 and 7470186 from the Délégation Générale à la Recherche Scientifique et Technique and the Centre National de la Recherche Scientifique.

The Specificity and Action of Animal, Bacterial and Plant Toxins
(*Receptors and Recognition*, series B, Volume 1)
Edited by P. Cuatrecasas
Published in 1976 by Chapman and Hall, 11 New Fetter Lane, London EC4P 4EE

6.1 INTRODUCTION

During the past decade there has been a rapid increase in research on the cytolytic effects and mode of action of various biological agents of low molecular weight such as polyene and macrocyclic antibiotics (Harold, 1970; Kinsky, 1970; Hamilton-Miller, 1973, 1974), protein substances such as complement–antibody systems (Kinsky, 1972; Müller-Eberhard, 1975), and a number of bacterial toxins endowed with potent hemolytic properties (Bernheimer, 1970, 1974).

Since these substances appear to act primarily on the cytoplasmic membrane of various eukaryote or prokaryote cells and to disrupt it, the study of their mode of action on susceptible cells has aimed essentially at using them as probes to elucidate problems of biomembrane organization and function.

The purpose of this chapter is to survey the recent developments in the study of bacterial cytolytic toxins, with emphasis on their biochemical characterisitics and mode of action on membrane integrity, permeability, and function.

For detailed information and references to the progress made prior to 1970 we refer the reader to the multi-volume treatise on microbial toxins (Ajl *et al.*, 1970; Kadis *et al.*, 1971; Montie *et al.*, 1970) and to the special chapter written by Bernheimer (1970). A recent review by this author (Bernheimer, 1974) with emphasis on some staphylococcal hemolysins, streptolysin O, and related toxins covers the literature up to 1973. We shall try to avoid as far as possible overlapping with the topic treated in this review and will deal in more detail with the advances in this field in the past three years. References to literature prior to 1950 with good historical background may be found in the general reviews by Prévot (1950) and Guillaumie (1950).

The term 'toxin' will be used in this chapter in its usual restrictive meaning employed by most bacteriologists to designate only those bacterial products of relatively large molecular size fully or partially protein in nature (including polypeptides) which possess toxic properties (Bonventre, 1970). Therefore non-protein bacterial substances endowed with cytolytic properties will be beyond the scope of this review. Bacterial toxins are known to occur as extracellular, intracytoplasmic, or cell-bound material (either loosely bound or as a moiety of the bacterial cell wall) as described in detail by Raynaud and Alouf (1970). They show wide heterogeneity of pharmacological effects and heterogeneity of host specificity which are both reflected at organ, tissue cellular, and subcellular levels.

A great number of bacterial toxins induce cellular damage in intoxicated hosts as shown by histological examination. The cytopathogenic effects of toxins were also assessed *in vitro* after the introduction of tissue culture techniques in the study of the mode of action of a number of them (Solotorovsky and Johnson, 1970).

Table 6.1 Cytolytic protein bacterial toxins

Producing bacterial species	Name of toxin	References*
Gram-positive bacteria		
Staphylococcus aureus		
	α-Toxin	Cassidy *et al.*, (1974)
	β-Toxin	Bernheimer *et al.*, (1974)
	γ-Toxin	Taylor and Bernheimer (1974)
	δ-Toxin	Kantor *et al.*, (1972)
	ε-Toxin (?)	Wadström *et al.*, (1974)
	Leukocidin	Woodin (1970)
Streptococcus pyogenes	Streptolysin O	Alouf and Raynaud (1973)
	Streptolysin S	Ginsburg (1970), Bernheimer (1972)
Streptococcus pneumoniae	Pneumolysin	Johnson (1972), Johnson and Allen (1975)
Corynebacterium ovis	Hemolysin	Soucék *et al.*, (1971)
Bacillus subtilis	Surfactin	Bernheimer and Avigad (1970)
Bacillus cereus	Cereolysin	Bernheimer and Grushoff (1967a)
Bacillus thuringiensis	Thuringiolysin	Pendleton *et al.*, (1973)
	Second thuringiolysin	*ibid.*
Bacillus alvei	Hemolysin	Bernheimer and Grushoff (1967b)
Bacillus laterosporus	Hemolysin	*ibid.*
Listeria monocytogenes	Listeriolysin	Jenkins and Watson (1971)
Clostridium chauvoei	α-Toxin	Moussa (1958)
Clostridium bifermantans	Hemolysin	Guillaumie (1950)
Clostridium botulinum	Hemolysin	Guillaumie and Kréguer (1951)
Clostridium histolyticum	ε-Toxin	Howard (1953)
Clostridium novyi type A	δ-Toxin	Rutter and Collee (1969)
	γ-Toxin	*ibid.*
Clostridium novyi types B,D	β-Toxin	*ibid.*
Clostridium perfringens	α-Toxin	Smyth and Arbuthnott (1974)
	θ-Toxin	Smyth (1975)
	δ-Toxin	Hauschild (1971)
Clostridium tetani	Tetanolysin	Cox *et al.*, (1974)
Clostridium septicum	α-Toxin	Moussa (1958)
	δ-Toxin	*ibid.*

Table 6.1 (continued)

Producing bacterial species	Name of toxin	References*
Gram-negative bacteria		
Escherichia coli	α-Hemolysin	Rennie and Arbuthnott (1974)
	β-Hemolysin	Smith (1963)
	γ-Hemolysin	Walton and Smith (1969)
Pseudomonas aeruginosa	Hemolysin	Liu (1974)
	Leukocidin	Frimmer and Scharmann (1975)
Pseudomonas fluorescens	Hemolysin	Gräf (1958)
Aeromonas hydrophila	Aerolysin	Bernheimer *et al.*, (1975)
Aeromonas proteolytica	Hemolysin	Foster and Hanna (1974)
Acinetobacter calcoaceticus	Hemolysin	Lehman (1971)
Vibrio parahemolyticus	Hemolysin	Sakurai *et al.*, (1973)

* The references are generally the most recent; they have been selected only to provide guidance to prior relevant literature.

Table 6.2 Cytolytic toxins bound to producing bacterial cells

Pneumolysin	(Cohen *et al.*, 1942).
E. coli β-hemolysin	(Smith, 1963).
Streptococcal hemolysin	(SLS ?) (Ginsburg, 1970)
P. aeruginosa leukocidin	(Frimmer and Scharmann, 1975)

However, among cell-damaging toxins only a more restricted number could be considered as cytolytic. The criteria which permit differentiation of cytolytic toxins *per se* from the other cell-damaging toxins are discussed in Section 6.3.

6.2 OUTLINES OF CYTOLYTIC TOXINS

Since the aim of this review is the discussion of the effects of these toxins on membranes, a survey of the chemical and physical characteristics of cytolytic toxins is beyond its scope. We shall briefly give an outline of these characterisitics and the references to recent purification and characterization of these toxins.

About forty cytolytic toxins listed in Table 6.1 are clearly defined. However, this list is not exhaustive. More than twenty toxins have still not been isolated in a pure state.

The vast majority of cytolytic toxins are extracellular. Only four have been found associated with producing bacterial cells (Table 6.2). No cytolytic toxin has been localized among intracellular protein toxins, except perhaps one form of streptolysin (Ginsburg, 1970).

Table 6.3 Bacterial species producing the immunologically and chemically related thiol-dependent cytolytic toxins

Family	Genus	Species	Toxin
Streptococcaceae	Streptococcus	*S. pyogenes*	Streptolysin O
		S. pneumoniae	Pneumolysin
Bacillaceae	Bacillus	*B. cereus*	Cereolysin
		B. thuringiensis	Thuringiolysin
		B. alvei	Hemolysin
		B. laterosporus	Hemolysin
	Clostridium	*C. bifermantans*	Hemolysin
		C. botulinum	Hemolysin
		C. histolyticum	ϵ-Toxin
		C. novyi type A	γ-Toxin
		C. perfringens	θ-Toxin
		C. septicum	Hemolysin
		C. tetani	Tetanolysin
Lactobacillaceae	Listeria	*L. monocytogenes*	Listeriolysin

6.2.1 Thiol-activated cytolytic toxins

A particular group comprising thirteen chemically and immunologically related cytolytic toxins can be distinguished (Table 6.3). The toxins of this group, the prototype of which is streptolysin O, are known as oxygen-labile or SH-dependent toxins since they lose their cytolytic effect in the oxidized state and are only active after reduction with hydrosulfite or thiols. These toxins have a common receptor on sensitive cells which is membrane cholesterol (see Bernheimer, 1974). They are produced by various strains from aerobic or anaerobic sporulating or non-sporulating Gram-positive species. Some of these toxins have been obtained in a pure state, namely, streptolysin O (Alouf and Raynaud, 1973), *Clostridium perfringens* θ-toxin (Mitsui *et al.*, 1973b; Möllby and Wadström, 1973; Hauschild *et al.*, 1973; Smyth, 1975), pneumolysin (Shumway and Klebanoff, 1971; Johnson, 1972). Cereolysin (Bernheimer and Grushoff, 1967a), and thuringiolysin (Pendleton *et al.*, 1973) have been partially purified.

The purification of tetanolysin and *B. alvei* lysin is in progress (Alouf, Pierrès, and Kiredjian, to be published). SH-toxins are proteins of molecular weight ranging from 47 000 to 68 000 daltons. Dimeric or polymeric forms are also found (Alouf and Raynaud, 1973; Smyth, 1975). Isoelectric points ranged around neutral pH except for pneumolysin (pI 4.9) as reported by Kreger and Bernheimer (1969). Multimolecular forms (as shown by pI) have been reported as discussed below. Some characteristics of these toxins are given in Table 6.4.

6.2.2 Other cytolytic toxins

Except for SH-dependent toxins, which constitute a homogeneous group, the other cytolytic toxins show no close relationship to each other. All of those which have been partially or completely purified appear to be proteins with molecular weights greater than 25 000 daltons except surfactin and streptolysin S. Surfactin produced by *B. subtilis* is a strong surface-active peptide lipid the structure of which has been established by Kakinuma *et al.*, (1969). It consists of a heptapeptide having an N-terminal glutamic acid in amide linkage with the carboxy group of 3-hydroxy-13-methyltetradecanoic acid. This agent has also been isolated and studied by Bernheimer and Avigad (1970).

Streptolysin S is a polypeptide of molecular weight about 12 000 daltons induced by a variety of molecules such as RNA, polyribonucleotides, serum albumin, and certain detergents (Ginsburg, 1970; Bernheimer, 1972). These substances act as carriers of the polypeptide. Only carrier–polypeptide complexes are cytolytic. This toxin possesses certain features common to surfactin and staphylococcal δ-toxin (Bernheimer, 1972, 1974), and a great deal of attention has been devoted in the past three years to the purification, molecular heterogeneity, and properties of the four staphylococcal toxins (Wadström *et al.*, 1974), on the one hand, and clostridial α-toxin on the other hand. Staphylococcal α-toxin has been purified by McNiven *et al.*, (1972), Six and Harshman (1973a,b), and Watanabe and Kato (1974). Staphylococcal β-toxin was isolated in a highly pure state by Wadström and Möllby (1971a,b) and Bernheimer *et al.*, (1974). It has a sphingomyelinase activity. Staphylococcal γ-toxin was purified by Taylor and Bernheimer (1974), and staphylococcal δ-toxin by Kreger *et al.*, (1971), Heatley (1971), and Kantor *et al.*, (1972).

γ-Toxin is composed of two separate proteins with molecular weights 29 000 and 26 000 daltons and showing pI at 9.8 and 9.9 which act synergistically. This situation is similar to that of staphylococcal leukocidin (Woodin, 1970) and *S. zymogenes* hemolysin (Granato and Jackson, 1971a,b) which are also two-component protein complexes acting synergistically.

Staphylococcal δ-toxin is heterogeneous with respect to charge and molecular size. It has a high content of hydrophobic amino-acids and lacks histidine, arginine, proline, tyrosine, and cysteine. A molecular weight of 100 000 has been reported. Subunits of 10 000 daltons appear to constitute the toxin (Bernheimer, 1974).

Purification of *Cl. perfringens* α-toxin has been a source of difficulties in the past. Its isolation in a highly pure state has been achieved by Möllby and Wadström (1973), Mitsui *et al.*, (1973a), and Smith and Arbuthnott (1974) by the conventional methods of protein purification used at present such as gel filtration and isoelectric focusing techniques. Since α-toxin is endowed with phospholipase C activity it has recently been purified by affinity chromatography on agarose-linked egg-yolk lipoprotein (Takahashi *et al.*, 1974) and by immunosorbent chromatography (Bird *et al.*, 1974).

Table 6.4 Some physical and biological properties of recently purified cytolytic toxins

Toxin	Reference	Molecular weight	pI	Specific activity* (hemolytic units/mg
SH-activated toxins				
Clostridial θ-toxin	Hauschild *et al.*, (1973)	74 000	–	768 000
	Mitsui *et al.*, (1973a)	53 000 (θ_A)	7.05	889 000
		50 000 (θ_B)	6.65	783 000
	Smyth (1975)	62 000 (θ_1)	6.8	043 000
		61 600 (θ_2)	6.5	1 151 000
		59 500 (θ_3)	6.1	609 000
		59 000 (θ_4)	5.7	412 000
Streptolysin O	Alouf and Raynaud (1973)	68 000 and 120 000	7.2 6.5	375 000 to 750 000
	Smyth and Fehrenbach (1974)	–	7.5 6.1	164 000
Cereolysin	Bernheimer and Grushoff (1967a)	52 000	6.5	2 000 000
	Bernheimer *et al.*, (1968)			
Pneumolysin	Shumway and Klebanoff (1971)	–	–	546 670
	Kreger and Bernheimer (1969)	63 000	4.9	
	Johnson (1972)	–	–	450 000
Listeriolysin	Jenkins and Watson (1971)	171 000	–	25 000
Thuringiolysin	Pendleton *et al.*, (1973)	47 000	6.2 6.5	2 000 000
Other toxins				
Streptolysin S	Ginsburg (1970)			
	Bernheimer (1972)	12 000	–	2 000 000
Staphylococcal α-toxin	Wadstrom *et al.*, (1974)	41 000 (Ia)	8.0–8.7	
	Wadström and Möllby (1972)	30 000 (Ib)	9.1–9.2	

Table 6.4 (continued)

Toxin	Reference	Molecular weight	pI	Specific activity* (hemolytic units/mg)
Staphylococcal	Wadstrom and Mollby (1972)	33 000 (II)	6.5–7.5	
		36 000 (III)	4.5–5.5	
	McNiven *et al.*, (1972)	36 000 (A)	8.55	
		36 000 (B)	9.15	
		(C)	7.36	
		(D)	6.28	
	Six and Harshman (1973 a, b)	28 000 (A)	7.2	20 000
		28 000 (B)	8.4	20 000
	Watanabe and Kato (1974)	36 000	7.98	51 000
	Wiseman *et al.*, (1975)	–	–	125 000
Staphylococcal β-toxin	Wadström and Möllby (1971a,b)	33 000	9.4	10^7–10^8
	Bernheimer (1974)	30 000	9.0	400 000 to 700 000
Staphylococcal γ-toxin	Taylor and Bernheimer (1974)			
Component I		29 000	9.8	
Component II		26 000	9.9	
Staphylococcal δ-toxin	Kreger *et al.*, (1971)			
Basic form		–	9.5	200
Acidic form		–	5.0	
	Heatley (1971)	–	–	300
	Kantor *et al.*, (1972)	103 000	4.65 6.7 9.0	
Clostridial α-toxin	Möllby and Wadström (1973)	30 000	5.7 4.6	– –
	Mitsui *et al.*, (1973b)	49 000	–	133 000
	Bird *et al.*, (1974)	48 000	4.75–5.5	35 520

Table 6.4 (continued)

Toxin	Reference	Molecular weight	pI	Specific activity* (hemolytic units/mg)
Clostridial α-toxin	Smyth and Arbuthnott (1974)	53 000	5.49 5.25	
	Takahashi *et al.*, (1974)	43 000 (α_0) (α_1) (α_2)	5.2 5.3 5.5	409 000
E. coli hemolysin	Rennie and Arbuthnott (1974)	580 000	4.6	1 000 000
Aeromonas hemolysin	Wretlind *et al.*, (1971)			
Component I			5.5	6 000
Component II			4.3	1 800
	Bernheimer *et al.*, (1975)	49 000	–	415 000

* The data are given for the erythrocytes which are the most sensitive to the toxin being considered. The values reported are not always comparable since assay systems vary greatly between laboratories.

The molecular heterogeneity of many cytolytic toxins either SH-dependent or not, particularly as regards multiple electrophoretic forms, has been shown. There is a survey and discussion of this topic in articles by Alouf and Raynaud (1973), Smyth and Fehrenbach (1974), Takahashi *et al.*, (1974), Smyth and Arbuthnott (1974), Wadström *et al.*, (1974), and Smyth (1975). Purified toxins are listed in Table 6.4.

6.3 CONCEPT OF CYTOLYTIC TOXIN

An obvious criterion of a cytolytic effect is the prompt physical dissolution or disruption of mammalian or other kinds of cells *in vitro* (Bernheimer, 1970, 1974). The best illustration of this disruption is the lysis and the release of hemoglobin of erythrocytes of appropriate animal species (hemolysis) which occur for almost all cytolytic toxins. However, two toxins, staphylococcal (Woodin, 1970) and *Pseudomonas aeruginosa* (Frimmer and Scharmann, 1975) leukocidins, are active on leukocytes but do not impair the permeability of erythrocytes.

Up to fifteen years ago the study of cytolytic toxins or other agents was indeed largely confined to erythrocyte suspensions in isotonic buffer and occasionally to the other kinds of blood cells (leukocytes and platelets).

Although erythrocytes remain a basic tool for the investigation of the mode of action as well as a convenient system for the assay of their lytic potency (see Section 6.8) the development since 1952 of cell culture technology has proved of great value in the study of mechanisms of cell lysis by cytolytic toxins as below.

Many human or animal cell lines, including tumor cells, have their permeability properties impaired by these toxins.

Some toxins have been shown to lyse prokaryote cells. *S. faecalis* var. *zymogenes* hemolysin behaves as a bacteriocin (Applebaum and Zimmerman, 1974) and lyses Gram-positive organisms. Some streptococcal, staphylococcal, or clostridial hemolysins lyse protoplasts, spheroplasts, and parasitic or saprophytic mycoplasma (see Bernheimer, 1970, 1974; Kreger *et al.*, 1971; and the monographs on various toxins in the Microbial Toxins treatise). A number of toxins, such as streptolysins S and O, staphylococcal β– and δ-hemolysins, and *Listeria* and *Cl. tetani* hemolysins (listeriolysin, tetanolysin), have been reported to lyse subcellular organelles from various mammalian cells particularly lysosomes and mitochondria (Bernheimer, 1970; Kingdom and Sword, 1970; Wadström and Möllby, 1972; Cox *et al.*, 1974). These organelles are surrounded by a membrane bilayer similar at least qualitatively to the cytoplasmic one. Lysis of these intracellular bodies has been shown by microscopic and biochemical technique to occur in living cells such as macrophages (Fauve *et al.*, 1966), leukocytes (Hirsch *et al.*, 1963; Zucker-Franklin, 1965), or platelets (Wadström and Möllby, 1972) as well as in the isolated state in suspension in a suitable menstrum. Concomitant solubilization and release of associated enzymes or pharmacological mediators takes place after lysis of these organelles (Keiser *et al.*, 1964; Wadström and Möllby, 1972; Cox *et al.*, 1974). According to the criterion of gross lysis of cells reflecting membrane disruption, many toxins such as diphtheria toxin (Collier, 1975), streptococcal erythrogenic (scarlatinal) toxin (Kim and Watson, 1972), staphylococcal dermoexfoliative toxins (Melish *et al.*, 1974; Wueper *et al.*, 1975) particularly *Cl. perfringens* enterotoxin (McDonel and Duncan, 1975), which damage or kill *in vivo* or *in vitro* various mammalian cells, cannot be classified as cytolytic toxins. However, some of these non-lytic toxins are known to interact and bind on to cell membranes, as studied in detail for diphtheria toxins (Gill *et al.*, 1973) and cholera enterotoxin (Cuatrecasas, 1973). These toxins affect various functions of the cell. They involve penetration of the hydrophobic lipid barrier of cytoplasmic membrane by a fragment of the toxin to reach the cytosol (Bonventre *et al.*, 1975; Gill, 1975).

A good delineation between cytolytic toxins *per se* and other cell-damaging toxins is indeed the criterion of gross lysis. However, this criterion is not always evident since the degree of cellular disruption may vary considerably with different kinds of cells and to some extent with cytolytic agents (Bernheimer, 1970). The primary criterion should therefore be based on the evidence that the damage is specifically confined to the cytoplasmic membrane accompanied by the loss of morphological and/or functional integrity of the osmotic barrier with concomitant leakage of intracellular material. According to this principle we may define a bacterial cytolytic toxin as any toxic protein or polypeptide of microbial origin which, upon reversible or irreversible binding to one or several components of the cytoplasmic membrane of any living animal, plant, or microbial cell, disorganizes the lipoprotein structure of the membrane, leading therefore to a loss of morphological and functional integrity of the cell.

When obvious lysis under the action of a toxin is not disclosed, or when sublytic levels of toxin are employed in order to detect subtle changes in membrane permeability for the study of the primary events which trigger membrane modifications, suitable techniques have to be devised for this type of investigation. The methods employed should be sufficiently sensitive.

6.4 ASSESSMENT OF CYTOLYTIC ACTIVITY

Several complementary experimental approaches have been employed for the assessment and study of membrane damage induced *in vitro* by cytolytic toxins on cultured animal or human cells as well as on cell (erythrocytes, platelets, leukocytes) or subcellular organelle suspensions.

The techniques are either the conventional ones devised for the study of cytotoxicity of toxins or the more specific ones based on permeability tests.

The former techniques involve the observation of morphological modification of cells by optical or electron microscopic techniques or several metabolic or physiological tests. The contribution and limitations of each of the techniques employed are developed in the following sections. However, it is to be noted that, whatever the technique used and its intrinsic value as a criterion of membrane damage, one should be cautious in the interpretation of the results analyzed below when the study is not undertaken with purified homogeneous toxin preparations free from other bacterial products of high or low molecular weight contaminating the preparation used. In a number of cases cytolytic and other effects attributed to specific toxins were due to the presence of other enzymes, toxins, or complex marcomolecular components. Thus the α– and β-toxins of *Staphylococcus aureus* were later shown to be contaminated with the highly cytolytic δ-toxin (Bernheimer *et al.*, 1968). *Clostridium perfringens* α-toxin (phospholipase C) was contaminated with θ-toxin (Möllby *et al.*, 1973; Smyth *et al.*, 1975a).

6.5 MORPHOLOGICAL EVALUATION OF MEMBRANE DAMAGE BY OPTICAL AND ELECTRON MICROSCOPY

Microscopical observation of cellular or membrane morphology has been a useful tool in studying the cytopathogenic effects of toxins (Solotorovsky and Johnson, 1970) including cytolytic ones. This morphological approach proved valuable for determining the scope of action of the latter particularly at the ultrastructural level studied by electron microscopic techniques in recent years.

6.5.1 Light microscopical studies

Most studies have been undertaken on streptolysin O and some of the other thiol-dependent immunologically and chemically related cytolytic toxins (listeriolysin, pneumolysin, tetanolysin, cereolysin, clostridial θ-toxin, etc.) listed in Table 6.3, streptolysin S, and staphylococcal toxins.

Listeriolysin effects were studied on rabbit and mouse phagocytic cells (Kingdon and Sword, 1970) and on L-M strain of mouse cells (Siddique, 1969). Striking membrane damage was observed. Pneumolysin tested on rabbit erythrocytes (Shumway and Klebanoff, 1971) was shown to promote marked alteration of cell morphology at the level of the surrounding membrane. A great deal of work was carried out with crude or purified streptolysin O preparations on rabbit leukocytes (Hirsch *et al.*, 1963), mouse macrophages (Fauve *et al.*, 1966), Ehrlich ascites tumor cells (Ginsburg and Grossowicz, 1960), rabbit platelets (Bernheimer and Schwartz, 1965), KB cells (Mastroeni *et al.*, 1969), mouse myocardium (Halpern *et al.*, 1969), and rat heart ventricular cells and rat kidney epithelium (Thompson *et al.*, 1970). Numerous spherical cytoplasmic blebs which extruded from the cell membranes appeared within seconds or minutes; cytoplasmic organelles became intensely granular in the former and collapsed in the latter. At the same time, the double walled nuclear envelope appeared swollen and the contents of the nuclei were disorganized. These findings supported other experimental indications that streptolysin O acts primarily through disruption of cell membrane integrity. Membrane damage of mouse macrophages is shown in Fig. 6.1.

Free or cell-bound streptolysin S (Ginsburg, 1970) also induced striking morphological modification of cell membrane and intracytoplasmic organelles on a variety of mammalian cells with swelling and bleb formation. Motion picture study of this toxin and of streptolysin O on leukocytes was a particular interest (Hirsch *et al.*, 1963). The cytolytic effects of staphylococcal α-, β-, γ-, δ-toxins on HeLa cells or human fibroblasts have been studied by phase-contrast microscopy (Wadström and Möllby, 1972). Similar investigation on cell cultures and white blood cells was reported earlier by Gladstone and Yoshida (1967). Highly purified β-toxin (Bernheimer *et al.*, 1974) incubated with erythrocytes has shown membrane deterioration and size reduction of cells and the appearance of phase-dense masses. More recently the study of phase-contrast microscopy of erythrocyte membrane from various species lysed by the hemolysin of *Aeromonas hydrophila* (Bernheimer *et al.*, 1975) has shown that phase-dense droplets developed in rabbit and human ghosts but not in sheep cell ghosts.

By immunofluorescence techniques Klainer *et al.*, (1964) detected α-toxin on the surface of rabbit erythrocytes during the period of maximal hemolysis followed by a decrease of fluorescence with time. This result has been interpreted as being indicative of release of toxin. It appears to be in accord with earlier experiments of

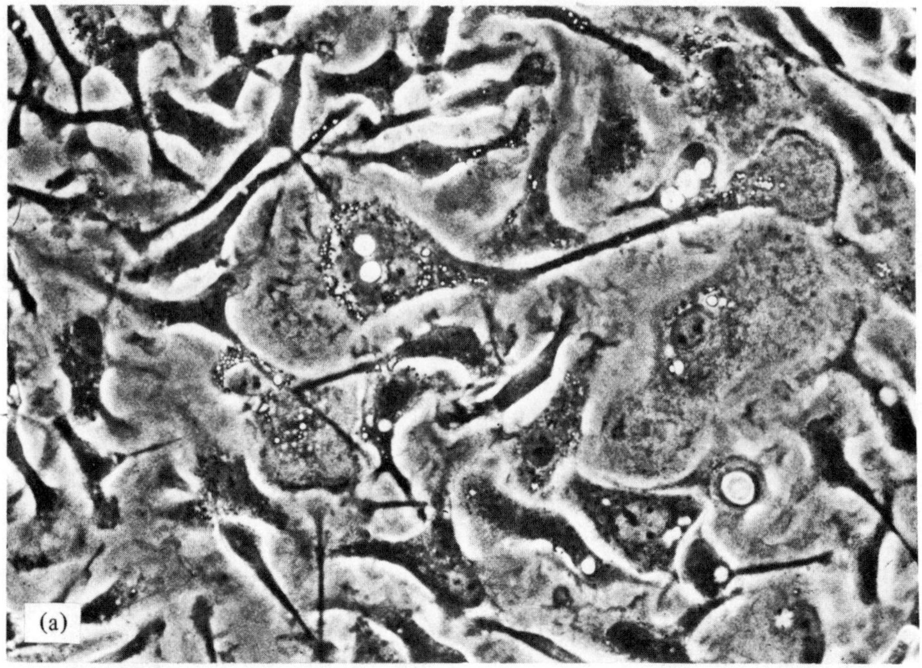

Fig. 6.1 Three-day-old culture of mouse macrophage in a serum-free medium (unstained, phase contrast x 1000).

(a) Control (untreated). (b) After 2 minutes exposure to streptolysin O (200 HU ml^{-1}). (c) After 5 minutes exposure to streptolysin O (200 HU ml^{-1}).

Note the swollen cells after treatment with toxins and blebs extruding from cell surface and lysis of many intracytoplasmic granules. These figures are from work by Fauve and Alouf (Institut Pasteur).

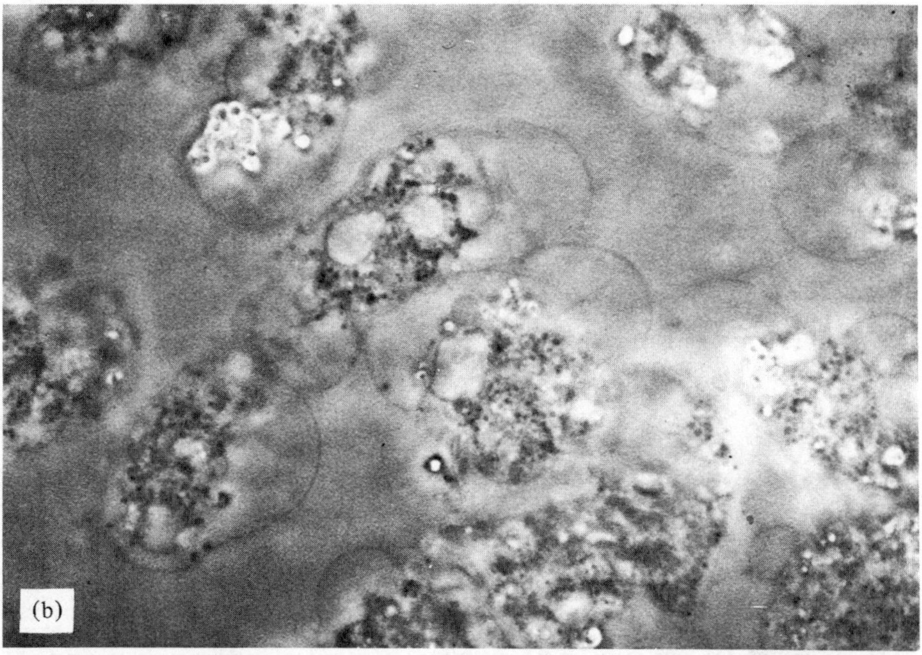
(b)

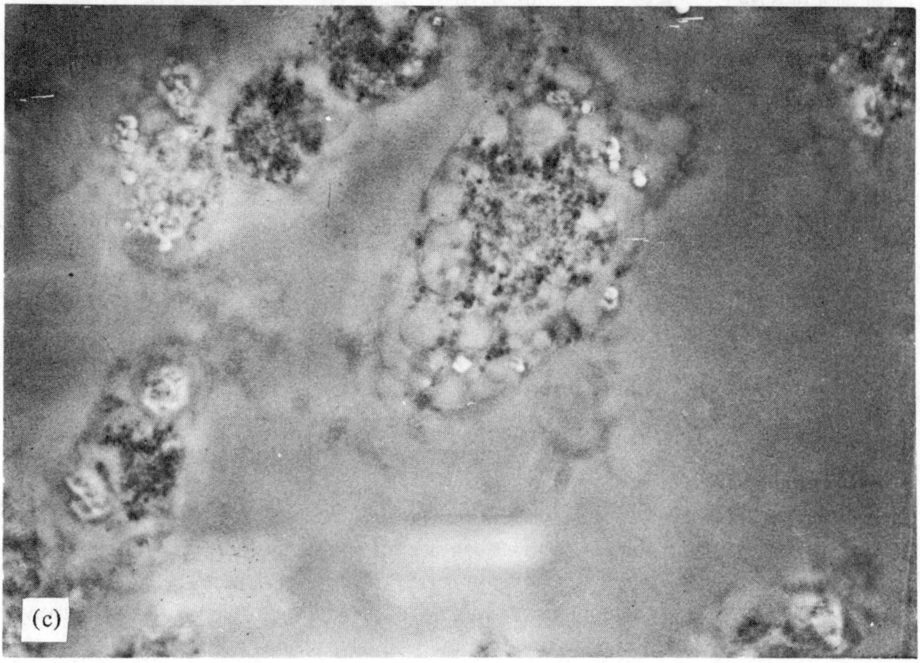
(c)

Lominski and Arbuthnott (1962) who presented evidence of recovery of α-toxin after hemolysis. More recently Cassidy and Harshman (1973) have shown that ^{125}I-labelled α-toxin bound very rapidly during the prelytic period and that part of the bound toxin was released just before the onset of visible lysis. It is to be noted that the same concentration of toxin which gave maximum fluorescence on rabbit cells produced no fluorescence on human cells which are known to be at least 100-fold less sensitive to lysis by α-toxin. Fluorescence occurred when toxin concentration was about 100 times that used for rabbit erythrocytes. It was also found that the capacity of erythrocytes or of ghosts of different species to bind α-toxin (Wiseman and Caird, 1972) correlated with their sensitvity to lysis. This finding leads to the inference (Bernheimer, 1974) that specific toxin receptors exist in the rabbit erythrocyte plasma membrane and that these receptors are present either in reduced amounts or have a different location in the erythrocyte plasma membrane of other species.

6.5.2 Electron microscopic studies

The effects of most of the toxins quoted above on a variety of cells were also studied by various electron microscopic techniques after the initial study reported by Dourmashkin and Rosse (1966) on the morphologic changes in erythrocyte membranes undergoing hemolysis by streptolysin O, *Clostridum perfringens* α-toxin (phospholipase C), saponin, and complement. These authors reported that streptolysin O produced what were described as 'holes' or 'pits' of about 50 nm in diameter in the erythrocyte membrane. This size is large enough to permit escape of the hemoglobin molecule. These holes appeared to be lined with material having different electron density, after negative staining, from that of the surrounding matrix. They found that clostridial α-toxin produce the same type of holes. Bernheimer (1972) suggested that these holes might be due to contaminating clostridial θ-toxin rather than α-toxin itself.

(a) *Thiol-dependent cytolytic toxins*

Erythrocyte membrane alteration has recently been investigated in detail by electron microscopy with three highly purified SH-dependent cytolytic toxins, streptolysin O (Pendleton *et al.*, 1972; Duncan and Schlegel 1975), cereolysin (Pendleton *et al.*, 1972; Shany *et al.*, 1974), and θ-toxin (Smyth *et al.*, 1975a). Ring or arc-shaped structures as well as 'holes' or 'pits' with a characteristic ring border appeared on the cell surface. Pendleton *et al.*, found that streptolysin O caused extensive damage to the structure of the membrane through the formation of holes of about 50 nm. The effect of cereolysin appeared less drastic; holes (of similar diameter) were less frequently obtained. The ring structures reported by Duncan and Schlegel for streptolysin O were of smaller size than those reported above. They had an electron-dense center of 24 nm diameter and a raised periphery about 17 nm in width. The C-shaped structures had the same width and were found to be 75 nm in length.

With θ-toxin, the outside diameter of the rings displayed some heterogeneity and varied from 24 to 35 nm with a border thickness of 4.1–7.8 nm. At high toxin concentration ghosts appeared frequently fragmented. Identical arc and ring-shaped structures were visible in the plane of the membrane and were seen peeling off at the edges of ghosts. Free rings and arcs were also visible (Smyth *et al.*, 1975a). According to Duncan and Schlegel (1975) rings are assumed to represent the C-circularized structure. These authors reported similar structures with lecithin–cholesterol dicetyl phosphate liposomes treated with streptolysin O as well as with cholesterol–toxin dispersion. An identical finding (Smyth *et al.*, 1975a) was reported for θ-toxin mixed with cholesterol-containing lipids. Cholesterol-free lipids or esterified cholesterol which in contrast to cholesterol neither inhibits the lytic properties of these toxins (Bernheimer, 1974) nor binds them (Prigent and Alouf, in the press) do not give such structures.

These findings give very strong support to the suggestion that membrane cholesterol is the cellular receptor of streptolysin) and related toxins (Alouf and Raynaud, 1968a; Bernheimer, 1974) as well as that of non-protein lytic agents such as polyene antibiotics and saponins. The toxins and these agents do not lyse prokaryote cells which are devoid of cholesterol whereas they do lyse eukaryote cells in which this sterol is a permanent and essential constituent of the cytoplasmic membrane (Nes, 1974).

The antibody–ferritin electron microscopic method was employed by Pendleton *et al.*, (1972) who demonstrated binding of streptolysin O and cereolysin to erythrocyte on cholesterol-containing mycoplasma membranes but not to bacterial membranes in which cholesterol is absent. Uneven distribution of ferritin in the erythrocyte membrane was shown, suggesting that the distribution of cholesterol (at least the molecules available to the toxin) may not be entirely random. Shany *et al.*, (1974) demonstrated that membrane cholesterol is the common binding site of streptolysin O, cereolysin, saponin, and the polyene antibiotic filipin. Reciprocal interference of binding to erythrocyte membrane or to cholesterol-containing liposomes between these agents was shown. The same investigators demonstrated that the topographic distribution of erythrocyte membrane cholesterol can be portrayed without the use of antibody by employing ferritin conjugated directly to cereolysin. The question as to whether the ring or arc-shaped structures formed upon interaction of streptolysin O and clostridial θ-toxin with erythrocyte membrane or at the surface of artificial liposomes are actually holes or pits extending through the full thickness of the membrane is still without a clear answer. It remains to be shown if there is any relationship between ring formation and permability changes leading to lysis of intact erythrocytes.

That these structures are not transverse holes at least in liposomes is suggested by the finding that streptolysin O did not release trapped internal markers whereas Triton X-100, saponin, sodium dodecyl sulfate, and streptolysin S release markers (Duncan and Schlegel, 1975). Similar structures have been observed on natural and artificial membranes treated with other lytic agents. Freeze-etching data suggest that

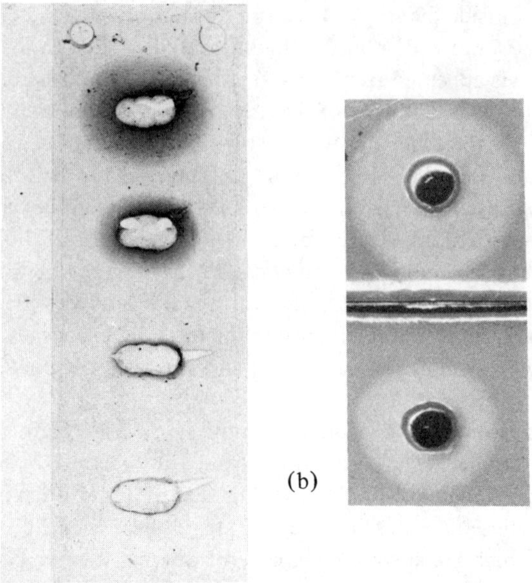

(a) (b)

Fig. 6.2 Complex formation in agar gel between cholesterol and (a) streptolysin O or (b) digitonin; 0.1 ml of cholesterol solution in acetone (1 mg ml^{-1}) mixed with 5 ml of melted agar (50°) and layered on glass sides (25 x 75 mm). Wells from top to bottom contain decreasing concentrations of streptolysin O and digitonin. These figures are from work by Prigent and Alouf (Institut Pasteur).

these features are not transverse holes in the case of filipin (Tillack and Kinsky, 1973; Verkleij *et al.*, 1973), immune antibody–complement lysis (Iles *et al.*, 1973; Seeman, 1974), and very likely saponins (Seeman *et al.*, 1973). Recent studies by De Kruijff *et al.*, (1974a, b) and De Kruijff and Demel (1974) with the polyene antibiotics filipin, amphotericin B, nystatin, etruscomycin, and pimaricin interacting with *Acholeplasma laidlawii* cells and lecithin liposomes led to the conclusion that amphotericin B, nystatin, and estruscomycin generate hydrophobic channels which traverse the membrane whereas pimaricin cannot. Filipin was visualized as an aggregate with cholesterol molecules of diameter 150–250 Å oriented in the hydrophobic core of the membrane. The structures seen with these antibiotics have been interpreted as molecular complexes with cholesterol. Space-filling models of these complexes have been constructed (De Kruijff and Demel, 1974). The nature of the ring and arc-shaped structures has still to be determined. It is very likely that they represent aggregates of toxins and cholesterol molecules or possibly stoichiometrically defined complexes of these molecules as appears to be the case for polyenes. In recent experiments with streptolysin O diffusion in agar gel containing cholesterol we have shown the specific insolubilization of toxin (Prigent and Alouf, to be published) (Fig. 6.2). Other, although less likely, interpretations of the structures observed could be a simple reorganization or packing of cholesterol molecules in the membrane. The release of ring structures from treated erythrocyte ghosts and cholesterol dispersions reported for θ-toxin makes isolation and investigation of their composition possible. It is to be noted that the occurrence of similar arc and ring structures in highly purified θ-toxin has been reported. Their presence, albeit in extremely small quantities (Smyth *et al.*, 1975a), suggests that they may be a polymeric form. High molecular weights forms have been reported for θ-toxin (Smyth, 1975), listeriolysin (Jenkins and Watson, 1971), and streptolysin O (Alouf and Raynaud, 1973). The possibility that traces of cholesterol present in the culture medium or during the purification process specifically induced toxin polymerization could not be ruled out. This situation bears an analogy to that reported for the 3S form of staphylococcal toxin which may be transformed into a 12S ring-structured polymer by contact with a considerable variety of lipids or to liposomes prepared from various lipids including cholesterol (Arbuthnott *et al.*, 1973; Bernheimer, 1974). Spontaneous polymerization of this toxin very likely catalyzed by lipids has also been shown. Polymerization of the F component of staphylococcal leukocidin by some phospholipids has been reported by Woodin (1970); this polymerization occurs at the cell surface upon interaction of leukocidin with leukocytes.

Other electron microscopic studies of this effect of streptolysin O have been reported on polymorphonuclear leukocytes (Zucker-Franklin, 1965) and myocardial cells (Thompson *et al.*, 1970). The observations reported displayed clear damage of cytoplasmic and lysosomal membrane in leukocytes. The study of Thompson *et al.*, (1970) on rat myocardial cells disclosed dramatic morphological changes within 1–2 minutes of contact with toxin. The blebs observed under phase contrast could be resolved by electron microscopy as protrusions of the plasmalemma into which there was comparatively sparse leakage of organelles from the cytoplasm. No evidence of cell membrane rupture was seen. However, the most dramatic and unusual early alteration of cardiac cells was strikingly manifested in the endoplasmic reticulum. The cisternae were considerably swollen and the contents became condensed. There was widespread vacuolation of the cytoplasm accompanied by similar changes in the mitochondria with some swelling of the cristae. Golgi complex was also vacuolated. The nuclear envelope remained intact and its apparent thickening observed by phase-contrast microscopy was seen to be due to the piling up of dense material (presumably chromatin) against the inner side of the nuclear envelope. Similar lesions were reported by Halpern *et al.*, (1969). The study of the effects of the toxin on kidney or endotheial cells has shown similar blebs but their number was much less than those observed on myocardial cells (Thompson *et al.*, 1970). It is conceivable that this is a reflection of the number of 'attack' sites on the cell membrane. Recent studies by Alouf and Chevance (unpublished data) on rabbit erythrocyte lysis by sublytic doses of streptolysin O disclosed discrete lesions of the membrane and leakage of intracellular material (Fig. 6.3) as well as granules on ghosts (Fig. 6.4) similar to those reported by Bernheimer *et al.*, (1974) for

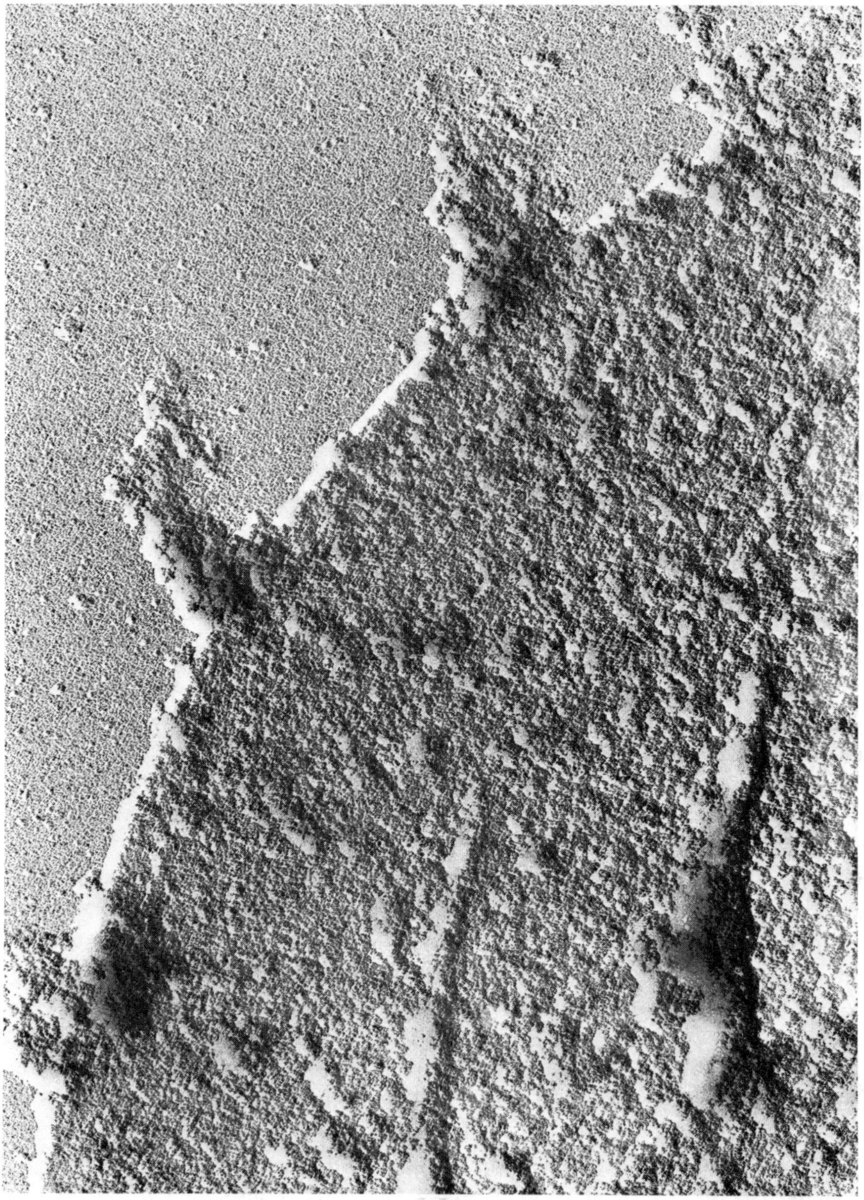

Fig. 6.3 Electron micrograph of rabbit erythrocytes exposed to streptolysin O (10 HU ml^{-1}); 0.15M-PBS buffer; palladium treatment at 15°C and shadowing (x 43 000); Edwards evaporator (5×10^{-5}). From work by Alouf, Prigent, and Chevance (Institut Pasteur).

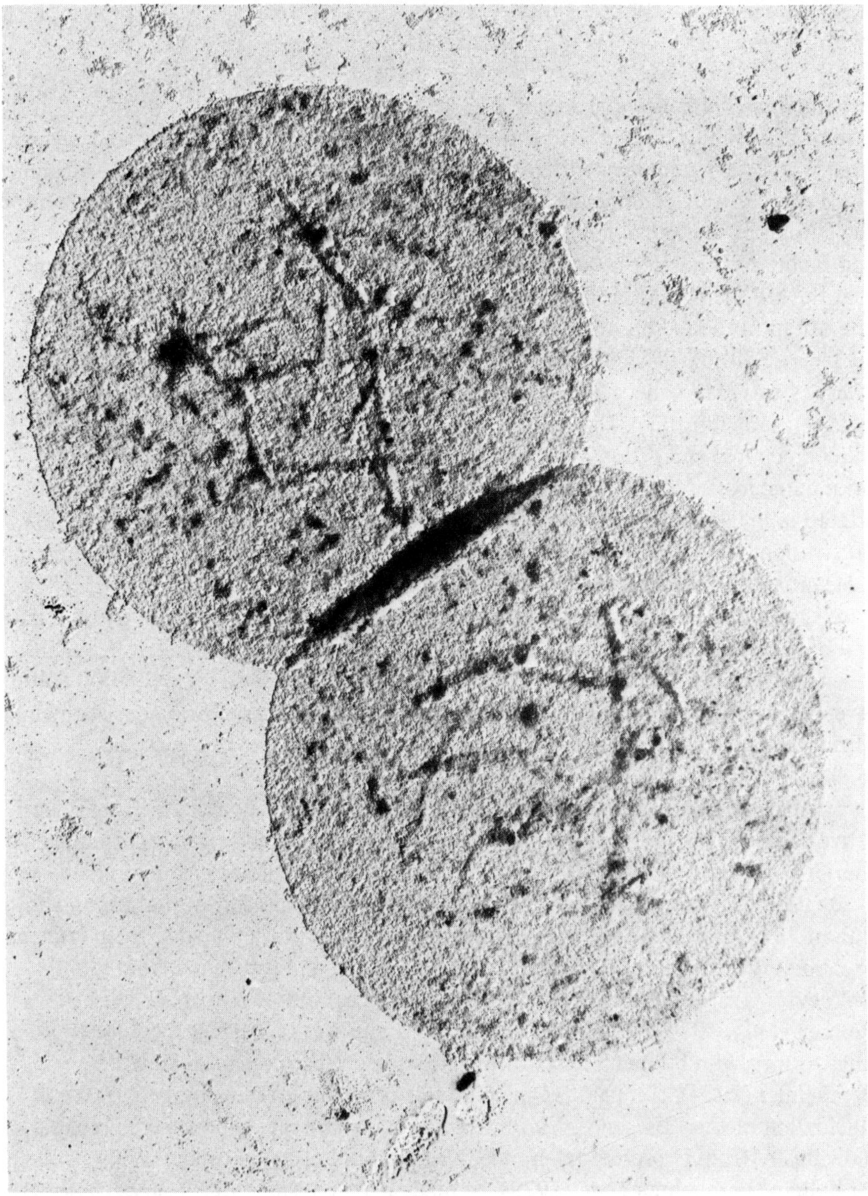

Fig. 6.4 Electron micrograph of rabbit erythrocytes exposed to streptolysin O (10 HU ml^{-1}); 0.15M-PBS buffer; palladium treatment at 15°C and shadowing (x 5750); Edwards evaporator (5 x 10^{-5}). From work by Alouf, Prigent, and Chevance (Institut Pasteur).

staphyloccal β-toxin. Whether these granules represent toxin aggregates or complexes (with cholesterol ?) remains to be established.

(b) *Staphylococcal α- and δ-toxins*

Rabbit neutrophils treated with α-toxin (Maheswaran *et al.*, (1969) showed aggregation of glycogen granules, extensive degranulation, and loss of cytoplasmic substance. Nuclear lobes were round and showed chromatin dissolution. Vesicles formed within the cytoplasm and outside the cells. The studies of Klainer *et al.*, (1972) on human and rabbit erythrocytes by scanning electron microscopy provided further evidence that the cell membrane is the primary site of action of α-toxin. Progressively severe alterations in the red blood cell membrane occurred with time. During the prelytic lag phase, multiple discrete blisters appeared on the surface of rabbit cells but cell integrity was maintained. Whether these small defects are related to potassium leakage characterisitic of this phase (Madoff *et al.*, 1964) is not known. During lysis cellular collapse and shoots were seen with separation of large fragments of cell membrane; loss of the self-sealing properties of the cells was evident. By contrast, alteration in less sensitive human erythrocytes was limited to occasional finger-like prostrusions during the period of accelerated lysis. Transmission electron microscopy substantiated these changes.

A systematic electron microscopic study of the 3S and 12S forms of α-toxin itself, and of toxin action on a variety of natural and artificial membranes, has been undertaken (see Bernheimer, 1974, for a review). The 3S-monomeric form (28 000 daltons) reveals amorphous finely granular material. The 12S polymeric form consists mainly of rings about 10 nm in diameter made up of six subunits per ring each 2.0–2.5 nm in diameter.

Similarly, to many mammalian cells, liposomes of various qualitative and quantitative composition are lysed by staphylococcal α-toxin, as shown by lipid fragmentation of these spherules and release of sequestered marker molecules.

By negative staining and freeze-etching techniques it was shown that, upon addition of 3S α-toxin to a variety of biomembranes osmotically prepared from human or rabbit erythrocytes and platelets, rat hepatocytes, and lysosomes from rabbit leukocytes, a large number of ring-shaped structures form (Freer *et al.*, 1968; Remsen *et al.*, 1970; Bernheimer *et al.*, 1972; Freer *et al.*, 1973). By contrast no rings formed with biomembranes from various bacterial species. As stated by Bernheimer (1974), comparative membranology would presumably benefit from elucidation of this difference. Various synthetic membranes (liposomes) prepared individually from phosphatidylcholine, phosphatidylserine, phosphatidylinositol, and cardiolipin (Bernheimer *et al.*, 1972), or made from phosphatidylcholine, cholesterol and diacetyl phosphate (Arbuthnott *et al.*, 1973), and chloroform–methanol extracts of brain white matter (Bernheimer *et al.*, 1972) gave similar ring structures upon treatment by α-toxin. By contrast no rings formed with liposomes made from phosphatidylethanolamine. Differences in ring pattern (ordered rectangular or random structure) were shown in both natural and synthetic membranes. The arrangement of the rings was affected by the negative stain used (molybdate or phosphotungstate) and by the eventual treatment of the membranes by δ-toxin added prior to α-toxin, or after this toxin, or simultaneously. It is to be noted that δ-toxin alone did not give any ring structure. It has also been shown that pre-treatment of erythrocyte membranes with phospholipase C abolished the capacity to yield the rectangular pattern. Several lines of evidence suggest that the structures formed are 12S toxin rings. Polymerization of the 3S form occurs therefore on biomembranes and lipid-containing artificial spherules as well as with individual lipids. A detailed study of this problem (Arbuthnott *et al.*, 1973) has shown that diglyceride was the most active lipid in inducing polymerization. This finding shows that the polar group of lecithin is not required for polymerization, and adds support to the concept that staphylococcal α-toxin is capable of hydrophobic interaction with lipids as suggested by the experiments of Freer *et al.*, (1968).

The relationship between toxin polymerization at the surface of the cytoplasmic membrane and the molecular mechanism of membrane disruption by this toxin still remains to be clarified. According to Wiseman and Caird (1972) and Wiseman *et al.*, (1975), the toxin is a symogen that degrades membrane protein when activitated by membrane protease. Freer *et al.*, (1973) were unable to find evidence supporting the proteolytic mechanism. By contrast they suggest high surface activity of the toxin. Their experiments are also supported by those of Buckelew and Colacicco (1971) and Colacicco and Buckelew (1971) who demonstrated that the toxin readily spreads as a film on aqueous media and that it is capable of penetrating into lipid monolayers. Film penetration was related to the structure of the lipid; it was greatest with cholesterol and least with ganglioside. It is to be noted that disruption of lipid liposomes by α-toxin leads to the inference that this toxin acts through surface-active properties. The capacity of α-toxin to penetrate and disrupt the hydrophobic regions of membranes may explain its lytic properties. Freeze-fractured replicas of normal and treated ghosts show drastic alteration in the membrane. Control ghosts appear smooth in outline with the membranes largely intact whereas treated ghosts show extremely irregular outlines due to fragmentation and vesiculation of the membrane and the appearance of plaques bearing relatively few intramembrane particles.

A mechanism involving both physical and enzymatic modes of action cannot however be ruled out. Further study is needed to solve this problem.

(c) *Staphylococcal β-toxin and clostridial α-toxin*

This toxin has a phospholipase C activity specific for sphingomyelin (sphingomyelin choline phosphohydrolase) which hydrolyzes either free or erythrocyte membrane sphingomyelin into N-acetylsphingosine and phosphorylcholine (see Bernheimer, 1974). Ultrastructural study of sheep or human ghost erythrocytes treated with β-toxin revealed large numbers of mebranous vesicles of various sizes within the ghosts (Berheimer *et al.*, 1974). This effect contrasted with that of clostridial α-toxin which has a phospholipase C activity for both phosphatidylcholine and sphingomyelin. This toxin induced osmophilic granules not observed with β-toxin.

The effect of β-toxin on bovine or human erythrocyte ghost membranes was also studied by the freeze-etching technique (Low *et al.*, 1974). Extensive alteration of the membrane was observed. Vesicles or invaginations appear in the interior of the ghosts as well as particle-free areas in the hydrophobic fracture plane distinct from the outer surface of etched erythrocyte ghosts. In addition large globules of particle-free solid material were frequently seen within ghosts, possibly the ceramide product of hydrolysis (N-acylsphingosine) accumulated in the hydrophobic region of the bilayer. These observations can be explained in terms of a membrane in which sphingomyelin is preferentially located in the outer half of the bilayer. This finding is in agreement with the increasing evidence for significant asymmetry between the inner and outer leaflets of the lipid bilayer of human erythrocytes and more generally for many other biomembranes (Bretscher and Raff, 1975). In erythrocytes, glycolipids and most choline-containing phospholipids (phosphatidylcholine and sphingomyelin) are found in the outer half of the bilayer while the aminophospholipids (phosphatidylserine and phosphatidylethanolamine) are largely confined to the inner half.

Staphylococcal β-toxin, clostridial α-toxin, and very likely other lytic toxins endowed with phospholipase C and D activity, such as those from *Acinetobacter calcoaceticus* and *Corynebacterium ovis* as well as non-toxic bacterial phospholipases such as that from *Bacillus cereus* (see Wadström *et al.*, 1974), should prove useful in further studies of the distribution of phospholipids in various membranes and their relationship to membrane function. Recently, clostridial α-toxin and *B. cereus* phospholipase have been used to demonstrate the asymmetry of influenza virus membrane bilayer (Tsai and Lenard, 1975).

6.6 MODIFICATION OF CELL METABOLISM AND PHYSIOLOGICAL FUNCTIONS BY CYTOLYTIC AGENTS

Obviously cell membrane damage by cytolytic toxins may lead to various, more or less severe impairment of cell function. The modification observed are however not specific to cytolytic toxins since the effects of other cytopathogenic toxins may be reflected similarly at the physiological or metabolic levels, such as cell growth (number of cells, total cell protein), pH of culture medium, intracellular enzyme activities, specific metabolic functions, incorporation of protein and nucleic acid precursors, etc. (see Thelestam, 1975; Thelestam *et al.*, 1973; Thelestam and Möllby, 1975a).

Cell viability of tissue culture mouse peritoneal macrophage monolayers was tested for listeriolysin by Watson and Lavizzo (1973). A temporary growth-enhancing effect was observed. Trypan blue uptake by KB and monkey kidney cells incubated with staphylococcal β-toxin was correlated with non-adherence to glass surfaces and absence of pH changes in the medium (Wiseman, 1968). The effect on intracellular enzymatic activities with staphylococcal toxins were studied by Korbecki and Jeljaszewicz (1965). Symington and Arbuthnott (1969) have shown that staphylococcal α-toxin and streptolysin S cause immediate increase of succinate oxidation upon incubation with ascites tumor cells. This effect was interpreted as resulting from irreversible damage of the cell membrane with consequently increased permeability to succinate. More recently it was established (Symington and Arbuthnott, 1973) that streptolysin S impaired succinic and cytochrome oxidase activity of isolated mouse liver mitochondria. This effect was attributed to the disruption of the electron transport system of these organelles between cytochrome C and oxygen.

Partially purified staphylococcal α-toxin preparation has been shown to inhibit the phosphorylation of ADP coupled to the electron transport system in rat liver mitochondria (Novak *et al.,* 1971a,b). This effect was attributed to a thermostable fraction in the preparation. It is unlikely that α-toxin is the active product since this hemolysin is thermolabile. Indeed, purified α-toxin free of δ-toxin was shown to have no effect on mitochondrial functions of rat or rabbit liver. Oxidation of succinate, glutamate, malate, or pyruvate was not affected (Rahal, 1972). More recently the purified toxin was shown to have no effect at high concentration on membrane ATPases and proton translocation in mitochondria (Cassidy *et al.*, 1974). By contrast the highly cytolytic thermostable δ-toxin, which has long been known to contaminate α-toxin preparations (Bernheimer *et al.*, 1968), was shown to cause mitochondrial lysis with inhibition of respiration and phosphorylation (Rahal, 1972). However, in the presence of succinate and cytochrome C respiration was stimulated, suggesting an intact electron-transport system except for loss of functional cytochrome C. Magnesium-dependent mitochrondrial adenosine triphosphatase was strimulated by δ-toxin. This toxin, purified and characterized by Kreger *et al.*, (1971), appears to behave on biomembranes or artificial membranes as a detergent-like agent similar to Triton X-100 and deoxycholate (Bernheimer, 1974). It is likely that the heat-stable factor of Novak and associates which uncouples oxidative phosphorylation is δ-toxin. It is also likely that the α-toxin preparation used by Symington and Arbuthnott (1969) on tumor cell mitochondria also contained this active factor.

A number of physiological studies have also been reported with various preparations of clostridial α-toxin (phospholipase C). However, since these preparations, particularly commercial ones, are very often contaminated by other proteins, the results should be interpreted with caution unless highly purified toxin is employed. This was the case for the preparations used by Strandberg *et al.*, (1974) who observed the release of histamine from rat mast cell suspensions treated with α-toxin. A similar result was obtained with purified θ-toxin which has no phospholipase activity. Differences between the conditions of histamine release by the two toxins were demonstrated especially as regards temperature and the effects of Ca^{2+}, Zn^{2+}, and N-ethylmaleimide. The histamine-releasing capacity of α-toxin has been shown to interfere with some electrophysiological properties of frog muscle membrane (Boethius *et al.*, 1973). Cuatrecasas (1971) reported that clostridial α-toxin and phospholipase A from *Vipera russeli* unmasked insulin receptors in fat cell membranes and stimulated glucose uptake. An irreversible aggregating effect of this toxin on human platelets was reported by Chap *et al.*, (1971); its mechanism remains to be elucidated. An insulin-like effect of α-toxin was reported on brown fat cells (Rosenthal and Fain, 1971). The toxin also appeared to suppress the responsiveness to thyroid stimulating hormone by producing a specific alteration in thyroid cell membrane (Macchia and Pastan, 1967). An interesting overall model of the modification of various cell functions upon treatment of leukocytes by highly purified staphylococcal leukocidin has been described by Woodin (1970). Fig. 6.5 shows the

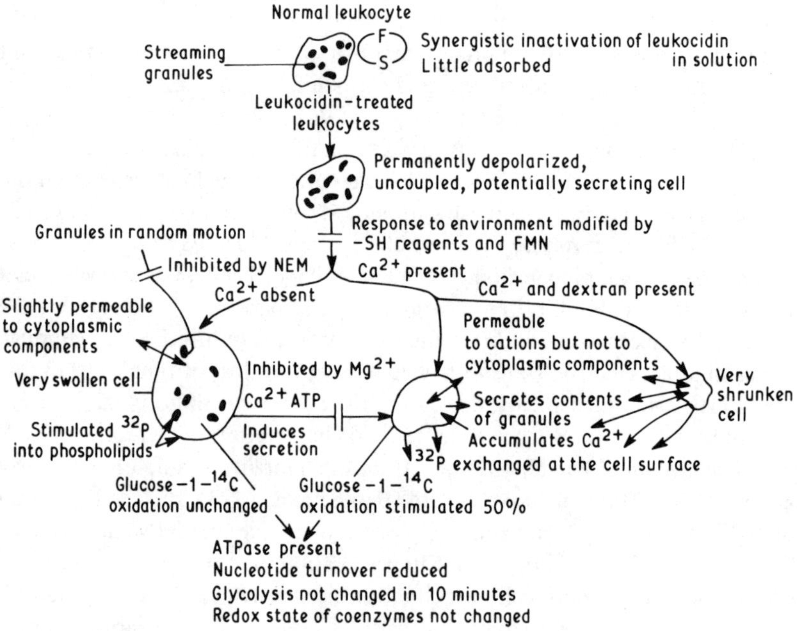

Fig. 6.5 The changes induced in leukocytes during 10 minutes of treatment with staphylococcal leukocidin. From Woodin (1970).

changes observed during the first 10 minutes of incubation with the toxin. The main features of leukocidin-treated cells were increased cell permeability to cations, secretion of protein, accumulation of calcium stimulated ATPase, and nucleotide–orthophosphate exchange. It is to be noted that all cytotoxic effects of the toxin stem from increased permeability to cations, and the site of action was unambiguously shown to be the cell membrane. The molecular mechanism of this toxin with leukocyte membrane will be discussed in another section. Most direct or indirect physiological tests showin metabolic or functional modification of cell life, although of much interest, are not very suitable for indicating the primary damage to the cytoplasmic membrane by cytolytic toxins and do not reflect rapid changes occurring at different subcellular levels. For such agents, measurement of changes in membrane permeability constitutes the better criterion and investigative approach to membrane damage. In additions, it permits quantitative rather than qualitative studies as is the case for morphological or metabolic evaluation of cellular lesions.

6.7 QUANTITATIVE STUDY OF PERMEABILITY CHANGES IN THE CELL MEMBRANE INDUCED BY CYTOLYTIC AGENTS

The normal cytoplasmic membrane acts as a selective permeability barrier protecting the cell from the outer environment. The structure of the membrane and the transport systems which regulate the active exchanges in both directions between the cell and its surrounding milieu in eukaryote or prokaryote organisms and particularly erythrocytes have been studied in great detail in recent years. The membrane appears as a mosaic fluid structure consisting of a bilayer of lipid array with intercalated proteins or glycoproteins (Singer, 1974; Bretscher and Raff, 1975). Exposure of any sensitive cell to a cytolytic agent impairs the osmotic barrier and, whatever its mechanism of action, transient or permanent rearrangement of membrane structure takes place. The modification of permeability is readily recognized by the release from the cells of small cytoplasmic metabolites which normally cannot leak out, such as K^+, phosphate, amino-acids, and sugars, or even macromolecules depending on the extent of membrane damage. At the same time internal enzymes may be rendered accessible to substrates which do not normally pass across the membrane; assay of β-galactosidase in the presence of toluene is a well known application. As noted by Harold (1970) the 'destruction' of the osmotic barrier frequently mentioned does not necessarily mean the physical disintegration of the membrane but any structural change in the orderly arrangement of both proteins and lipids such as to produce discontinuities in the barrier which are often designated as 'channels' or 'holes'. The terms should not be considered as reflecting physically fixed structures but rather as an operational concept meaning a more or less extended area of membrane disorganization or reorientation of one or several constituents of the bilayer.

As stated by Thelestam (1975), depending on the type of membrane lesion and the type of cell a slight change in membrane permeability may have more or less serious consequences. If the cell has little capacity for membrane repair it will eventually undergo lysis, whereas a high repair capacity will infer better chances of survival. Regardless of the degree of repair the altered membrane permeability will give rise to an increased bidirectional flow of matter through the membrane.

Therefore, permeability changes can be directly assessed either by detecting penetration of external substances into the cell or by measuring leakage of intracellular cell metabolites or markers into the extracellular medium. The first method involves the dye exclusion test as an index of cell injury or cell death (Holmberg, 1961; Phillips, 1973). The techniques based on the uptake of vital stains such as trypan blue, neutral red, or lissamine green have been employed in complement studies (Saijo, 1973) and applied to toxins by Wiseman (1968) and more recently by Thelestam and Mollby (1976). However, the second approach based on the measurement of the leakage of intracellular metabolites or markers (radioactively labelled in most cases) is much more valuable and of greater interest. It allows a very accurate study of membrane damage owing to its sensitivity, reproducibility, suitability for investigation of cytoplasmic damage without morphological changes, and the possibility of quantification of the cell damaging effect (Thelestam *et al.*, 1973).

6.7.1 Methodology of leakage tests on intracellular markers

It is assumed that the molecular size of the material leaking from damaged cells is indicative of membrane damage in terms of the 'size' of the functional holes caused by the lytic agent. Small functional holes induced in the membrane will give rise to a primary leakage of only small molecules. A strongly altered membrane permeability implies large holes which may permit the escape of high molecular weight substances as well. Although there are many kinds of initial reactions possible at the cell membrane level, the final event of membrane disruption is very often a simple colloid osmotic lysis (Seeman, 1974).

The material leaking from damaged cells and designated as 'intracellular marker' may be composed of naturally occurring components of the cytosol such as ions, small molecular weight metabolites, and macromolecules. In most studies this material is radioactively labelled by prior incubation of target cells with radioactive ions or metabolites which are incorporated in the cytoplasm. Other markers not occurring naturally in cells have also been used, such as $^{86}Rb^+$, ^{51}Cr, or the non-metabolizable amino-acid 1-^{14}C α-aminoisobutyric acid.

Whatever the substance taken as a reference to reflect cell damage, it is to be stressed (Thelestam and Möllby, 1976) that cytoplasmic markers leak as a result of a complicated interaction between test agent and cell membrane. Many factors are important for the outcome of this interaction: the potency of the test agent and the number of active molecules present, the time course of interaction between agent and membrane, the nature of the cell system used and its stage of growth, the nature of the toxin-induced holes, the possible formation of secondary holes, and not least important the capacity of the cell to repair its membrane. The first study on leakage of intracellular material from a lysed cell was reported by Green *et al.*, (1959) and Goldberg and Green (1960) for antibody-complement induced lysis of Krebs ascites tumor. It was observed that most of the smaller intracellular ions and molecules such as potassium, free amino-acids, and ribonucleotides were lost from the cells within a few minutes. These substances escaped through a cell membrane which appeared to be altered but unruptured, as judged by electron miscroscopy.

The release of low or high molecular weight markers was thereafter studied in a great number of investigations on the mechanism of immune lysis or cytotoxicity (Klein and Perlman, 1963; Forbes, 1963; Sanderson, 1964; Wigzell, 1965, 1967; Hingson *et al.*, 1969; Henney 1973) and osmotic hemolysis (McGregor and Tobias, 1972). The same method was applied for a number of cytolytic toxins and other cytolytic agents such as polyene antibiotics or detergents on living cells or on model artificial membrane systems. Most studies involved the use of radioactive markers. The measurements of the release of these markers was undertaken on prelabelled target cells.

We shall describe respectively the conventional or standard markers employed in a number of studies of toxin-induced or immune lysis and the new markers recently introduced in the systematic investigation of the effects of cytolytic toxins and other agents on cell membrane permeability of human fibroblasts (Thelestam *et al.*, 1973; Möllby *et al.*, 1974; Thelestam *et al.*, 1975a, b, 1976).

6.7.2 Release of conventional markers through the damaged membrane

The release of ^{35}S-methionine precipitable by trichloroacetic acid from rabbit kidney, Ehrlich ascites tumor, and human amnion cells treated by staphylococcal α-toxin was studied by Madoff *et al.*, (1963). The same toxin was shown to release ^{32}P from calf kidney cells but not from insensitive human or monkey kidney cells (Hallander and Bengtsson, 1967). ^{32}P was also used by Forbes (1963). The leakage of K^+ ions for cytotoxicity tests from rabbit erythrocytes treated with α-toxin was rapid and massive during the prelytic period (Madoff *et al.*, 1964). Bangham *et al.*, (1965b) have shown the release of Na^+ and K^+ from artificial phospholipid granules (in which these ions were trapped) upon treatment with streptolysin S. The same toxin was also shown to release alkaline phosphatase from a granule fraction of rabbit liver suspended in 0.25 M-sucrose.

Leakage of ^{42}K as well as that of adenylate kinase and hemoglobin was shown during gradual osmotic hemolysis (McGregor and Tobias, 1972). The release of K^+ or other cations was also studied in the action of polyene antibiotics on *Candida albicans* (Gale, 1974; Hammond *et al.*, 1974), *Acholeplasma laidlawii* or liposomes (De Kruijff *et al.*, 1974a), red blood cells (Cass and Dalmark, 1973), and lipid bilayer membranes (Andreoli, 1974). ^{82}Rb release has also been used either in membrane model systems such as liposomes for the study of the effects of sterols on permeability (Demel *et al.*, 1972) or in investigations of the lysis of red blood cells. The kinetics of release of $^{86}Rb^+$ and hemoglobin from erythrocytes damaged by antibody and complement was studied by Hingson *et al.*, (1969) and more recently by Duncan (1974) for the lysis of rabbit erythrocytes by streptolysin O. In the last case both markers were released at the same rate. The addition of serum to suspending medium did not retard the escape of hemoglobin whereas the efflux of $^{86}Rb^+$ only slightly preceded the loss of this protein. Addition of sucrose to the cell delayed the release of both markers. The author inferred that a colloid-osmotic process is not involved in the streptolysin O lytic effect, and suggested that the functional lesion produced by the toxin has an effective radius greater than 3.55 nm.

It has also been shown by Bernheimer (1970) that erythrocytes treated by streptolysin O did not undergo swelling prior to their disruption. This finding also rules out a colloidosmotic process. As discussed by Duncan (1974), by contrast, lysis by *Clostrodium speticum* hemolysin and streptolysin S involved osmotic mechanisms (Bernheimer, 1972).

A highly purified preparation of *S. aureus* α-toxin was tested by Cassidy *et al.*, (1974) on liposomes composed of lecithin cholesterol and dicetyl phosphate and with liposomes composed of erythrocyte membrane lipids. The interaction was measured by the release of glucose as a low molecular weight marker and hexokinase as a macromolecular one. The latter was released at high toxin concentrations. In spite of these findings it was however established that the action of α-toxin towards biomembranes may not be fully explained by postulating interaction with membrane lipids. ^{86}Rb and ^{42}K are probably the most sensitive indicators of membrane damage hitherto described (Hingsom *et al.*, 1969; McGregor and Tobias, 1972; Duncan, 1974). As noted by Thelestam (1975) these isotopes are not ideal for routine work, owing to their γ-emission and short half-lives. However, sensitive potassium electrodes are now available and may be used for the detection of release of unlabelled K^+ ions (Gale, 1974; Hammond *et al.*, 1974). Furthermore, released potassium may be measured by atomic absorption spectrophotometry. The latter method was employed by Thelestam in some preliminary experiments with fibroblasts. The results indicated a sensitivity comparable to that found with aminoisobutyric acid as marker as discussed later. A spontaneous release approaching 60 per cent in 30 minutes was however observed.

The practical limitations with Rb^+ and K^+ have initiated a search for other low molecular weight cytoplasmic markers. Release of cellular ATP (Nungester *et al.*, 1969; Henney, 1973) has been monitored in immune lysis but its measurement is not very simple and this restricts its use as a marker for large series of samples. Its sensitivity appears similar to that of nucleotide labels used by Thelestam and Möllby (1975a, b, 1976).

^{14}C-Nicotinamide has been used as marker in the study of immune lysis of mastocytoma cells, assuming that it is released as NAD (mol.wt. 664) (Martz *et al.*, 1974). Uptake of this substance in mastocytoma cells was considerably less efficient than uptake of uridine in fibroblasts, implicating a lower sensitivity of the NAD-leakage test than of the nucleotide leakage test (Thelestam, 1975).

Leakage of ^{51}Cr as a criterion of cell damage has been widely used in immunological studies (Perlman and Holm, 1969; Wigzell, 1967; Brunner *et al.*, 1968; Wagner and Feldman, 1972; Henney, 1973; Stulting and Berke, 1973). The release of this isotope was also used with other new markers in a comparative study of the action of clostridial θ-toxin and other cytolytic agents such as the non-ionic detergent Triton X-100, melittin, and the polyene antibiotic amphotericin B on cultured human diploid fibrobalsts (Thelestam and Mollby, 1976). However, the use of

^{51}Cr poses some problems. In spite of its wide use, the fate of this isotope in different types of nucleated target cells was not satisfactorily elucidated until recently although the isotope was early postulated to bind unspecifically to hemoglobin in erythrocytes (Gray and Sterling, 1950). Once it is bound intracellularly ^{51}Cr does not interchange between cells (Bunting *et al.*, 1963). It was recently shown (Henney, 1973) that in the murine mastocytoma cell line P815 ^{51}Cr became bound to proteins of molecular weights in the range 80 000–250 000. In the fibroblast system of Thelestam and Möllby, ^{51}Cr label proved of lower molecular weight (3000 daltons with some variation around this value) than postulated or found for other systems. The size found is approximately that of a nucleotide label.

The sensitivity of ^{51}Cr release was not satisfactory for detection of subtle changes in membrane permeability (Thelestam and Möllby, 1975c). In addition non-specific release of this marker and the others reported above which may occur present experimental disadvantages. Moreover, as ^{51}Cr becomes non-specifically bound (to various degrees depending on the type of cell system and stage of growth) to intracellular protein of different sizes, the release of this label gives little information on the nature of membrane damage (Bunting *et al.*, 1963; Henney, 1973). Formulas for expressing the relative leakage, taking into consideration non-specific release, have been reported by several authors (Henney, 1973; Poon and Cauchi, 1973; Stulting and Berke, 1973; Martz *et al.*, 1974).

6.7.3 Release of new markers through the damaged membrane

An assay system attempting at creating more sophisticated leakage test as regards sensitivity and specificity under well standardized conditions has been devised in recent years by the Swedish group of Wadström, Möllby, and Thelestam who greatly contributed with C. Smyth to the purification and the study of the mode of action of a number of cytolytic toxins.

The human diploid fibrobalst system was chosen because it is well characterized and easy to cultivate under reproducible conditions. In addition, it was of great interest with such a system to extend to nucleated metabolizing cells previous studies made on toxin interaction with red blood cells.

The following cytoplasmic markers of different molecular sizes were developed and employed in leakage tests: (1) ^{3}H-labelled uridine nucleotides referred to as nucleotide label (mol.wt. < 1000); (2) ^{3}H-labelled RNA referred to as RNA-label (mol.wt. $> 200\,000$); (3) ^{14}C-labelled α-aminoisobutyric acid referred to as AIB-label (mol.wt. 103). To evaluate these leakage tests they were compared with the three hitherto most widely used cytotoxicity tests, i.e. observation of cell morphology, a dye exclusion test using trypan blue, and a leakage test with ^{51}Cr.

^{3}H-Uridine is suitable for labelling most types of cells because it is efficiently taken up, and trapped within the cell owing to immediate phosphorylation (Prescott, 1964). Actively growing cells rapidly incorporate uridine nucleotides into different species of RNA.

Therefore, labelling of the fibroblasts in the early growth phase, followed by a long non-radioactive chase, resulted predominantly in labelled rRNA. The incorporation of uridine into macromolecules was substantially slowed down in the stationary growth phase. In the stationary phase uptake was as efficient into the cytoplasm but substantially slowed down. Under these conditions nucleotide labelling resulted in only a low molecular weight cytoplasmic label, with uridine (mol. wt. 244) as the smallest component. Measurement of membrane damage by release of nucleotide label is thus as sensitive a method as the release of ATP or NAD. A good marker should be readily accumulated by the cell, should not be incorporated into macromolecules, and should be treated as non-foreign material in the cell. The non-metabolizable AIB offers such qualities (Thelestam, 1975).

The release of the three new markers was compared with release of ^{51}Cr after treatment with cytolytic agents. The sensitivity of the leakage tests was inversely related to the molecular size of the marker. The most sensitive detection of subtle changes in membrane permeability was obtained with the AIB-label. By contrast, high concentrations of highly cytolytic agents were required to permit release of the RNA (Thelestam and Möllby, 1976). A significant release of both AIB-label and nucleotide label was frequently detected before any morphological changes were evident, as seen in the light microscope. By contrast, release of the ^{51}Cr label correlated well with uptake of trypan blue and morphological signs of cell damage. The following order of sensitivity for detection of membrane damage was obtained with the tests used by Thelestam and Möllby (1975a, b, 1976): leakage of AIB label > nucleotide label > ^{51}Cr label ~ uptake of trypan blue ~ morphological changes > leakage of RNA label.

The leakage tests and their results have been reported in several articles (Thelestam *et al.*, 1973; Möllby *et al.*, 1974; Thelestam, 1975; Thelestam and Möllby, 1975a, b,1976). These tests were applied with the following cytolytic toxins: staphylococcal α-, β-, γ-, and δ-hemolysins; *Clostridium perfringens* α-toxin; and θ-toxin. The other cytolytic test substances were: the non-ionic detergent Triton X-100; bee venom mellitin; and two polyene antibiotics, filipin and amphotericin B.

(a) *Triton X-100, mellitin, and polyenes*

Triton X-100 is a surfactant which solubilizes membrane proteins (Simons *et al.*, 1973). Melittin is a cytolytic polypeptide of 2840 daltons having a characteristic arrangement of basic hydrophilic residues at the carboxy-terminal part (position 21–26) and of hydrophobic residues at the amino-terminal part of the peptide (1–20). It has strong surface activity and reacts with non-polar fatty acids of membrane phospholipids (Williams and Bell, 1973; Habermann, 1972; Mebs, 1973). The polyene antibiotics combine with membrane cholesterol as discussed in Section 6.5.2(a). Amphotericin appears to form transverse channels of about 8 Å in diameter, and it is very likely that leakage of low molecular substances takes place through these pores (De Kruijff *et al.*, 1974a,b; De Kruijff and Demel, 1974). Filipin–cholesterol complexes did not seem to cause transverse channels but very likely lateral disordering in the bilayer allowing the escape of both low and high molecular weight substances. Leakage in experiments with these four reference substances gave rise to different types of leakage patterns. Low concentrations of Triton X-100 caused an identical release of all four markers used, consistent with the formation of large functional 'holes'. The other three substances all induced a certain release of AIB label in concentrations that did not permit the release of other markers. On the basis of the release of RNA label it appeared that the holes induced by melittin were enlarged at high concentrations or upon prolonged incubation. The holes produced by filipin seemed to be of a smaller size since only small amounts of the RNA label were released even at high concentrations or after prolonged incubation. Treatment with amphotericin B induced a very significant release of AIB label in concentrations far below those permitting release of any other marker. The other markers were released only at comparatively high concentrations and to a much lower extent than was found with filipin. The leakage pattern induced with amphotericin B was thus consistent with the formation of the postulated small pores with a diameter of 8 Å (De Kruijff and Demel, 1974).

(b) *Staphylococcal toxins*

Crude preparations of α-, β-, and δ-toxin induced a significant release of the nucleotide label, while highly purified preparations of α-toxin seemed to be without effect on the fibroblasts. Neither crude nor purified γ-toxin permitted leakage of any nucleotide label. The effects found with the crude preparations were due to contaminating δ-toxin. This toxin is strongly surface-active as shown by several investigators (see Bernheimer, 1974). Highly purified toxin was shown to give rise to almost instantaneous nucleotide release similarly to Triton X-100 (Thelestam *et al.*, 1973). However, RNA label was less easily released by δ-toxin, than by Triton X-100. The release pattern induced by δ-toxin more closely resembled that found with melittin. Thus, the use of the high molecular weight RNA label in addition to the small nucleotide label permitted a distinction to be made between the functional 'holes' caused by δ-toxin and Triton X-100.

The effects of the staphylococcal toxins on membrane permeability were investigated in experiments with the AIB label as a more sensitive criterion of membrane damage. The α-, β-, and γ-toxins induced a concentration-dependent release of the AIB label, indicating that these hemolytic toxins also affected the permeability of the fibroblast membrane. Since no nucleotide label was released it seems that the induced functional holes were considerably smaller than those produced by the δ-toxin (Thelestam and Möllby, 1975a).

(c) *Clostridial α- and θ-toxins*

A crude preparation containing both the α- and the θ-toxin induced significant nucleotide release in concentrations which were low both in terms of phospholipase C enzymatic activity and hemolytic activity. Most of this release was shown to be due to θ-toxin. However, purified α-toxin at fairly high concentrations also induced

Table 6.5 ED_{50} Ratios (from Thelestam, 1975)

Cytolytic agent	Nucleotide/AIB	RNA/AIB
Triton X-100	1	1
Melittin	4	8
δ-Toxin	10	60
Filipin	3	200
θ-Toxin	8	> 31
Streptolysin S	100	not determined
Amphotericin B	320	> 2000

ED_{50} values are the ratios between the concentrations needed to release 50 per cent nucleotide, RNA, and α-aminoisobutyric acid.

nucleotide release. That this release was actually associated with enzymatic splitting of membrane phospholipids by toxin was indicated by the release of radioactivity from ^{3}H-choline labelled cells.

Increased release of nucleotide label took place when incubation with α-toxin was performed in hypotonic solution or in the presence of sublytic concentrations of Triton X-100. This effect is very likely due to improved accessibility of membrane phospholipids to phospholipase C.

The nature of the functional holes caused in the fibroblast membrane by highly purified θ-toxin was investigated by comparing θ-toxin induced leakage patterns with those induced by staphylococcal δ-toxin and by the reference substances. The nucleotide and RNA labels were released in a manner resembling that found with filipin. The release of RNA label induced by θ-toxin or filipin never exceeded 50 per cent, although high concentrations and prolonged incubation were used. By contrast, with melittin and δ-toxin the release of all four markers converged at high concentrations and after prolonged incubation. The mechanism of action of θ-toxin and of the related toxin streptolysin O has been discussed as well as their similarity, in some respects at least, to polyene antibiotics.

(d) *Comparative sizes of functional holes*

A comparison of the sizes of the induced functional holes was made (Table 6.5) by calculating the ratios between the concentrations needed to release 50 per cent of the respective markers in 30 minutes (Thelestam, 1975).

A low ED_{50} ratio should indicate relatively large functional holes whereas the ratio should be higher for a small hole. Furthermore, for small holes the comparison between nucleotide and AIB release should give lower ratios than the comparison between RNA and AIB release, because the difference in molecular size of the markers is much greater in the latter case. It is obvious that the absolute sizes cannot be given by this technique. However, if by other methods the physical dimensions of the hole could be determined, a correlation becomes feasible.

In conclusion, the leakage tests performed by the markers used show that staphylococcal α-, β-, and δ-toxin induced subtle permeability changes in the fibroblast membrane, permitting release of only the smallest marker (AIB). Staphylococcal δ-toxin induced the release of all four markers in a leakage pattern resembling that found with melittin from bee venon. Clostridial θ-toxin caused relatively smaller functional holes since leakage of the RNA label was largely restricted. The polyene antibiotic filipin induced a similar leakage pattern. Clostridial phospholipase C (α-toxin) permitted a moderate release of the AIB and nucleotide labels, indicating accessibility of susceptible phospholipids in the fibrobalst membrane to attack by this enzyme. The detergent Triton X-100 induced release of all four markers to the same extent over the concentration range tested, indicating large functional holes. By contrast, the polyene antibiotic amphotericin B permitted significant release of only the smallest marker, indicating considerably smaller functional holes.

It is concluded that a combination of several cytoplasmic markers of different molecular sizes is therefore useful for characterizing induced membrane lesions. The assay methods may be applied in the study of alterations in membrane permeability caused by various cytolytic toxins, enzymes, and membrane active antibiotics. Furthermore, they may be useful in the study of immunologically induced cytotoxic events.

6.8 HEMOLYTIC PROPERTIES OF CYTOLYTIC TOXINS

Most of the studies on the mechanisms of membrane disruption by cytolytic toxins have been performed on erythrocytes since most toxins are hemolytic except leukocidins produced by *S. aureus* (Woodin, 1970) and *P. aeruginosa* (Frimmer and Scharmann, 1975). Erythrocytes provide particularly suitable experimental material owing to the simplicity of the cell compared with others and to the tremendous development of our knowledge in the past ten years concerning the qualitative and quantitative lipid and protein composition and localization in the bilayer leaflet of cell membrane especially in human red blood cells (see Bretscher and Raff, 1975, for a review). The plasma membrane of erythrocytes is the only membrane that can be readily isolated intact and free from contamination by other cells or organelles. Erythrocyte membrane 'ghosts' are prepared by hypotonic lysis of red cells which release the hemoglobin. By studying intact cells, permeable (unsealed) ghosts, or inside-out resealed ghosts, one can study both the external and cytoplasmic faces of the membrane.

Erythrocytes are also suitable for titration purposes. The estimation of the degree of lysis of standardized suspensions of erythrocytes of appropriate species is still the simplest method for the quantitative assay of cytolytic agents. The hemolytic assay involves the measurement of hemoglobin released under defined conditions. Unfortunately the unit of hemolytic activity, at least for a given toxin or group of related toxins, is as yet not standardized amongst the different laboratories as regards

Table 6.6 Relative sensitivity of erythrocytes from various mammalian species to some cytolytic toxins

Toxin	Sensitivity of erythrocytes compared to those of rabbit (%)										
	Rabbit	Sheep	Horse	Rat	Swine	Goat	Cat	Dog	Guinea-pig	Mouse	Human
Staphylococcal α-toxin (Bernheimer, 1974; Möllby and Wadström, 1971)	100	1	0.6	10	–	1	5	25	0.1	9	0.8
Staphylococcal β-toxin (Bernheimer *et al.*, 1974)	0.1	100	–	–	–	20	–	0.1	0.1	–	0.5
Staphylococcal γ-toxin (Wadström *et al.*, 1974)	100	–	–	–	–	–	–	–	–	–	0.1
Staphylococcal δ-toxin (Kreger *et al.*, 1971)	100	20	100	–	20	10	10	–	25	–	50
(Möllby and Wadström, 1971)	100	25	400	–	–	50	–	400	–	–	200
Streptolysin S (Ginsburg, 1970)	100	90	–	240	–	100	100	–	–	258	83
Streptolysin O (Alouf, unpublished)	100	90	95	–	–	–	90	80	100	2	100
Thuringiolysin (Pendleton *et al.*, 1973)	100	12.5	100	–	–	–	–	–	–	–	–
Aeromonas hemolysin (Bernheimer *et al.*, 1974)	100	10	20	3000	30	12	500	600	600	2500	53
E. coli hemolysin (Rennie and Arbuthnott, 1974)	100	100	59	50	–	–	75	66	–	21	7

incubation time, temperature, pH, origin and concentration of erythrocytes, and estimation of the end-point. Therefore it is often difficult to compare units from different sources.

6.8.1 Erythrocyte spectrum

Differences in susceptibility of red blood cells obtained from different animal species (mammals, birds, amphibians) have been described for many cytolytic toxins. Recent data obtained in this respect are shown in Table 6.6. References to earlier work may be found in the various monographs of the Bacterial Toxins treatise quoted in Section 6.1 and in the reviews of Halbert (1970) on streptolysin O and Bernheimer (1970, 1974) on staphylococcal and other toxins. It is to be noted that, except for highly purified toxins, the data obtained with other preparations should be considered with caution, since contamination with as low as a few per cent of highly potent toxin may considerably change the lysis spectrum.

Some toxins such as *E. coli* hemolysin or streptolysin O have a broad hemolytic spectrum at least as regards mammalian red blood cells, whereas staphylococcal α- and β-toxins show striking differences in the sensitivity of erythrocytes from different animal species.

In most cases no relationship between sensitivity and phylogenetic position is evident, and there does not appear to be any correlation with quantitative differences in the chemical constituents of the erythrocyte membrane. For example, attempts to correlate the sensitivity of erythrocytes to *Aeromonas* hemolysin with quantitative differences in particular membrane constituents were not notably successful (Bernheimer *et al.*, 1975). Similarly, although mouse erythrocytes are 30–50 times less sensitive than those of human or rabbit to streptolysin O (Table 6.6; and Howard and Wallace, 1953), and gave hemolysis titration curves much flatter than that with rabbit erythrocytes (see Halbert, 1970), similar fixation of the toxin on both cells was identical (Petersen *et al.*, 1965). These authors have also shown that there is little difference in cholesterol and total lipid composition between human and mouse erythrocytes. It is likely that the resistance of mouse cells to streptolysin O is linked to the second phase of the lytic process which follows the specific fixation of the toxin on the membrane receptor (Alouf and Raynaud, 1968a,b,c; Kanbayashi, 1972). As regards the striking differences in erythrocyte sensitivity to staphylococcal α-toxin, Wiseman *et al.*, (1975) suggested that the level of membrane protease activity in red cells of various species explains these differences since in their opinion α-toxin is a protoxin activated by membrane protease. This hypothesis has been discussed in a previous section. In the case of staphylococcal β-toxin the impressive differences in capacity to lyse erythrocytes from different animal species has been shown to correlate with the sphingomyelin content of red cells (Wiseman and Caird, 1967; Bernheimer *et al.*, 1974). This toxin is known to be a sphingomyelinase. Sensitivity decreases with decreasing sphingomyelin content. Several lines of evidence suggest that sphingomyelin degradation is the membrane lesion that leads to hemolysis when the cells are chilled (hot-cold lysis discussed below). As noted by Bernheimer (1974), β-toxin is the cytolytic toxin *par excellence.* It is both an enzyme and a toxin at the same time, and evidence for this fact is more straightforward than for α-toxin (phospholipase C) of *Clostridium perfringens.*

Finally, it is to be noted that the erythrocyte spectrum has been the basis for the differentiation of multiple hemolysins excreted by a single organism such as for *Staphylococcus aureus* (Wadström *et al.,* 1974) and *E.coli* (Walton and Smith, 1969).

6.8.2 Lytic potency

Except for a few cytolytic toxins such as staphylococcal δ-toxin (Kreger *et al.,* 1971) and the second thuringiolysin hemolysin (Pendleton *et al.,* 1973), most cytolytic toxins have very high specific activities (Table 6.4). They are several orders of magnitude greater than those of hemolytic agents of non-bacterial origin including polyene antibiotics (see Bernheimer, 1974). The calculation on molar bases has shown that slightly more than 100 molecules of streptolysin O are needed to lyse one erythrocyte whereas 10^6-10^8 molecules of detergents, vitamin A, sterols, and polyenes are required (Alouf and Raynaud, 1968b).

Special mention should be made of the hemolytic potency of staphylococcal β-toxin. The data of Wadström and Möllby (1971a,b) and Gow and Robinson (1969) gave extremely high values for the specific activity (10^8-10^{10} HU/mg) which are higher by several orders of magnitude than that of any known hemolysin. The recent purification procedure described by Bernheimer *et al.,* (1974) gave a product with roughly 400 000–700 000 HU/mg protein. These contradictory results have not been satisfactorily explained.

6.8.3 Hot-cold hemolysis

The 'hot-cold' hemolytic phenomenon reported 50 years ago is a characteristic of staphylococcal β-toxin (see Wiseman, 1970, for a review), *Cl. perfringens* α -toxin (van Heyningen, 1941), and *Acinetobacter calcoaceticus* hemolysin (Lehman and Wiig, 1972). Incubation at 37°C of these toxins with susceptible cells and suitable cations (sheep or other ruminant erythrocytes and Mg^{2+} for β-toxin) results in no or incomplete (at high toxin concentration) hemolysis. Treated cells cooled thereafter at 0–4°C rapidly lyse.

In spite of a number of investigations this phenomenon remains unsatisfactorily explained. Several hypotheses have been put forward (Bernheimer, 1974; Smyth *et al.,* 1975b). Bernheimer suggested that extensive splitting by the toxin of its sphingomyelin substrate in the outer leaflet of sheep erythrocyte membrane could leave areas of the membrane that are essentially lipid monolayers and that these regions are sufficiently stable to persist at 37°C but lowering of the temperature may be accompanied by thermodynamic instability leading to multiple areas of membrane collapse.

It was shown by Součková and Soucék (1972) that pretreatment of sheep erythrocytes with corynebacterial hemolytic toxin (phospholipase D) renders the cell insensitive to lysis by β-toxin.

Very recently the hot-cold phenomenon was studied in detail by Smyth *et al.*, (1975b) with highly purified staphylococcal β-hemolysin; they established the link between this phenomenon and cation involvement in membrane stabilization. Treatment by β-toxin of ruminant erythrocytes (sheep, ox, goat), in which sphingomyelin comprises about 50 per cent of the total membrane phospholipids, displays the hot-cold phenomenon. The addition of ethylenediaminetetraacetate (EDTA) to sheep erythrocytes preincubated with sphingomyelinase C was found to induce rapid hemolysis at 37°C. The treated cells became susceptible to chelator-induced hemolysis and to hot-cold hemolysis simultaneously, and the degree of lysis by both mechanisms increased equally with prolonged preincubation with sphingomyelinase C. Erythrocytes of species not readily susceptible to hot-cold hemolysis were equally insusceptible to chelator-induced lysis. Chelators of the EDTA series were the most effective, whereas chelators more specific for Ca^{2+}, Zn^{2+}, Fe^{2+}, Cu^{2+}, and Mg^{2+} were without effect. The rate of chelator-induced lysis was dependent on the preincubation period with β-hemolysin and on the concentration of chelator added. The optimal concentration of EDTA was found to equal the amount of exogenously added Mg^{2+}, a cation necessary for sphingomyelinase C activity. Hypotonicity increased the rate of chelator-induced hemolysis, whereas increasing the osmotic pressure to twice isotonic completely inhibited chelator-induced lysis. The data suggest that exogenously added and/or membrane-bound divalent cations are important for the stability of spingomyelin-depleted membranes. The phenomenon of hot-cold hemolysis may be a consequence of the temperature dependence of divalent ion stabilization.

6.8.4 Synergistic potentiation of hemolytic activity

A synergistic effect on sheep erythrocyte lysis was shown by mixing staphylococcal β- and δ-toxins (Kreger *et al.*, 1971) and α- and δ-toxins (Wadström *et al.*, 1974). For the latter system, stimulation of β-toxin sphingomyelinase activity by the detergent-like effect of δ-toxin is suggested. This would be similar to the stimulating effect of sublytic doses of Triton X-100 on membrane damage of human fibroblasts by clostridial phospholipase C (Möllby *et al.*, 1974).

Pretreatment of rabbit erythrocytes with clostridial α-toxin induces more rapid lysis of the cells upon the addition of staphylococcal α-toxin probably by promoting better binding of the toxins to diglycerides in the membrane (Wadström *et al.*, 1974). Enzymes endowed with non-hemolytic properties have also been shown to induce lysis with cytolytic toxins. Staphylococcal β-toxin is known to hydrolyze as much as 80 per cent of sphingomyelin in human and pig erythrocytes under conditions in which no hemolysis takes place. As shown by Colley *et al.*, (1973) the non-hemolytic phospholipase C from *B. cereus* induced hemolysis of the cells previously treated with β-toxin. It is very likely that in intact erythrocytes phosphoglyceride molecules are not accessible to phospholipase C of *B. cereus* but become available to its hydrolytic activity through prior removal of choline phosphate residues from sphingomeylin by β-toxin. These examples show how purified enzymes and toxins may be used as probes in the study of biomembrane organization.

A similar situation is that of the CAMP reaction which is the rapid hemolysis occurring when sheep erythrocytes are exposed to the combined effects of staphylococcal β-toxin and a diffusable product of group B streptococci known as CAMP factor. This factor is thermostable and most likely protein in nature (Brown *et al.*, 1974). Its mechanism of action is still unknown.

6.9 MECHANISMS OF ACTION OF CYTOLYTIC TOXINS

Cytolytic agents modify membrane permeability by interaction with one or more of its components. This interaction may occur through different mechanisms: (1) general surfactant activity involving reaction with bilayer lipids or proteins and solubilization of these components; (2) enzymatic activity involving the hydrolysis of lipids or proteins; (3) binding to specific molecules of the bilayer causing a transient or permanent rearrangement of the membrane framework.

The first mechanism is a detergent-like action which, as discussed before, is that by which staphylococcal δ-toxin, streptolysin S, and bacillus surfactin disrupts the cytoplasmic membrane leading to cells lysis.

Lysis due to enzymatic hydrolysis of membrane component(s) is the case with the cytolytic toxins endowed with lecithinase C activity, such as staphylococcal β-toxin, *Cl. perfringens* α-toxin, *A. calcoaceticus* hemolysin (Bernheimer, 1974; Wadström *et al.*, 1974), and very likely other clostridial toxins (Rutter and Collee, 1969), as well as lecithinase D activity in the case of *C. ovis* hemolysin (Soucék *et al.*, 1971). Staphylococcal α-toxin has been reported to be a protease by Wiseman *et al.*, (1975).

The third mechanism, involving rearrangement of membrane structure upon toxin binding on a specific component of the bilayer acting as a receptor, and triggering more or less complex biochemical or physical events leading to the destruction of the osmotic barrier, appears to operate for a number of toxins. We shall briefly summarize here our knowledge in this respect about the mode of action of two toxins, streptolysin O and staphylococcal leukocidin.

6.9.1 Streptolysin O

The biochemical and biological properties of streptolysin O have been reviewed in recent years (Halbert, 1970; Bernheimer, 1974; Duncan, 1975). The physical and biological properties of streptolysin O are given in Table 6.4. This toxin is the prototype of immunologically and chemically related (often termed oxygen-labile) hemolysins which lose their cytolytic activity by oxidation. This activity is restored by reduction with hydrosulfite or thiols. SLO and related toxins are inhibited by cholesterol and some other sterols which fulfil strict chemical and stereochemical requirements (Watson and Kerr, 1974). A detailed quantitative study of these requirements and of the interaction of streptolysin O *in vitro* with cholesterol has been undertaken in our laboratory (Prigent and Alouf, 1976).

Inhibitory and non-inhibitory sterols are shown in Fig. 6.6. The presence of a 3β (equatorial) hydroxy group on the steroid nucleus is a strict structural requirement illustrated by the failure of epicholesterol, which has a 3α (axially oriented) hydroxy group, to inhibit streptolysin O activity.

Another essential structural condition is the presence of a lateral aliphatic side-chain since dehydroepiandrosterone which has a steroid nucleus as in cholesterol but lacks the side-chain is ineffective. By contrast the stereochemical relationships of rings A and B to each other in the steroid nucleus are not critical since both dihydrocholesterol (5α cis non-planar steroid) and coprostanol (5β trans planar steroid) inhibit streptolysin O activity. Similar requirements are also necessary to inhibit the lytic activity of polyene antibiotics which are known to form complexes in a stereochemically and stoichiometrically defined manner with sterols (De Kruijff *et al.*, 1974a,b). We have demonstrated the formation of complexes between streptolysin O and inhibitory sterols (Fig. 6.2) by diffusion and immunodiffusion techniques on agar gel plates containing cholesterol or other sterols. A halo of hydrophobic toxin–sterol complex appears around a well of toxin. A similar pattern occurs with digitonin which is known to bind and precipitate 3β-hydroxy-sterols by forming a hydrophobic complex.

We also found that, upon interaction with inhibitory steroids, streptolysin O loses its immunoreactive properties toward homologous neutralizing and precipitating antibodies. It is to be noted that Kaplan and Wannamaker (1974) reported the decrease or loss of immunogenicity of streptolysin mixed with cholesterol. Streptolysin O appears therefore as a steroid-binding agent which through at least two stereospecific binding sites recognizes *in vitro* the 3β-OH group and aliphatic side-chain of cholesterol or steroid molecule.

Cholesterol is known to be a critical constituent of the biomembranes of eukaryote cells, playing a major part in the regulation of its stability and permeability because of its potent condensing effect on phospholipids in the bilayer (increase of molecular rigidity and order) (Papahadjopoulos, 1974; Bretscher and Raff, 1975; Newman and Huang, 1975).

Several lines of evidence suggest that membrane cholesterol is the binding site of streptolysin O, related toxins, polyenes, and saponins (Alouf and Raynaud, 1968a; Bernheimer, 1974; Shany *et al.*, 1974). The strong affinity of these agents for cholesterol shown *in vitro* suggests that their binding by this sterol at the surface of eukaryote membranes inhibits its stabilizing interaction with surrounding phospholipids. Considerable weakening and increase of membrane fluidity should ensue with formation of 'holes' or 'channels' lining cholesterol–toxin complexes in the

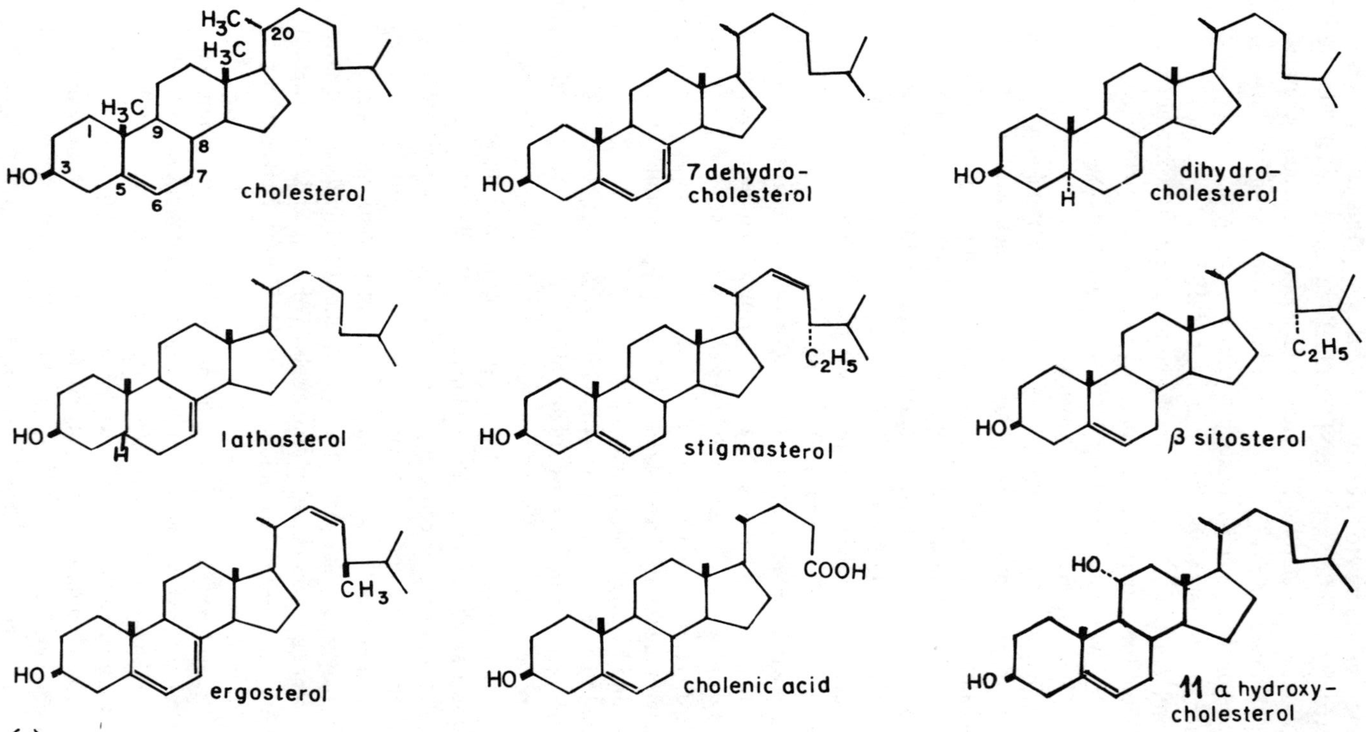

(a)

Fig. 6.6 Effect of chemical structure of steroids on the hemolytic activity of streptolysin O (a) inhibitory steroids, (b) non-inhibitory steroids. From work by Prigent and Alouf (Institut Pasteur).

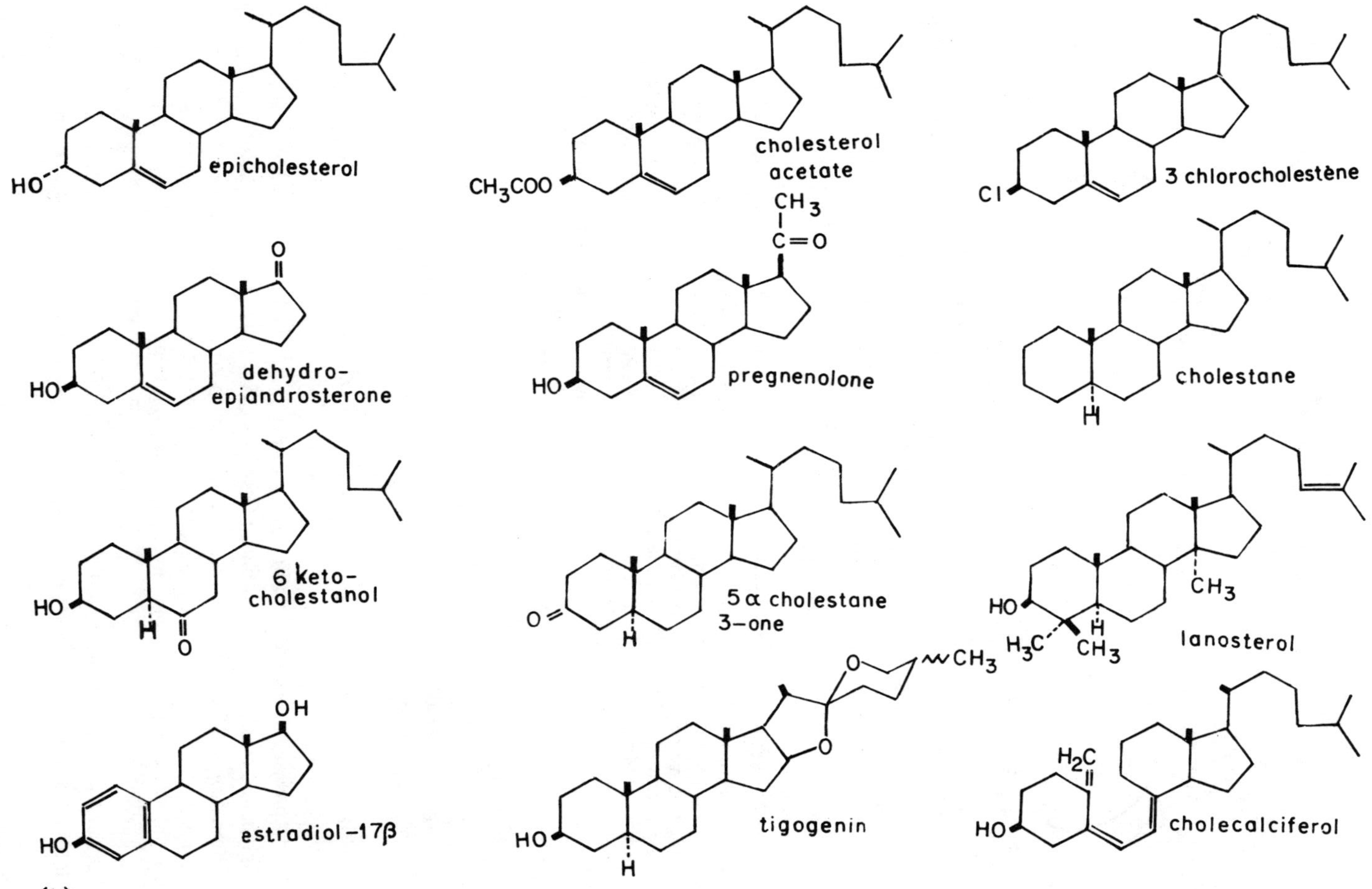
epicholesterol
cholesterol acetate
3 chlorocholestène
dehydro-epiandrosterone
pregnenolone
cholestane
6 keto-cholestanol
5α cholestane 3-one
lanosterol
estradiol-17β
tigogenin
cholecalciferol

(b)

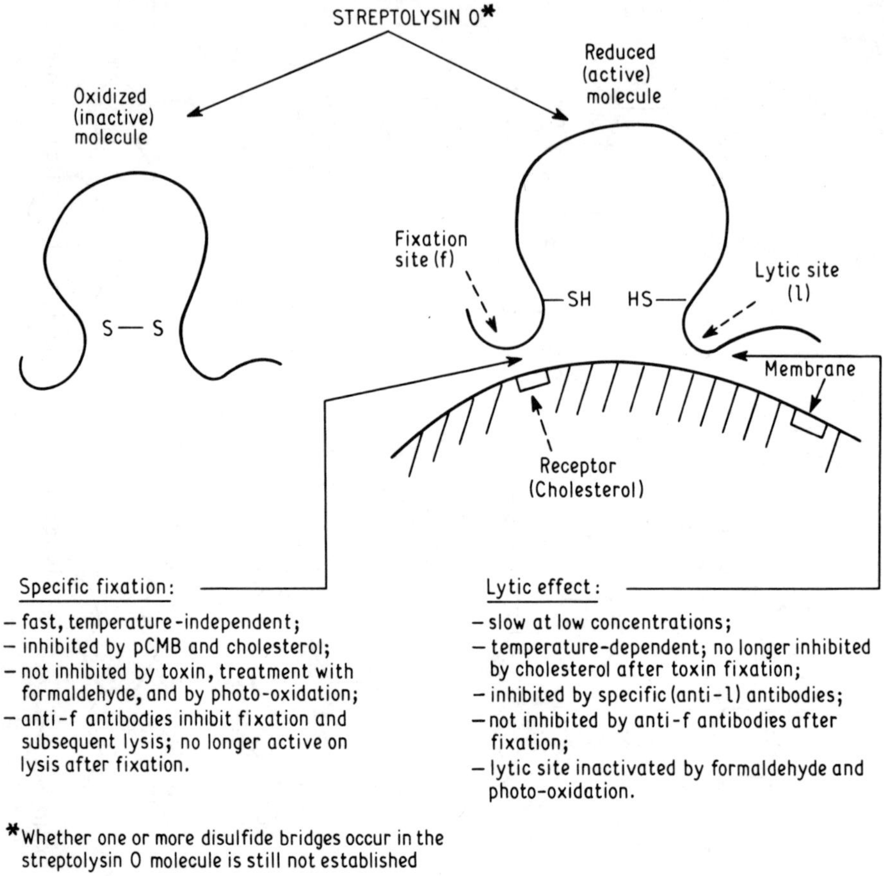

Fig. 6.7 Speculative model of streptolysin O molecule as regards interactions with erythrocyte membrane.

broad plane of the bilayer, allowing therefore hemoglobin escape and membrane collapse.

In previous work (Alouf and Raynaud, 1968a,b,c; Prigent *et al.*, 1974), using unlabelled and labelled streptolysin O and its toxoid, we studied the kinetics and interaction of toxin in the cold and at 37°C with erythrocytes. We found that lysis of erythrocytes involves two sequential steps. The first is temperature-independent and consists of toxin binding to erythrocyte, while the second is marked by release of hemoglobin. The first step can be dissociated from the lytic one at low temperature. Under these conditions bound toxin is no longer inhibited by sterols or by SH-inhibitors when the cells are incubated later at permissive temperature. Firm binding to the cell surface was shown to take place immediately at 0°C. Only reduced streptolysin O or toxoid can bind whereas oxidized forms of these molecules do not.

Once attachment has occurred hemolysis could take place at 37°C but not at 0°C (at least for moderate quantitites of toxin). Hemolysis may be prevented by various anti-streptolysin O antibodies after toxin binding. These results have been confirmed by Oberley and Duncan (1971). Binding appeared irreversible (Alouf and Raynaud, 1968a) or partly reversible under other experimental conditions (Kanbayashi *et al.*, 1972) during the prelytic period.

These data may be interpreted by a speculative model (Fig. 6.7) involving two topologically distinct sites on the molecule of streptolysin O: (1) a fixation site (f) which recognizes the receptor (cholesterol) at the cell surface and binds it; (2) a lytic site (l) which is necessary to the physical-chemical events following binding to trigger cell damage irreversibly and which can be blocked by specific antibodies or by toxoidation with formaldehyde or photo-toxidation. This model received support from the finding of sera or anti-streptolysin O monoclonal proteins which have either anti-f antibodies or anti-l antibodies (Prigent *et al.*, 1974). The former completely block toxin fixation whereas the latter do not. Neutralization of SLO activity by antibodies occurs therefore through two distinct mechanisms.

6.9.2 Staphylococcal leukocidin

This toxin is a very interesting model of a specific interaction of a cytotoxic protein with polymorphonuclear leukocyte and macrophage membrane with no effect on other cell types (Woodin, 1970). This toxin is a two-component complex formed by two proteins, designated F and S, of molecular weights 32 000 and 38 000. The primary response of sensitive cells is an increased permeability due to alteration in the potassium pump (Fig. 6.5). Leukocidin is not completely absorbed by the cell but is mainly converted into an inactive form which remains in solution. Inactivation is due to the interaction with a defined phospholipid, triphosphoinositide. The presence of this component in leukocytes was shown and its conformational alteration by leukocidin was demonstrated. The lytic events are supposed to occur as follows. (1) Entry of part of the F protein into a region of low dielectric constant in the membrane. (2) Entry of an esterified fatty acid into the hydrophobic regions of F, producing a change in the surface of this molecule. (3) Adsorption of S protein on to the altered surface of F and interaction of S with the exposed inositol triphosphate moiety of triphosphoinositide, altering the conformation of the latter. (4) Creation of a channel and reversible return of F to a compact state, and a sequence of events with restoration of the system.

6.10 CONCLUSION

The chemistry and biology of cytolytic toxins constitute a telling chapter of biology. They prove to be very interesting agents with which to probe biomembrane organization. Future development should aim at a better understanding of the molecular mode of action of these toxins, the nature of relevant receptors, the forces involved in interaction, and the primary and three-dimensional structure in relation to their cytolytic and other biological properties.

REFERENCES

The literature search was completed on November 30, 1975.

Ajl, S.J., Kadis, S. and Montie, T.C. (Eds.) (1970) *Microbial Toxins,* Vol. 1, Academic Press, New York.

Alouf, J.E. and Raynaud, M. (1968a) in *Current Research on Group A Streptococcus* (Caravano, R., Ed.), pp. 192–206. Excerpta Medica Foundation, Amsterdam.

Alouf, J.E. and Raynaud, M. (1968b) *Ann. Inst. Pasteur,* **114**, 812–827.

Alouf, J.E. and Raynaud, M. (1968c) *Ann. Inst. Pasteur,* **115**, 97–121.

Alouf, J.E. and Raynaud, M. (1973) *Biochimie,* **55**, 1187–1193.

Andreoli, T.E. (1974) *Ann. N.Y. Acad. Sci.,* **235**, 448–468.

Appelbaum, B. and Zimmerman, L.N. (1974) *Inf. Immun.,* **10**, 991–995.

Arbuthnott, J.P., Freer, J.H. and Billcliffe, B. (1973) *J. Gen. Microbiol.,* **75**, 309–320.

Bangham, A.D., Standish, M.M. and Watkins, J.C. (1965a) *J. Mol. Biol.,* **13**, 238–252.

Bangham, A.D., Standish, M.M. and Weissmann, G.C. (1965b) *J. Mol. Biol.,* **13**, 253–259.

Bernheimer, A.W. (1970) in *Microbial Toxins* (Ajl., S.J., Kadis, S. and Montie, T.C., Eds.), Vol. 1, pp. 183–212, Academic Press, New York.

Bernheimer, A.W. (1972) in *Streptococci and Streptococcal Diseases* (Wannamaker, L.W. and Matsen, J.M., Eds.), pp. 19–31, Academic Press, New York.

Bernheimer, A.W. (1974) *Biochim. Biophys. Acta,* **344**, 27–50.

Bernheimer, A.W. and Schwartz, L.L. (1965) *J. Pathol. Bacteriol.,* **89**, 209–223.

Bernheimer, A.W. and Grushoff, P. (1967a) *J. Gen. Microbiol.,* **46**, 143–150.

Bernheimer, A.W. and Grushoff, P. (1967b) *J. Bacteriol.,* **93,** 1541–1543.

Bernheimer, A.W. and Avigad, L.S. (1970) *J. Gen. Microbiol.,* **61**, 361–369.

Bernheimer, A.W., Grushoff, P. and Avigad, L.S. (1968) *J. Bacteriol.,* **95**, 2439–2441.

Bernheimer, A.W., Kim, K.S., Remsen, C.C., Antanavage, J. and Watson, S.W. (1972) *Inf. Immun.,* **6**, 636–642.

Bernheimer, A.W., Avigad, L.S. and Kim, S.S. (1974) *Ann. N.Y. Acad. Sci.,* **236**, 292–305.

Bernheimer, A.W., Avigad, L.S. and Avigad, G. (1975) *Inf. Immun.,* **11**, 1312–1319.

Bird, R.A., Low, M.G. and Stephen, J. (1974) *F.E.B.S. Letters,* **44**, 279–281.

Boethius, J., Rydqvist, B., Möllby, R. and Wadström, T. (1973) *Life Sci.*, **13**, 171–176.

Bonventre, P.F. (1970) in *Microbial Toxins* (Ajl, S.J., Kadis, S. and Montie, T.C., Eds.), Vol. 1, pp. 29–66, Academic Press, New York.

Bonventre, P.F., Saelinger, C.B., Ivins, B., Woscinski, C. and Amorini, M. (1975) *Inf. Immun.*, **11**, 675–684.

Bretscher, M.S. and Raff, M.C. (1975) *Nature*, **258**, 43–49.

Brown, J., Carnsworth, R., Wannamaker, L.W. and Johnson, D.W. (1974) *Inf. Immun.*, **9**, 377–383.

Brunner, K.T., Manuel, J., Cerottini, J.C. and Chapuis, B. (1968) *Immunology*, **14**, 181–196.

Buckelew, A.R. and Colacicco, G. (1971) *Biochim. Biophys. Acta*, **233**, 7–16.

Bunting, W.L., Kiely, J.M. and Owen, C.A. (1963) *Proc. Soc. Exp. Biol. Med.*, **113**, 370–374.

Cass, A. and Dalmark, M. (1973) *Nature New Biol.*, **244**, 47–79.

Cassidy, P.S. and Harshman, S. (1973) *J. Biol. Chem.*, **248**, 5545–5546.

Cassidy, P.S., Six, H.R. and Harshman, S. (1974) *Biochim. Biophys. Acta*, **332**, 413–423.

Chap, H., Lloveras, J. and Douste-Blazy, L. (1971) *Compt. Rend. Acad. Sci.*, **273**, 1452–1455.

Cohen, B., Halbert, S.P. and Perkins, M.E. (1942) *J. Bacteriol.*, **43**, 607–627.

Colacicco, C. and Buckelew, A.R. (1971) *Lipids*, **6**, 546–553.

Colley, C.M., Zwaal, R.F.A., Roelofsen, B. and Van Deenen, L.L.M. (1973) *Biochim. Biophys. Acta*, **307**, 74–82.

Collier, R.J. (1975) *Bacteriol. Rev.*, **39**, 54–85.

Cox, C.B., Hardegree, C. and Fornwald, R. (1974) *Inf. Immun.*, **9**, 696–701.

Cuatrecasas, P. (1971) *J. Biol. Chem.*, **246**, 6532–6542.

Cuatrecasas, P. (1973) *Biochemistry*, **12**, 3547–3558.

De Kruijff, B. and Demel, R.A. (1974) *Biochim. Biophys. Acta*, **339**, 57–70.

De Kruijff, B., Gerritsen, W.J., Oerlemans, A., Demel, R.A. and Van Deenen, L.L.M. (1974a) *Biochim. Biophys. Acta*, **339**, 30–43.

De Kruijff, B., Gerristen, W.J., Oerlmans, A., Van Dijck, P.W.M., Demel, R.A. and Van Deenen, L.L.M. (1974b) *Biochim. Biophys. Acta*, **339**, 44–56.

Demel, R.A., Bruckdorfer, K.R. and Van Deenen, L.L.M. (1972) *Biochim. Biophys. Acta*, **255**, 321–330.

Dourmashkin, R.R. and Rosse, W.F. (1966) *Am. J. Med.*, **41**, 699–710.

Duncan, J.L. (1974) *Inf. Immun.*, **9**, 1022–1027.

Duncan, J.L. (1975) *Microbiology 1975*, pp. 257–262, Amer. Soc. Microb.

Duncan, J.L. and Schlegel, R. (1975) *J. Cell Biol.*, **67**, 160–173.

Fauve, R.M., Alouf, J.E., Delaunay, A. and Raynaud, M. (1966) *J. Bacteriol.*, **92**, 1150–1153.

Forbes, I.J. (1963) *Austral. J. Exp. Biol.*, **41**, 255–264.

Foster, B.G. and Hanna, M.O. (1974) *Can. J. Microbiol.*, **20**, 1403–1409.

Freer, J.H., Arbuthnott, J.P. and Bernheimer, A.W. (1968) *J. Bacteriol.*, **95**, 1153–1168.

Freer, J.H., Arbuthnott, J.P. and Billcliffe, B. (1973) *J. Gen. Microbiol.*, **75**, 321–332.

Frimmer, M. and Scharmann, W. (1975) *Naunyn-Schmiedeberg's Arch. Pharmacol.*, **281**, 123–132.
Gale, E.F. (1974) *J. Gen. Microbiol.*, **80**, 459–465.
Gill, D.M. (1975) *Proc. Natl. Acad. Sci. U.S.*, **72**, 2064–2068.
Gill, D.M., Pappenheimer, A.M., Jr. and Uchida, T. (1973) *Fed. Proc.*, **32**, 1508–1515.
Ginsburg, I. (1970) in *Microbial Toxins* (Montie, T.C., Kadis, K. and Ajl, S.J., Eds.), Vol. 3, pp. 99–171, Academic Press, New York.
Ginsburg, I. and Grossowicz, N. (1960) *J. Pathol. Bacteriol.*, **80**, 111–119.
Gladstone, G.P. and Yoshida, A. (1967) *Brit. J. Exp. Pathol.*, **48**, 11–19.
Goldberg, B. and Green, H. (1960) *J. Biophys. Biochem. Cytol.*, **7**, 645–650.
Gow, J.A. and Robinson, J. (1969) *J. Bacteriol.*, **97**, 1026–1032.
Gräf, W. (1958) *Arch. Hyg. Bakteriol.*, **142**, 267–275.
Granato, P.A. and Jackson, R.W. (1971a) *J. Bacteriol.*, **107**, 551–556.
Granato, P.A. and Jackson, R.W. (1971b) *J. Bacteriol.*, **108**, 804–808.
Gray, S.J. and Sterling, K. (1950) *J. Clin. Invest.*, **29**, 1604–1613.
Green, H., Fleischer, R.A., Barrow, P. and Goldberg, B. (1959) *J. Exp. Med.*, **109**, 511–521.
Guillaumie, M. (1950) *Ann. Inst. Pasteur*, **79**, 661–671.
Guillaumie, M. and Kréguer, A. (1951) *Compt. Rend. Soc. Biol.*, **145**, 179–182.
Habermann, E. (1972) *Science*, **177**, 314–322.
Halbert, S.P. (1970) in *Microbial Toxins* (Montie, J.C., Kadis, J. and Ajl, S.K., Eds.), Vol. 3, pp. 69–98, Academic Press, New York.
Hallander, H.O. and Bengtsson, S. (1967) *Acta Path. Microbiol. Scand.*, **70**, 107–119.
Halpern, B., Hollmann, K.H., Rahman, S. and Verley, J.M. (1969) *Isr. J. Med. Sci.*, **5**, 1138–1148.
Hamilton-Miller, J.M.T. (1973) *Bacteriol. Rev.*, **37**, 166–196.
Hamilton-Miller, J.M.T. (1974) *Adv. Appl. Microbiol.*, **17**, 109–134.
Hammond, S.M., Lambert, P.A. and Kliger, B.N. (1974) *J. Gen. Microbiol.*, **81**, 325–330.
Harold, F.M. (1970), *Adv. Microb. Physiol.*, **4**, 46–104.
Hauschild, A.H.W. (1971) in *Microbial Toxins* (Kadis, S., Montie, T.C. and Ajl, S.J., Eds.), Vol. 2, pp. 159–188, Academic Press, New York.
Hauschild, A.H.W., Lecroisey, A. and Alouf, J.E. (1973) *Can. J. Microbiol.*, **19**, 881–885.
Heatley, N.G. (1971) *J. Gen. Microbiol.*, **69**, 269–278.
Henney, C.S. (1973) *J. Immunol.*, **110**, 73–84.
Hingson, D.J., Massengill, R.K. and Mayer, M.M. (1969) *Immunochem.*, **6**, 295–307.
Hirsch, J.G., Bernheimer, A.W. and Weissmann, G. (1963) *J. Exp. Med.*, **118**, 223–228.
Holmberg, B. (1961) *Exptl. Cell Res.*, **22**, 406–414.
Howard, J.G. (1953) *Brit. J. Exp. Pathol.*, **34**, 564–567.
Howard, J.G. and Wallace, K.R. (1953) *Brit. J. Exp. Pathol.*, **34**, 181–184.
Iles, G.H., Seeman, P., Naylor, D. and Cinader, B. (1973) *J. Cell Biol.*, **56**, 528–539.
Jenkins, E.M. and Watson, B.B. (1971) *Inf. Immun.*, **3**, 589–594.
Johnson, M.K. (1972) *Inf. Immun.*, **6**, 755–760.

Johnson, M.K. and Allen, J.H. (1975) *Amer. J. Ophthalm.*, **80**, 518–521.

Kadis, S., Montie, T.C. and Ajl, S.J. (Eds.) (1971) *Microbial Toxins*, Vol. 2A, Academic Press, New York.

Kakinuma, A., Sugino, H., Isuno, M., Tamura, G. and Arima, X. (1969), **33**, 971–976.

Kanbayashi, Y., Hotta, M. and Koyama, J. (1972) *J. Biochem.* (Japan), **71**, 227–237.

Kantor, H.S., Temples, B. and Shaw, W.W. (1972) *Arch. Biochem. Biophys.*, **151**, 142–156.

Kaplan, E.L. and Wannamaker, L.W. (1974) *Proc. Soc. Expt. Biol. Med.*, **146**, 205–208.

Keiser, H., Weissmann, G. and Bernheimer, A.W. (1964) *J. Cell Biol.*, **22**, 101–113.

Keusch, G.T. and Donta, S.T. (1975) *J. Inf. Dis.*, **131**, 58–63.

Kim, Y.B. and Watson, D.B. (1972) in *Streptococci and Streptococcal Diseases* (Wannamaker, L.W. and Matsen, J.M., Eds.), pp. 33–59, Academic Press, New York.

Kingdon, G.C. and Sword, C.P. (1970) *Inf. Immun.*, **1**, 356–360.

Kinsky, S.C. (1970) *Ann. Rev. Pharmacol.*, **10**, 119–142.

Kinsky, S.C. (1972) *Biochim. Biophys. Acta*, **265**, 1–23.

Klainer, A.S., Madoff, M.A., Cooper, L.Z. and Weinstein, L. (1964) *Science*, **145**, 714–715.

Klainer, A.S., Chang, T.W. and Weinstein, L. (1972) *Inf. Immun.*, **5**, 808–813.

Klein, G. and Perlmann, P. (1963) *Nature*, **199**, 451–453.

Korbecki, M. and Jeljaszewicz, J. (1965) *J. Inf. Dis.*, **155**, 205–213.

Kreger, A.S. and Bernheimer, A.W. (1969) *J. Bacteriol.*, **98**, 306–307.

Kreger, A.S., Kim, K.S., Zaboretzky, F. and Bernheimer, A.W. (1971) *Inf. Immun.*, **3**, 449–465.

Lehman, V. (1971) *Acta Path. Microbiol. Scand.*, **79B**, 372–376.

Lehman, V. and Wiig, N. (1972) *Acta Path. Microbiol. Scand.*, **80B**, 827–834.

Liu, P.V. (1974) *J. Inf. Dis.*, **130**, Suppl. Nov., S94–99.

Lominski, I. and Arbuthnott, J.P. (1962) *J. Path. Bacteriol.*, **83**, 515–520.

Low, D.K.R., Freer, J.H., Arbuthnott, J.P., Möllby, R. and Wadström, T. (1974) *Toxicon.*, **12**, 279–285.

Macchia, V. and Pastan, I. (1967) *J. Biol. Chem.*, **242**, 1864–1869.

Madoff, M.A., Artenstein, M.S. and Weinstein, L. (1963) *Yale J. Biol. Med.*, **35**, 382–389.

Madoff, M.A., Cooper, L.Z. and Weinstein, L. (1964) *J. Bacteriol.*, **87**, 145–149.

Maheswaran, S.K., Frommes, S.P. and Lindorfer, R.K. (1969) *Can. J. Microbiol.*, **15**, 128–129.

Martz, E., Burakoff, S.J. and Benacerraf, B. (1974) *Proc. Nat. Acad. Sci. U.S.*, **71**, 177–181.

Mastroeni, P., Misefari, A. and Nacci, A. (1969) *Appl. Microbiol.*, **17**, 650–651.

McDonel, J.L. and Duncan, C.L. (1975) *Inf. Immun.*, **12**, 1214–1218.

McGregor II, R.D. and Tobias, C.A. (1972) *J. Membrane Biol.*, **10**, 345–356.

McNiven, A.C., Owen, P. and Arbuthnott, J.P. (1972) *J. Med. Microbiol.*, **5**, 113–122.

Mebs, D. (1973) *Experientia*, **29**, 1328–1334.

Melish, M.E., Glasgow, L.A., Turner, M.D. and Lillibridge, C.B. (1974) *Ann. N.Y. Acad. Sci.*, **236**, 317–341.
Mitsui, K., Mitsui, N. and Hase, J. (1973a) *Jap. J. Exp. Med.*, **43**, 65–80.
Mitsui, K., Mitsui, N. and Hase, J. (1973b) *Jap. J. Exp. Med.*, **43**, 377–391.
Möllby, R. and Wadström, T. (1971) *Inf. Immun.*, **3**, 633–635.
Möllby, R. and Wadström, T. (1973) *Biochim. Biophys. Acta*, **321**, 569–584.
Möllby, R., Nord, C.E. and Wadström, T. (1973) *Toxicon.*, **11**, 139–147.
Möllby, R., Thelestam, M. and Wadström, T. (1974) *J. Membrane Biol.*, **16**, 313–330.
Montie, T.C., Kadis, S. and Ajl, S.J. (Eds.) (1970) *Microbial Toxins*, Vol. 3, Academic Press, New York.
Moussa, R.F. (1958) *J. Bacteriol.*, **76**, 538–545.
Müller-Eberhard, H.J. (1975) *Ann. Rev. Biochem.*, **44**, 697–724.
Nes, W.R. (1974) *Lipids*, **9**, 596–612.
Newman, C.G. and Huang, C. (1975) *Biochem.*, **14**, 3363–3370.
Novak, E., Seifert, J., Buchar, E. and Raskova, H. (1971a) *Toxicon.*, **9**, 211–218.
Novak, E., Seifert, J., Buchar, E. and Raskova, H. (1971b) *Toxicon.*, **9**, 361–366.
Nungester, W.J., Pardise, L.J. and Adair, J.A. (1969) *Proc. Soc. Exp. Biol. Med.*, **132**, 582–586.
Oberley, J.D. and Duncan, J.L. (1971) *Inf. Immun.*, **4**, 683–689.
Papahadjopoulos, D. (1974) *J. Theor. Biol.*, **43**, 329–337.
Pendleton, I.R., Kim, K.S. and Bernheimer, A.W. (1972) *J. Bacteriol.*, **110**, 722–730.
Pendleton, I.R., Bernheimer, A.W. and Grushoff, P. (1973) *J. Invert. Pathol.*, **21**, 131–135.
Perlmann, P. and Holm, G. (1969) *Adv. Immunol.*, **11**, 117–185.
Petersen, K.F., Novak, P., Thiele, O.W. and Urbaschek, B. (1966) *Int. Arch. Allergy*, **29**, 69–81.
Phillips, H.J. (1973) in *Tissue Culture; Methods and Application* (Kruse, P.F., Jr and Patterson M.K., Jr., Eds.), pp. 406–408, Academic Press, New York.
Poon, S.K. and Cauchi, M.N. (1973) *Transplantation*, **16**, 522–526.
Prescott, D. (1964) in *Progress in Nucleic Acid Research and Molecular Biology* (Davidson, J.N. and Cohn, W.E., Eds.), Vol. 3, pp. 33–57, Academic Press, New York.
Prévot, A.R. (1950) *Sang*, **21**, 565–587.
Prigent, D. and Alouf, J.E. (1976)
Prigent, D., Alouf, J.E. and Raynaud, M. (1974) *Compt. Rend. Acad. Sci.*, **278**, 651–653.
Rahal, J.J., Jr. (1972) *J. Inf. Dis.*, **126**, 96–103.
Rahal, J.J., Jr. Plaut, M.E., Rosen, H. and Weinstein, L. (1967) *Science*, **155**, 1118–1120.
Raynaud, M. and Alouf, J.E. (1970) in *Microbial Toxins* Ajl, S.J., Kadis, S. and Montie, J.C., Eds.), Vol. 1, pp. 67–117, Academic Press, New York.
Remsen, C.C., Watson, S.W. and Bernheimer, A.W. (1970) *Biochem. Biophys. Res. Commun.*, 1297–1304.
Rennie, R.P. and Arbuthnott, J.P. (1974) *J. Med. Microbiol.*, **7**, 179–188.
Rosenthal, J.W. and Fain, J.N. (1971) *J. Biol. Chem.*, **246**, 5888–5895.
Rutter, J.M. and Collee, J.G. (1969) *J. Med. Microbiol.*, **2**, 395–417.

Saijo, N. (1973) *Immunology*, **24**, 683–690.
Sakurai, J., Matsuzaki, A. and Miwatani, T. (1973) *Inf. Immun.*, **8**, 775–780.
Sanderson, A.R. (1964), *Nature*, **204**, 250–253.
Seeman, P. (1974) *Fed. Proc.*, **31**, 2116–2124.
Seeman, P., Cheng, D. and Iles, G.H. (1973) *J. Cell Biol.*, **56**, 519–527.
Shany, S., Grushoff, P. and Bernheimer, A.W. (1973) *Inf. Immun.*, **7**, 731–734.
Shany, S., Bernheimer, A.W., Grushoff, P.S. and Kim, K.W. (1974) *Molec. Cellul. Biochem.*, **3**, 179–186.
Shumway, C.N. and Klebanoff, S.J. (1971) *Inf. Immun.*, **4**, 388–392.
Siddique, I.H. (1969) *Can. J. Microbiol.*, **15**, 955–957.
Simons, K., Helenius, A. and Garoff, H. (1973) *J. Mol. Biol.*, **80**, 119–133.
Singer, S.J. (1974) *Ann. Rev. Biochem.*, **43**, 805–833.
Six, H.R. and Harshman, S. (1973a) *Biochem.*, **12**, 2672–2677.
Six, H.R. and Harshman, S. (1973b) *Biochem.*, **12**, 2677–2683.
Smith, H.W. (1963) *J. Pathol. Bacteriol.*, **85**, 197–211.
Smyth, C. (1975) *J. Gen. Microbiol.*, **87**, 219–238.
Smyth, C. and Arbuthnott, J.P. (1974) *J. Med. Microbiol.*, **7**, 41–65.
Smyth, C. and Fehrenbach, F.J. (1974) *Acta Path. Microbiol. Scand.*, **82B**, 860–870.
Smyth, C.J., Freer, J.H. and Arbuthnott, J.P. (1975a) *Biochim. Biophys. Acta.* **382**, 479–493.
Smyth, C.J., Möllby, R. and Wadström, T. (1975b) *Inf. Immun.*, **12**, 1104–1111.
Solotorovsky, M. and Johnson, W. (1970) in *Microbial Toxins* (Ajl, S.J., Kadis, S. and Montie, T.C., Eds), Vol. 1, pp. 277–327, Academic Press, New York.
Soucék, A., Michalex, Č. and Součková, A. (1971) *Biochim. Biophys. Acta*, **227**, 116–128.
Součková, A. and Soucék, A. (1972) *Toxicon.*, **10**, 501–509.
Strandberg, K., Möllby, R. and Wadström, T. (1974) *Toxicon.*, **12**, 199–208.
Stulting, R.D. and Berke, G. (1973) *Cell. Immunol.*, **9**, 474–476.
Symington, D.A. and Arbuthnott, J.P. (1969) *J. Med. Microbiol.*, **2**, 495–505.
Symington, D.A. and Arbuthnott, J.P. (1973) *J. Med. Microbiol.*, **6**, 225–234.
Takahashi, T., Sugahara, T. and Ohsaka, A. (1974) *Biochim. Biophys. Acta*, **351**, 155–171.
Taylor, A.G. and Bernheimer, A.W. (1974) *Inf. Immun.*, **10**, 54–59.
Thelestam, M. (1975) Doctoral Dissertation, Karolinska Instituted (Stockholm).
Thelestam, M. and Möllby, R. (1975a) *Inf. Immun.*, **11**, 640–648.
Thelestam, M. and Möllby, R. (1975b) *Inf. Immun.*, **12**, 225–232.
Thelestam, M. and Möllby, R. (1976). *Med. Biol.*, **54**, 39–45.
Thelestam, M., Möllby, R. and Wadström, T. (1973) *Inf. Immun.*, **8**, 938–946.
Thompson, A., Halbert, S.P. and Smith, U. (1970) *J. Exp. Med.*, **131**, 745–763.
Tillack, T.W. and Kinsky, S.C. (1973) *Biochim. Biophys. Acta*, **323**, 43–54.
Tsai, K.H. and Lenard, J. (1975) *Nature*, **253**, 554–555.
van Heyningen, W.E. (1941) *Biochem. J.*, **35**, 1257–1269.
Verkleij, A.J., De Kruijff, B., Gerrijsen, W.F., Demel, R.A., Van Deenen, L.L.M. and Ververgaert, P.H.J. (1973) *Biochim. Biophys. Acta*, **291**, 577–581.
Wadström, T. and Möllby, R. (1971a) *Biochim. Biophys. Acta*, **242**, 288–307.
Wadström, T. and Möllby, R. (1971b) *Biochim. Biophys. Acta*, **242**, 308–320.

Wadstrom, and Möllby, R. (1972) *Toxicon.*, **10**, 511–519.
Wadstrom, T., Thelestam, M. and Mollby, R. (1974) *Ann. N.Y. Acad. Sci.*, **236**, 343–361.
Wagner, H. and Feldman, M. (1972) *Cellular Immunol.*, **3**, 405–420.
Walton, J.R. and Smith, D.H. (1969) *J. Bacteriol.*, **98**, 304–305.
Watanabe, M. and Kato, I. (1974) *Jap. J. Exp. Med.*, **44**, 165–178.
Watson, B.B. and Lavizzo, J.C. (1973) *Inf. Immun.*, **7**, 753–758.
Watson, K.C. and Kerr, E.J.C. (1974), *Biochem. J.*, **140**, 95–98.
Wigzell, H. (1965) *Transplantation*, **3**, 423–431.
Wigzell, H. (1967) *Intern. Sympos. Immunol. Meth. Biol. Standardization*, **4**, 251–260.
Williams, J.C. and Bell, R.M. (1973) *Biochim. Biophys. Acta*, **288**, 255–262.
Wiseman, G.M. (1968) *Can. J. Microbiol.*, **14**, 179–181.
Wiseman, G.M. (1970) in *Microbial Toxins* (Montie, T.C., Kadis, S. and Ajl, S.J., Eds.), Vol. 3, pp. 237–263, Academic Press, New York.
Wiseman, G.M. and Caird, J.D. (1967) *Can. J. Microbiol.*, **13**, 369–376.
Wiseman, G.M. and Caird, J.D. (1972) *Can. J. Microbiol.*, **18**, 987–992.
Wiseman, G.M., Caird, J.D. and Fackrell, H.B. (1975) *J. Med. Microbiol.*, **8**, 29–38.
Woodin, A.M. (1970) in *Microbial Toxins* (Montie, J.C., Kadis, S. and Ajl, S.J., Eds.), Vol. 3, pp. 327–355, Academic Press, New York.
Wretlind, B., Möllby, R. and Wadström, T. (1971) *Inf. Immun.*, **4**, 503–505.
Wuepper, K.D., Dimond, R.L. and Knutson, D.D. (1975) *J. Invest. Dermatol.*, **65**, 191–200.
Zucker-Franklin, D. (1965) *Ann. J. Pathol.*, **47**, 419–433.

NOTE ADDED IN PROOF

Since the completion of the literature search the following reviews or articles have appeared: on the four hemolysins of *S. aureus* (Wiseman G.M. (1975) *Bact. Rev.*, **39**, 317–344); γ-hemolysin of *S. aureus* (Fackrell, H.B. and Wiseman G.M. (1976) *J. gen. Microbiol.*, **92**, 1–10 and 11–24.); hemolysin of *V. parahemolyticus* (Honda T. *et al.*, (1976) *Inf. Immun.*, **13**, 133–139); Cellular Streptolysin S (Calandra, G.B. and Oginsky E.L. (1975) *Inf. Immun.*, **12**, 13–28) and Streptolysin O (Calandra, G.B. and Theodore, T.S. (1975) *Inf. Immun.*, **12**, 13–28).

7 Presynaptic Actions of Botulinum Toxin and β-Bungarotoxin

L. L. SIMPSON

Acknowledgement
Supported in part by National Heart and Lung Institute Program Project Grant HL 12738.

The Specificity and Action of Animal, Bacterial and Plant Toxins
(Receptors and Recognition, series B, Volume 1)
Edited by P. Cuatrecasas
Published in 1976 by Chapman and Hall, 11 New Fetter Lane, London EC4P 4EE

7.1 INTRODUCTION

Throughout the history of medicine, toxins have been the subject both of awe and of alarm. Because of the natural occurrence of so many biological poisons, man – during all of his existence – has been vulnerable to accidental poisoning. Even worse, there have been those unfortunate persons who have fallen out of favor with their compatriots, and as a result been the victims of poisoning that was not the least bit accidental! And, as an added unhappy note, civilized man, in his zest to create new drugs, has synthesized a stunning array of toxic substances. It is not too much an exaggeration to claim that man lives in a milieu (albeit dilute) of poisons.

Given the fact that poisons are so widespread and so numerous, medical science has clear mandate: It is imperative that quick and efficacious techniques be developed for the diagnosis, prophylaxis, and treatment of poisoned victims. In some cases this mandate has been achieved. As a result, medical research has brought several forms of poisoning largely under control. One of the byproducts of this research has been a re-shaping of the way we envisage toxic substances. No longer do we view poisons just as sources of human malady. Instead, there is a growing appreciation that some poisons, when used carefully and imaginatively, can aid the study of life processes.

To some extent, the notion that poisons can be used as investigational tools is mere common sense. In order to be poisonous, a substance must be active in miniscule quantities. For this to be true, the substance must be highly selective both in its site and in its mechanism of action. Furthermore, the target organ or receptor upon which the poison acts must be crucial to the function of an isolated tissue or to an entire organ. A poison, by its very nature, is a substance that seeks out and impairs a process vital to life. Therefore, in the hands of a thoughtful investigator, a poison can be used to locate and analyze essential functions of cells and organs.

Much of the current interest in toxins is secondary to a belief that these substances can be used to aid medical research. But as a prerequisite to using the substances as research tools, there must exist at least a minimal understanding of the toxins themselves and of the target organs upon which they act. Accordingly, this chapter will deal with three related topics. First, there is a brief description of data on the isolation and characterization of botulinum toxin. Next, there is a discussion of the action of botulinum toxin in relation to its target organ, i.e. the cholinergic nerve ending. Finally, there is a comment on the ways in which botulinum toxin has aided research on the nervous system.

7.2 ISOLATION AND CHARACTERIZATION OF BOTULINUM TOXIN

Botulinum toxin is produced by the organism *Clostridum botulinum.* The toxin occurs in at least six different types that are designated A, B, C, D, E, and F. The individual toxins vary somewhat in their potency and in their antigenicity, but they are similar in the sense that all of them cause paralysis of the nervous system.

Although substantive research has been conducted on all six botulinal toxins, this review will emphasize the type A substance. This emphasis is based on the facts that: (a) type A toxin was the first botulinal toxin to be crystallized; (b) type A toxin is the most potent of the botulinal toxins; and (c) research on the mechanism of action of botulinum toxin has focussed on the type A substance. In addition to these facts, recent work suggests that the different botilinal toxins have structural similarities (DasGupta and Sugiyama, 1972). Therefore, the type A toxin can be viewed as a prototype for the group.

Botulinum toxin type A was first crytallized by Lamanna *et al.*, (1946) and Abrams *et al.*, (1946). These two groups were working independently, and their methods of crystallization were slightly different. However, in both cases the major purification step involved precipitation of toxin with ammonium sulfate. The crystalline molecule was found to be a globular protein of molecule weight *ca.* 900 000. The molecule was composed entirely of amino acids, but neither the number nor the nature of the amino acids in the molecule gave any hint as to why the substance should be so extraordinarily lethal.

Although early work suggested that the crystalline molecule was a homogenous neurotoxin, later studies revealed hetergeneity. More precisely, it was found that the crystalline molecule contained at least two biologically active components. One component was a neurotoxin that paralyzed synaptic and neutromuscular transmission; the other component was a hemagglutinin that agglutinated red blood cells. The two components could be separated by exposing crystalline toxin to isolated chicken erythrocytes (Lamanna, 1948; Lowenthal and Lamanna, 1953). When this mixture was centrifuged, neurotoxin remained in the suoernatant whereas hemagglutinin and red blood cells sedimented in a pellet. Removal of hemagglutinin by selective adsorption on red blood cells did not produce loss of neurotoxin activity. Therefore, the data were interpreted to mean that crystalline botulinum toxin type A is composed of separable and independently acting neurotoxin and hemagglutinin molecules.

The studies on red blood cells stimulated interest in the isolation and characterization of pure neurotoxin. Drs. Boroff, DasGupta, and their associates were the first to develop a successful biochemical technique for fractionating botulinum toxin. Their original report (DasGupta, Boroff, and Rothstein, 1966) described chromatographic separation of crystalline toxin into two major fractions, which they labelled α and β. The α fraction weighed 128 000–150 000, and it contained the neurotoxin activity; the β fraction weighted 500 000, and it contained the hemagglutinin activity. The molecular weights of α and β do not equal the reported molecular weight of the crystalline material (*ca.* 900 000). However, this does not necessarily represent a contradiction, because the crystalline molecule may contain polypeptides not related to neurotoxicity or hemagglutination. Furthermore, the hemagglutinin can exist in different forms of self-aggregation. And finally, in the hands of DasGupta *et al.*, (1966), the crystalline molecule actually weighed less than 900 000.

The 150 000 molecular weight fraction has been shown to be homogeneous by the techniques of ion exchange chromatography, gel filtration, immunoelectrophoresis, Ouchterlony gel double-diffusion, and sedimentation equilibrium (Boroff and DasGupta, 1971). These findings suggest, but do not prove, that the α-fraction is pure neurotoxin. Efforts have been made to cleave the α-molecule into smaller fragments. For example, DasGupta and Sugiyama (1972) have reported that type A neurotoxin can be reduced (cleavage of disulfide bond) to two smaller components with molecular weights of about 97 000 and 53 000. Unfortunately, neither of these components is neurotoxic, so it remains to be determined whether there is a botulinum neurotoxin of small molecular weight.

Simultaneously with the work on isolation of the neurotoxin, some attention has been directed toward identification of the 'active site' in the molecule. This work has been done almost exclusively with the crystalline toxin rather than with the neurotoxin (α fraction). The major research tactic has been to expose the crystalline toxin to a chemically reactive compound, and then to determine whether the toxin has lost activity. The underlying rationale is that loss of activity could mean that the active site in the molecule has been altered or destroyed. In experiments of this nature, crystalline toxin has been exposed to ketene, nitrous acid, methylene blue, 2-hydroxy-5-nitrobenzyl bromide, and sodium *p*-chloromercuribenzoate, to name only a few. In all cases, the activity of crystalline toxin was diminished. Indeed, it has been difficult to find a reactive substance that does not reduce the neuroparalytic activity of crystalline botulinum toxin. The conclusions to be drawn from this work are: (1) the active site responsible for neurotoxicity has not been identified; and (2) the crystalline molecule is rather labile, and thus susceptible to many forms of chemical inactivation.

Relatively little work has been done on the amino acid content or on active site determination of the isolated neurotoxin. Dr. Gerwing and her associates have published data on the amino acid content of a presumed neurotoxin derived from botulinum toxin. Sadly enough, the substance on which their analyses were performed (see Gerwing, Dolman, and Bains, 1965) was later shown not to be true neurotoxin (Boroff, DasGupta, and Fleck, 1968; Hauschild and Hilsheimer, 1968; Knox, Brown, and Spero, 1970). An amino acid analysis of type A neurotoxin has been published (Boroff, Meloche, and DasGupta, 1970), but the work remains to be confirmed. As yet, there is only anecdotal evidence on the active site in the isolated neurotoxin. It appears that the neurotoxin is, if anything, even more labile than the crystalline molecule. Any of a variety of site-directed, chemically reactive substances will inactivate the neurotoxin. Consequently, the identity of the active site remains undetermined.

7.3 SITE OF ACTION OF BOTULINUM TOXIN

When investigators first sought to determine the target organ upon which botulinum toxin exerts its effect, much of their attention was directed at the central nervous system. In many respects, the brain or spinal cord appeared to be the 'logical' choice as a site of action. Perhaps the most compelling evidence implicating the central nervous system was clinical in nature. Patients who had become poisoned with botulinum toxin seemed to have a massive neurological defect, the chief symptom of which was paralysis of all neuromuscular and autonomic activity. This clinical finding was suggestive of severe toxin-induced pathology of the central nervous system.

At the turn of the century, several investigators tried to demonstrate lesions in the brains of poisoned animals and humans (autopsy specimens). Their data were not convincing. In fact, as the pace of research quickened, it became more and more obvious that the brain was not seriously affected. Then, in rapid succession, three independent groups of investigators published data which showed that the peripheral nervous system rather than the brain is the major site of toxin action (Dickson and Shevky, 1923a, b; Edmunds and Long, 1923; Schubel, 1923). They reported that botulinum toxin could paralyze motor nerves and autonomic nerves in a wide variety of animal species. In addition, they found that the paralytic action of the toxin was circumscribed to nerve endings. As these and later studies would demonstrate, the toxin is highly preferential; it tends to act only upon those peripheral nerve endings which release the substance acetylcholine (Ach).

A substantial body of evidence indicates that Ach is active at three peripheral sites. These are: (a) nerve endings of preganglionic fibers from both the parasympathetic and sympathetic divisions of the autonomic nervous system; (b) nerve endings of postganglionic fibers from the parasympathetic nervous system; and (c) nerve endings of motor fibers which innervate striate muscle. When one considers how dependent the body is upon autonomic and neuromuscular transmission, one can easily understand how botulinum poisoning can produce a widespread and debilitating paralysis. Furthermore, one can understand, in hindsight, why early investigators assumed that the brain was the target organ for the toxin. Prior to the discovery of peripheral cholinergic (i.e., acetylcholine-mediated) transmission, it was difficult to imagine anything other than gross morphological damage to the brain as a basis for the symptoms of botulism.

Although there is nearly universal agreement that botulinum toxin exerts its paralytic action in the periphery, the possibility of a central action has had moments of resurgence. Moreover, many Russian workers have always believed that the toxin does have a significant central effect (Abrosimov, 1956; Pak, 1959; Pak and Bulatova, 1962; Ado and Abrosimov, 1964; Mikhadilov and Mikhailova, 1964).

By contrast, western investigators have been largely skeptical about whether the toxin affects the brain or spinal cord. Experiments which purport to show an action on the central nervous system are usually subject to some criticism. As an example, the author can cite his own experience (Simpson, deBalbian Verster, and Tapp, 1967). In an attempt to determine whether the crystalline toxin could act on the brain, we administered the substance via the sublingual artery of laboratory animals. These animals had previously been implanted with cerebral electrodes for the recording of electroencephalograms (EEG). At various times after administration of the toxin, we observed profound changes in the EEG. These findings were similar to those that had been reported by another group (Polley *et al.*, 1964). However, at the same time that our studies were being conducted, other workers (see above) were fractionating botulinum toxin into its neurotoxic and hemagglutinating components. Therefore, in a collaborative study (Simpson, Boroff, and Fleck, 1968) we examined the effects of neurotoxin (α fraction) and hemagglutinin (β fraction) on the EEG of experimental animals. We found that neither of the substances produces central effects. The latter result could mean that our original results were not due to neurotoxin, but instead to some contaminant or small molecular weight substance that had dissociated from the crystalline molecule. It may well be that other investigators, in their efforts to find a central action for the toxin, have likewise documented nonspecific effects unrelated to blockade of cholinergic transmission.

Although there is no sound evidence that the toxin can affect the brain, the problem is one that deserves continued interest. If indeed the toxin does not act centrally, that in itself is a matter of some importance. The importance can best be understood when one considers two related facts. Firstly, botulinum toxin paralyzes peripheral cholinergic nerve endings in species as diverse as earthworms, snails, goldfish, dogface baboons, and man. Secondly, both the central nervous system and the peripheral nervous system of all species that have been examined contain cholinergic nerve endings. Thus, if it were true that the toxin acts only on peripheral nerves, then we would be left with a curious state of affairs. Namely, peripheral cholinergic nerves in earthworms, man, and other creatures may have more in common than do peripheral and central cholinergic nerves in any one species. Such a conclusion would not be taken lightly by neurobiologists. It may be judicious to reserve final judgement on the matter until more compelling positive or negative data are forthcoming.

7.4 MECHANISM OF ACTION OF BOTULINUM TOXIN

For more than half a century we have known that botulinum toxin acts on acetylcholine-containing nerve endings. Nevertheless, the precise mechanism of action of the toxin remains only poorly understood. The major reason for slow progress is that we have only limited knowledge about the function of nerve endings. Needless to say, it is difficult to understand the pathology of an organ until one understands the normal physiology of that organ. As a result, advances in our knowledge about botulinum toxin activity have trailed closely behind our advancing knowledge of the nerve ending. In deference to that fact, a discussion of toxin activity should begin with a brief explanation of nerve ending physiology.

A nerve cell is composed of three main elements – a cell body, an axon, and a nerve ending. A nerve impulse normally arises in or near the cell body and is conducted distally by the axon. The nerve ending (or terminal) is a specialized structure that transmits impulses from itself to the structure that it innervates, i.e. another nerve, a muscle, or a gland. The nerve ending is not ordinarily coupled electrically to the organ it innervates. Therefore, a nerve action potential that is transmitted down the axon into the terminal cannot electrically excite the organ to which it is connected. There are several reasons for this, the most important of which is that there is a small anatomical gap between nerve endings and the structures they innervate. This gap is referred to as the synapse. The synaptic space prevents nerve endings (presynaptic structures) from being electrically coupled to innervated organs (postsynaptic structures).

In man and most other animals, synaptic transmission is a chemical process. Chemical mediators, known as synaptic transmitters, are stored in vesicles within nerve terminals. When an action potential invades the nerve terminal, it causes a transient inward flux of calcium ions. This in turn, by a process that is not fully understood, causes vesicles to migrate toward the nerve membrane and discharge their transmitter substance into the synaptic space. The transmitter diffuses across the synapse and interacts with receptors on the postsynaptic cell. The result of this drug–receptor interaction is the initiation of a new action potential.

There are many different synaptic transmitters in the body. Interestingly enough, botulinum toxin appears to paralyze only those synapses at which Ach is a transmitter. This specificity of action is something of an advantage to investigators, because there is a sizable body of knowledge about cholinergic transmission. Of the various cholinergic synapses that have been studied, the neuromuscular junction between motor neurons and striate muscle is the one that is best understood. Because of this, it is the neuromuscular junction that most investigators use in attempts to describe the mechanism of botulinum toxin activity.

A simple scheme depicting the neuromuscular junction and the metabolism of Ach is shown in Fig. 7.1. As the figure illustrates, Ach is synthesized within the presynaptic nerve ending. Acetate derived from acetyl coenzyme A, and choline of dietary origin (but see below), are enzymatically coupled by the enzyme choline acetyltransferase. After its synthesis, Ach is taken up and stored by vesicles. When a nerve impulse invades the nerve ending, it causes an initial influx of calcium, and this in turn causes an efflux of vesicular Ach. The prevailing view is that Ach is released from vesicles via the process of exocytosis, or reverse pinocytosis. [For a contrasting view, see the article by Marchbanks (1975).]. Released Ach diffuses across the synapse to interact with receptors. After receptor activation, Ach either diffuses away or is enzymatically degraded (esteratic cleavage) by acetylcholinesterase

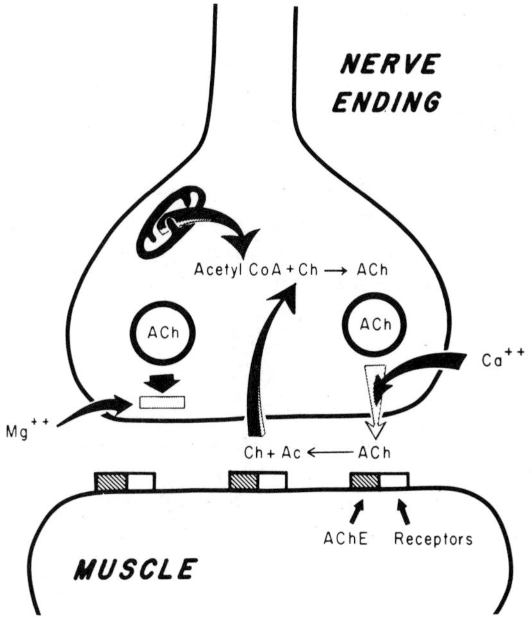

Fig. 7.1 The metabolism of acetylcholine (Ach) at the neuromuscular junction. Acetate, derived from acetyl coenzyme A, and choline (Ch) are linked enzymatically in the cytoplasm to form Ach. The latter substance is taken up and stored in vesicles. When a nerve impulse is transmitted into the nerve ending, it evokes an influx of calcium (Ca^{++}); this ion acts to trigger exocytosis of Ach. When released into the synapse, Ach can interact with receptors or with acetylcholinesterase (AchE). After interaction with AchE, Ach is converted into acetate (Ac) and Ch. Ac diffuses away from the synapse, but Ch is pumped back into the neuron.

Magnesium (Mg^{++}) is the only substance at the neuromuscular junction that might potentially block transmission. It accomplishes this by virtue of being a Ca^{++} antagonist. However, the concentration of Ca^{++} at the neuromuscular junction is greater than that of Mg^{++}, so transmission is not ordinarily impaired.

(AchE). In the latter case, acetate diffuses away, but choline is actively pumped into the nerve ending for re-utilization.

Given this scheme of events, one can hypothesize several sites at which botulinum toxin might act to impair cholinergic transmission. For example, it might inhibit the synthesis of Ach, either by blocking uptake of choline or by interfering with choline acetyltransferase. As another possibility, it might block the release of Ach from nerve endings. Or a third alternative is that botulinum toxin might bind to acetylcholine receptors, thus preventing the transmitter from exerting its physiological action. In point of fact, all of these possibilities have been explored.

7.4.1 Synthesis of acetylcholine

The synthesis of transmitter depends upon the activity of the enzyme choline acetyltransferase and the availability of its substrates. Choline acetyltransferase is known to be present in high concentrations at the nerve ending; when motor nerves are cut and allowed to degenerate, there is nearly complete disappearance of the enzyme (Hebb, Krnjević, and Silver, 1964). Both classical (Burt, 1970) and antibody-based (Eng *et al.*, 1974) techniques have been developed for the histochemical localization of choline acetyltransferase. These techniques, in combination with cell fractionation techniques, have shown that the enzyme is in the cytoplasm (Hebb, 1972).

Of the various possible mechanisms of botulinum toxin activity, the first to receive active consideration was that the toxin inhibited choline acetyltransferase in cholinergic nerves. In an early study, Torda and Wolf (1947) reported marked enzyme inhibition by the toxin. Subsequent studies did not confirm the original observation. Burgen, Dickens, and Zatman (1949) found that the toxin did not alter enzyme activity in nervous tissue. In a related study, Stevenson (1951) found that the toxin did not inhibit acetylation of choline by the bacterium *Lactobacillus plantarum,* an organism that synthesizes Ach in a manner similar to synthesis by nervous tissue.

The possibility that the toxin limits availability of substrates to choline acetyltransferase has not been directly studied. However, there is presumptive evidence that the toxin does not possess such action. With respect to acetyl coenzyme A, this substance is involved in many cellular functions in many cell types. If botulinum toxin were to act by limiting availability of acetyl coenzyme A, then one might expect the toxin to be a general or nonspecific poison. In reality, the toxin is highly specific to cholinergic nerve endings.

The presumptive evidence relating to choline is more compelling. It has been demonstrated that motor nerves possess an uptake mechanism for choline, and that a substantial amount of the choline taken up is converted into Ach (Potter, 1970). Experiments on cholinergic nerve endings have shown that the uptake mechanism has a very high affinity, and that the rate of nerve impulse flow regulates the rate of uptake (Haga and Noda, 1973; Yamamura and Snyder, 1973; Simon and Kuhar, 1975). The latter observation may mean that uptake of choline is the rate-limiting step in the synthesis of Ach.

Conceivably, a drug that blocks uptake of choline might block transmission. In other words, as the available stores of transmitter were released, there would be no new transmitter synthesized to replenish that which was lost. An example of a drug that exerts this effect is hemicholinium-3 (HC-3). One interesting property of this drug is that it has no paralytic effect on inactive nerves, or even on nerves that are firing slowly. The explanation is that an inactive nerve does not exhaust its supplies of Ach, and thus blockade of choline uptake is not harmful.

It is highly unlikely that botulinum toxin exerts its paralytic effect by blocking choline uptake. This conclusion is based on the fact that the toxin will block neuromuscular transmission even when motor nerves are totally quiescent (Hughes and Whaler, 1962; Simpson, 1973). In such nerves, blockade of choline uptake, even if it did occur, could not lead to paralysis.

7.4.2 Release of acetylcholine

It is currently believed that botulinum toxin causes paralysis by preventing release of Ach from cholinergic nerve endings. Although a number of investigators had provided suggestive evidence in support of this hypothesis, the first group to provide direct evidence was that of Burgen, Dickens, and Zatman (1949). They demonstrated that release of Ach from rat phrenic nerve-hemidiaphragm preparations was markedly diminished by botulinum toxin. Their data have been confirmed by the electrophysiological studies of Brooks (1956), Harris and Miledi (1971), Spitzer (1972), and Boroff *et al.*, (1974). All of the groups listed above have found that botulinum toxin produces a sharp reduction in the amplitude and in the frequency of Ach-induced postsynaptic potentials.

The precise mechanism by which the toxin blocks transmitter release is poorly understood. However, one of the key observations in the search for a mechanism was published by Hughes and Whaler (1962). These investigators showed that the rate of onset of toxin-induced paralysis of neuromuscular transmission was related to the rate at which nerves were firing. That is, paralysis was more rapid in onset when nerves were stimulated frequently. This result was interpreted to mean that toxin activity is dependent upon one of two processes, either membrane changes associated with the nerve action potential, or membrane changes associated with the release of Ach. Subsequent research has demonstrated that the second alternative is true (Simpson, 1971, 1973).

Recent findings indicate that there is a rather complex interaction between the toxin and the nerve ending (Simpson, 1974a). More precisely, toxin-induced paralysis of neuromuscular transmission is at least a two-step process. The initial step involves binding of the toxin to the nerve ending. This binding has a low dependence on temperature, it is not dependent upon transmitter release, it is poorly reversible, but it can be antagonized by antitoxin. The second step has a high dependence on temperature, it is dependent upon the process of transmitter release, it is irreversible, and it cannot be antagonized by antitoxin. Both steps must occur before paralysis will develop.

There are at least two plausible ways to interpret the aforementioned data. According to one scheme, bound toxin might move into or through the membrane during the process of transmitter release. As a result, the toxin might become lodged at a critical membrane site and prevent vesicles from undergoing exocytosis. Another possibility (see Boroff *et al.*, 1974) is that the toxin binds to the vesicles themselves and impairs their function. It is likely that high resolution electron microscopy with suitably labelled toxin will permit a choice between these possible mechanism;

studies of this nature are underway in several laboratories.

Very little work has been done in an attempt to identify the receptor(s) for botulinum toxin. A major difficulty is that the toxin is extraordinarily potent, so the number of molecules necessary to paralyze a synapse is quite small. By extrapolation, one can assume that the number of receptors must also be small. The problem of isolating or identifying receptors is common to research on all poisons of extreme potency. Hence, an indirect method has frequently been used to try to gain insight into receptor identification. According to this technique, known constituents of nerve membranes are incubated with toxin. Thereafter, the potency of the toxin is assayed. If potency has been diminished, that could mean that the constituent tested has the properties of a receptor. In experiments of this nature, it has been found that gangliosides will interact with toxin to produce a loss of potency (Simpson and Rapport, 1971a,b). It is the sialic acid moiety of gangliosides that accounts for their ability to inactive the toxin. This could be an indication that a molecule containing sialic acid is the toxin receptor. On the other hand, it is possible that the highly reactive sialic acid merely produces nonspecific losses in toxin potency. In view of the indirectness of the techniques, strong conclusions are not warranted.

7.4.3 Acetylcholine receptors

Botulinum toxin has no ability to block the Ach receptor (AchR), as evidenced by the fact that it does not diminish responsiveness of tissues to exogenous Ach (Guyton and MacDonald, 1947; Masland and Gammon, 1949; Burgen, Dickens, and Zatman, 1949). Neither does the toxin compete with specific ligands (α-bungarotoxin) for AchR; on the contrary, chronic botulinum toxin treatment leads to an increase in the number of AchR and to an increase in α-bungarotoxin binding (see below).

It should be noted that there is another mechanism by which the toxin might impair AchR function. When acetylcholinesterase (AchE) is inhibited, there is a brief potentiation of neuromuscular transmission that is followed by neuromuscular blockade. The late occurring blockade is due to excessive accumulation of Ach that binds to and inactivates (desensitizes) AchR. Therefore, if the toxin were an inhibitor of AchE, it might produce a secondary blockade of AchR.

One group has reported that botulinum toxin, at least *in vitro*, can inhibit AchE (Marshall and Quinn, 1967). Attempts to reproduce the finding have been unsuccessful (Sumyk and Yocum, 1968). Indeed, some question has been raised about the merit of the report claiming inhibition (Simpson and Morimoto, 1969). At present, there is no reason to believe that botulinum toxin is an AchE inhibitor.

7.4.4 Other nerves and other transmitters

The foregoing discussion has focused on neuromuscular transmission. This focus is warranted, because it is the isolated neuromuscular preparation that has most commonly been used in studies on toxin activity. For the sake of completeness, it must be reported that botulinum toxin does paralyze autonomic nerves (Dickson and Shevky, 1923a; Edmunds and Long, 1923; Ambache, 1949, 1951a, 1954; Kupfer, 1958; Winbury, 1959; Eccles and Libet, 1961; Harry, 1962). Both preganglionic and postganglionic nerves are affected, whenever the transmitter being release is Ach.

A more vexing question is that of whether botulinum toxin acts at synapses where transmitters other than Ach are released. Early (e.g. Ambache, 1949, 1951a) as well as more recent work (Vincenzi, 1967) indicated that the toxin did not affect autonomic fibers that are non-cholinergic. However, a number of investigators have reported that botulinum toxin can block transmission at sites at which norepinephrine is being released (Rand and Whaler, 1965; Westwood and Whaler, 1968; Holman and Spitzer, 1973). In all of these cases, the amount of toxin needed and/or the time necessary for onset of blockade was much greater than would be the case of cholinergic junctions. At least one group (Holman and Spitzer, 1973) has tentatively concluded that botulinum toxin has different actions on cholinergic and noradrenergic nerves. The significance of reports that botulinum toxin at high doses can impair non-cholinergic transmission is unsettled.

7.5 USE OF BOTULINUM TOXIN AS AN INVESTIGATIONAL TOOL

T he major attribute of botulinum toxin is that it is highly selective both in its site and in its mechanism of action. A further advantage is that the toxin has a long duration of action. These features of the toxin suggest that it might have utility in three areas of research: (1) the localization of cholinergic nerve endings; (2) the study of subcellular processes involved in Ach release; and (3) an analysis of the trophic effects of nerve on muscle.

Of these three areas, the first two have received only minor attention. Botulinum toxin has been used to help establish the cholinergic nature of certain ganglia or autonomic fibers of both mammalian (Ambache, 1951b; Ambach and Lessin, 1955: Hilton and Lewis, 1955) and non-mammalian origin (Shankland, Rose, and Donniger, 1971). The strategy inherent in this research is that blockade of transmission by botulinum toxin and other cholinergic blocking agents might be taken as suggestive evidence that the neurons under study release Ach. The technique has merit, but it has been largely replaced by more direct techniques. The availability of micro-assays for Ach, AchE, and choline acetyltransferase, as well as histochemical techniques for localizing the latter two substances, make the use of botulinum toxin a non-preferred method for identifying cholinergic fibers.

Perhaps the greatest putative value of botulinum toxin lies in the study of subcellular mechanisms governing transmitter release. The data that are currently available demonstrate that the toxin acts specifically on macromolecules that regulate

Ach release. Hence, an understanding of toxin activity may simultaneously be a way of gaining insight into exocytosis. As a related possibility, isolation of the toxin receptor(s) may mean the isolation of a membrane component that is in or near to the releasing site. If these possibilities bear true, they would make botulinum toxin one of the most valuable pharmacological tools available to neurobiologists. The limitations that have thus far precluded using botulinum toxin as a 'ligand for the release site' are twofold. Firstly, the entire mechanism of toxin action has not been unraveled, so it is not clear whether there is one or more than one receptor. Secondly, the toxin is remarkably potent, and there are no assay techniques sensitive enough to monitor the number of toxin molecules bound to a single synapse. Fortunately, neither of these limitations is insurmountable; it is highly likely that botulinum toxin will eventually be used in studies on transmitter release.

An area of research that has profited substantially from botulinum toxin is that dealing with trophic effects of nerve on muscle (Guth, 1968). [This area of research has been reviewed twice in recent years (Drachman, 1971; Simpson, 1974).] It is well known that efferent nerves exert a number of trophic influences on the structures that they innervate. These influences can best be demonstrated by cutting nerves; when the nerves have degenerated, the denervated structures undergo many changes. A question that inevitably arises in such studies is whether it is necessary for the nerve to degenerate before trophic influences are lost. Botulinum toxin is well suited to answering this question. The toxin can be used to produce complete blockade of neuromuscular transmission, while at the same time producing no morphological damage to nerves (Thesleff, 1960; Harris and Miledi, 1971). By comparing the effects of nerve section and of botulinum toxin, one can deduce whether the nerve itself, or merely the release of Ach and other substances is the trophic factor.

The most serious consequence of nerve section is that denervated muscle will ultimately atrophy. In the absence of tomely re-innervation, muscle function will deteriorate and may fail altogether. There is convincing evidence that Ach is the trophic factor that supports normal muscle integrity. This conclusion is based on the repeated observation that neuromuscular blockade due to botulinum toxin will produce muscle atrophy similar to that caused by nerve section (Jirmanova *et al.*, 1964; Drachman, 1964, 1967; Duchen, 1971b; Giacobini-Robecchi *et al.*, 1975). Further evidence in support of a role for Ach has been adduced by Drachman (1971). He has demonstrated that chronic treatment with HC-3, an agent that blocks Ach synthesis, or with d-tubocurarine, an agent that blocks AchR, will lead to muscle atrophy. All of these data point to the importance of Ach as trophic factor. On the other hand, it should not be assumed that the release of Ach is an adequate stimulus to prevent atrophy. It may be that the end result of transmitter release, i.e. evoked muscular activity, is the major factor.

Another trophic influence that the nerve exerts on muscle relates to neurotization. In most mammals, there is a one-to-one relationship between muscle fibers and the nerves that innervate them. Muscle fibers will not accept additional innervation, even if an investigator implants accessory nerves. This contrasts with the situation in denervated muscle; in the latter case, implanted nerves can make functional contact with underlying muscle. These findings suggest that intact nerves govern the receptivity of muscle to the formation of neuromuscular junctions.

It has been demonstrated that botulinum-poisoned muscle *in situ* will accept implanted nerves (Fex *et al.*, 1966). Even more interesting, it has been shown that poisoned nerves themselves will sprout new nerve endings that will innervate the adjacent muscle. The process of sprouting occurs more rapidly in slow muscle (soleus) than in fast muscle (gastrocnemius), but in either case the process requires several weeks (Duchen, 1970, 1971a). Upon recovery from poisoning, the nerve retracts the sprouts and once again relies upon transmission at the original neuromuscular junctions. These data show that nerve section is not a prerequisite for the formation of nerve–muscle contacts.

A phenomenon that may be related to neurotization is the distribution of AchR. In the normally innervated muscle, AchR is confined alsmost exclusively to the junctional region. Ach that is iontophertically applied to the junctional region will evoke an electrophysiological response, whereas Ach applied outside the junctional region evokes little or no response. The situation changes dramatically following denervation. When deprived of its neural input, muscle fibers become sensitive to Ach along their entire length (Axelsson and Thesleff, 1959; Miledi, 1960). The spread of Ach sensitivity is due, at least in part, to a spread of AchR outward from the junctional region. As might be expected, the spread of AchR is accompanied by an increase in tissue sensitivity, which is referred to as denervation supersensitivity, and by an increase in the binding of ligands for AchR, such as α-bungarotoxin (Lee, Tseng, and Chiu, 1967; Barnard, Wieckowski, and Chiu, 1971; Miledi and Potter, 1971; Berg *et al.*, 1972; Fambrough and Hartzell, 1972; Hartzell and Fambrough, 1972; Chang, Chen, and Chuang, 1973; Porter *et al.*, 1973; Anderson and Cohen, 1974; Chiu *et al.*, 1974; McConnell and Simpson, in press).

In a classic manuscript, Thesleff (1960) demonstrated that denervation supersensitivity is not dependent upon nerve section; botulinum toxin treatment is adequate to elicit the phenomenon. Muscles that have been poisoned with the toxin are sensitive to Ach along their entire fiber length. Moreover, toxin-treated muscles undergo a marked increase in their binding of α-bungarotoxin (Simpson, in press). The binding of α-bungarotoxin is relatively uniform across the surface of muscles that have been treated with botulinum toxin (Fig. 7.2). The original, electrophysiological studies of Thesleff (1960), coupled with the more recent studies on the distribution of binding of AchR ligands (Fig. 7.2), demonstrate that denervation supersensitivity and the spread of AchR can occur even in the presence of a morphologically intact nerve.

The data discussed above seem to indicate that nerve section and botulinum poisoning are essentially indistinguishable in their postsynaptic effects. However, there is at least one exception. When motor nerves are cut, there is substantial loss of cholinesterase activity in the junctional region (Hall, 1973). Curiously enough botulinum toxin does not cause such marked loss of cholinesterases (Duchen, 1970, 1971a; Stromblad, 1960; Drachman, 1972). This finding implies that something other

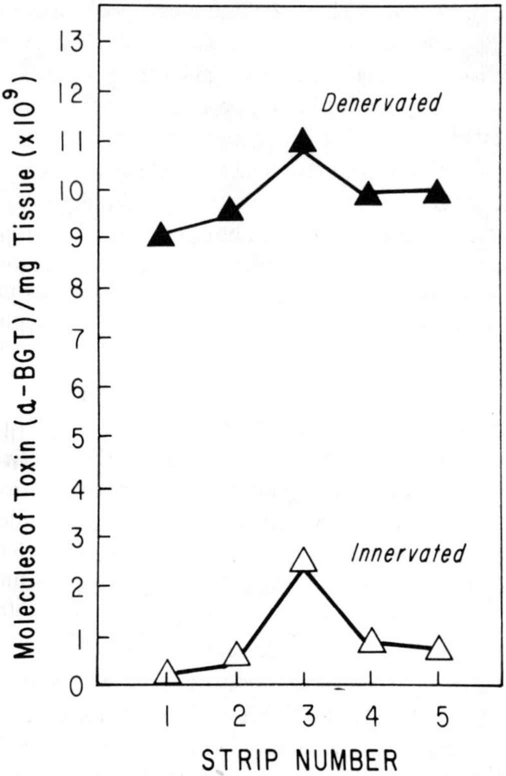

Fig. 7.2 The binding of α-bungarotoxin to strips of innervated and functionally denervated, i.e. botulinum toxin treated, hemidiaphragm. Hemidiaphragms were cut into five longitudinal strips, strip 1 being that closest to the rib and strip 5 that closest to the central tendon. In the innervated muscle, most of the α-bungarotoxin is bound to the middle region of the muscle (strip 3); this is the expected outcome, because neuromuscular junctions are mainly in the middle region. In functionally denervated muscle (botulinum toxin treatment *in situ*), α-bungarotoxin binding is markedly increased across the entire muscle surface. The increase in α-bungarotoxin binding is indicative of an increase in acetylcholine receptors.

than Ach is the trophic factor that governs muscle cholinesterase activity.

7.6 COMPARISON OF BOTULINUM TOXIN AND β-BUNGAROTOXIN

In addition to botulinum toxin, there are a number of poisons of biological origin that act presynaptically. Of these various poisons, the one that has been most thoroughly studied is β-bungarotoxin, a substance derived from the venom of *Bungarus multicinctus.* Although early work suggested that botulinum toxin and β-bungarotoxin might have much in common, more recent data indicate that the two toxins differ in many important ways.

7.6.1 Isolation and characterization of β-bungarotoxin

When the venom of *Bungarus multicinctus* is chromatographically fractionated, a large number of pharmacologically active substances is obtained (Lee *et al.*, 1972; Dryden, Harvey, and Marshall, 1974a). Most of these substances are enzymes or toxins (Lee, 1972). Among the toxins are two groups of neuroparalytic agents that display notable specificity of action. One of these, referred to as α-bungarotoxin, produces a non-depolarizing block of neuromuscular transmission. α-Bungarotoxin is believed to be a reasonably specific and irreversible ligand for AchR. The other toxin(s), referred to as β-bungarotoxin, acts presynaptically to paralyze neuromuscular transmission. β-Bungarotoxin, like botulinum toxin, is essentially devoid of postsynaptic activity.

The β-bungarotoxin originally isolated by Lee and his associates (1972) had an apparent molecular weight of 28 000, and it was composed of about 180 amino acid residues. A more recent study by Kelly and Brown (1974), in which β-bungarotoxin was purified, reported a molecular weight of about 22 000. The amino acid analyses given in both reports were quite similar. The β-bungarotoxin isolated and purified by Kelly and Brown (1974) could be reduced (cleavage of disulfide bond) to yield two subunits whose molecular weights were 8800 and 12 400. The pharmacological activity of these subunits has not been described. As yet, there are no data on the active site in the β-bungarotoxin molecule.

7.6.2 Site of action and mechanism of action of β-bungarotoxin

In the periphery, β-bungarotoxin acts mainly at the neuromuscular junction (Chang and Lee, 1963; Chen and Lee, 1970; Chang, Chen, and Lee, 1973; Kelly and Brown, 1974). There is no evidence that the toxin blocks propagation of the nerve action potential (Chang, Chen, and Lee, 1973; Kelly and Brown, 1974); instead, it acts to block transmission from nerve to muscle. Like botulinum toxin, β-bungarotoxin does not inhibit the acetylation of choline (unpublished data of Lee and Chiou, cited in Chen and Lee, 1970), it does not deplete the nerve ending of Ach stores (Chang, Chen, and Lee, 1973), and it does not block AchR (Chang, Chen, and Lee, 1973; Dryden, Harvey, and Marshall, 1974b). β-Bungarotoxin is also similar to botulinum toxin in the sense that paralysis is a two-stage phenomenon, i.e. an initial binding process and a secondary lytic process (Chang and Huang, 1974).

In spite of the many similarities between the two toxins, it is clear that they have different mechanisms of action. Some of the major differences were first noted by Chang, Lee, and their colleagues (Chang, Chen, and Lee, 1973; Chang and Huang, 1974;

Chang, Huang, and Lee, 1973). As reported by these investigators, β-bungarotoxin causes an initial facilitation of transmitter release, after which it causes neuromuscular blockade. By contrast, botulinum toxin causes no facilitation whatever; it acts solely to block transmitter release. In addition, these seems to be a mutual antagonism between the two toxins. To some extent, botulinum toxin is able to antagonize the facilitation of transmitter release caused by β-bungarotoxin.

Another major difference between the toxins relates to enzyme activity. To date, no one has described any enzymatic activity associated with botulinum toxin. It was likewise reported that β-bungarotoxin did not exert enzymatic effects (Lee *et al.*, 1972; Wernicke, Oberjat, and Howard, 1974). However, in the latter case there have now emerged data that show β-bungarotoxin to possess significant phospholipase activity (Kelly *et al.*, in press; Strong *et al.*, in press). Moreover, the phospholipase activity is an integral, but not complete, explanation for the neuromuscular effects of β-bungarotoxin.

A partial model to explain the action of β-bungarotoxin has been suggested (Kelly *et al.*, in press). According to this model, β-bungarotoxin binds to the external surface of the nerve terminal membrane. This binding has two effects that bear on transmitter release. Firstly, it alters calcium flux in such a way as to increase the likelihood of transmitter release. Secondly, it alters the release mechanism itself in such a way that exocytosis is facilitated. Jointly, these two suggested actions (altered calcium flux and altered release process) account for the ability of β-bungarotoxin to enhance Ach release. The model goes on to suggest that ultimate blockade of transmitter release is due to an as yet unexplained inactivation of the release mechanism.

7.6.3 Use of β-bungarotoxin as an investigational tool

The duration of action of β-bungarotoxin has not been studied in detail, but the substance may share with botulinum toxin a potential utility in the study of trophic effects of nerve on muscle (Giacobini *et al.*, 1973). Likewise, if β-bungarotoxin acts directly on the transmitter release process, then it may share with botulinum toxin a possible usefulness in isolating and characterizing macromolecules that govern Ach release.

The only study in which β-bungarotoxin has actually been used as an investigational tool was published by Dryden, Harvey, and Marshall (1974). These authors were interested in establishing whether nicotinic agonists act prejunctionally to release Ach. Therefore, the activity of nicotinic agonists was compared in neuromuscular preparations before and after treatment with β-bungarotoxin. Such treatment did not alter responses to the agonists. Presumably this demonstrates that nicotinic agonists do not act prejunctionally to release Ach.

7.7 CONCLUDING REMARKS

It has become common knowledge that botulinum toxin is the most poisonous substance known. Furthermore, poisoning due to the toxin, i.e. botulinum dreaded form of illness. In view of these unhappy facts, it is gratifying to report that the incidence of botulism is quite low. In addition, methods of diagnosis and treatment have improved steadily. As a result, death due to poisoning by botulism toxin is rather uncommon.

Since botulism is not a major health problem, at least in terms of its incidence, research on botulinum toxin has undergone a transition. Whereas early work centered on the ability of the toxin to produce certain types of pathology, more recent work has emphasized the potential value of the toxin as an investigational tool. If botulinum toxin does become a useful tool, this substance would join the ranks of several other toxins already being exploited by medical investigators.

The idea that poisons can be valuable is not novel. As with so many other important ideas, this one was first stated by Claude Bernard. In his classic work entitled *Experimental Science,* Bernard (1875) made the following comments:

> 'Poisons can be employed as means for the destruction of life or as agents for the treatment of the sick, but in addition to these two well recognized uses there is a third of particular interest to the physiologist. For him, the poison becomes an instrument which dissociates and analyzes the most delicate phenomena of living structures, and by attending carefully to their mechanism in causing death, he can learn indirectly much about the physiological processes of life. Such is the way in which I have long regarded the actions of toxic substances.'

It would appear that the vision of Claude Bernard is now being realized with many poisons, including botulinum toxin.

REFERENCES

Abrams, A., Kegeles, G. and Hottle, G.A. (1946), The purification of toxin from *Clostridium botulinum* type A. *J. Biol. Chem.,* **164**, 63–79.

Abrosimov, V.N. (1956), Mechanism of action of botulinum toxin on respiration. *Arkhiv. Patol.,* **18**, 86–91.

Ado, A.D. and Abrosimov, V.N. (1964), The specific effect of bacterial toxins on the nervous regulation of respiration. *J. Hyg. Epidemiol. Microbiol. Immunol.,* **8**, 433–441.

Ambache, N. (1949), The peripheral action of *Cl. botulinum* toxin. *J. Physiol.* (Lond.), **108**, 127–141.

Ambache, N. (1951a), A further survey of the action of *Clostridium botulinum* toxin upon different types of autonomic nerve fibre. *J. Physiol.* (Lond.), **113**, 1–17.

Ambache, N. (1951b), Unmasking, after cholinergic paralysis by botulinum toxin, of a reversed action of nicotine on the mammalian intestine, revealing the probable presence of local inhibitory ganglion cells in the enteric plexuses. *Brit. J. Pharmacol.,* **6**, 51–67.

Ambache, N. (1954), Autonomic ganglion stimulants. *Arch. Int. Pharmacodyn.*, **97**, 427–446.

Ambache, N. and Lessin, A.W. (1955), Classification of intestinomotor drugs by means of type D botulinum toxin. *J. Physiol.* (Lond.), **127**, 449–478.

Anderson, M.J. and Cohen, M.W. (1974), Fluorescent staining of acetylcholine receptors in vertebrate skeletal muscle. *J. Physiol.* (Lond.), **237**, 385–400.

Axelsson, J. and Thesleff, S. (1959), A study of supersensitivity in denervated mammalian skeletal muscle. *J. Physiol.* (Lond.), **147**, 178–193.

Barnard, E.A., Wieckowski, J. and Chiu, T.H. (1971), Cholinergic receptor molecules and cholinesterase molecules at mouse skeletal muscle junctions. *Nature*, **234**, 207–209.

Berg, D.K., Kelly, R.B., Sargent, P.B., Williamson, P. and Hall, Z.W. (1972), Binding of α-bungarotoxin to acetylcholine receptors in mammalian muscle. *Proc. Nat. Acad. Sci.*, **69**, 147–151.

Bernard, C. (1875), *La Science Experimentale*, Bailliere, Paris.

Boroff, D.A. and DasGupta, B.R. (1971), Botulinum toxin. In S.J. Ajl (Ed.) *Microbial Toxins*, Vol. 2A, pp. 1–68, Academic Press, New York and London.

Boroff, D.A., DasGupta, R.R. and Fleck, U.S. (1968), Homogeneity and molecular weight of toxin of *Clostridium botulinum* type B. *J. Bacteriol.*, **95**, 1738–1744.

Boroff, D.A. del Castillo, J., Evoy, W.H. and Steinhardt, R.A. (1974), Observations on the action of type A botulinum toxin on frog neuromosucular junctions. *J. Physiol.* (Lond.), **240**, 227–253.

Boroff, D.A., Meloche, H.P. and DasGupta, B.R. (1970), Amino acid analysis of the isolated and purified components from crystalline toxin of *Clostridium botulinum* type A. *Infect. Immun.*, **2**, 679–680.

Brooks, V.B. (1956), An intracellular study of the action of repetitive nerve volleys and of botulinum toxin on miniature end-plate potentials. *J. Physiol.* (Lond.), **134**, 264–277.

Burgen, A.S.V., Dickens, F. and Zatman, L.J. (1949), The action of botulinum toxin on the neuromuscular junction. *J. Physiol.* (Lond.), **109**, 10–24.

Burt, A.M. (1970), A histochemical procedure for the localization of choline acetyltransferase activity. *J. Histochem. Cytochem.*, **18**, 408–415.

Chang, C.C., Chen, T.F. and Chuang, S.T. (1973), N, O-di- and N,N,O-tri(^{3}H)acetyl α-bungarotoxins as specific labelling agents of cholinergic receptors. *Brit. J. Pharmacol.*, **47**, 147–160.

Chang, C.C., Chen, T.F. and Lee, C.Y. (1973), Studies on the presynaptic effect of β-bungarotoxin on neuromuscular transmission. *J. Pharmacol. Exp. Ther.*, **184**, 339–345.

Chang, C.C. and Huang, M.C. (1974), Comparison of the presynaptic actions of botulinum toxin and β-bungarotoxin on neuromuscular junction. *Naunyn-Schmied. Arch. Pharmacol.*, **282**, 129–142.

Chang, C.C., Huang, M.C. and Lee, C.Y. (1973), Mutual antagonism between botulinum toxin and β-bungarotoxin. *Nature*, **243**, 166–167.

Chang, C.C. and Lee, C.Y. (1963), Isolation of neurotoxin from the venom of *Bungarus multicinctus* and their modes of neuromuscular blocking action. *Arch. Int. Pharmacodyn. Ther.*, **144**, 241–257.

Chen, I.L. and Lee, C.Y. (1970), Ultrastructural changes in the motor nerve terminals caused by β-bungarotoxin. *Virchows Arch. Abt. B. Zellpath.*, **6**, 318–325.

Chiu, T.H., Lapa, A.J., Barnard, E.A. and Albuquerque, E.X. (1974), Binding of d-tubocurarine and α-bungarotoxin in normal and denervated mouse muscles. *Exper. Neurol.*, **43**, 399–413.

DasGupta, B.R., Boroff, D.A. and Rothstein, E. (1966), Chromatographic fractionation of the crystalline toxin of *Clostridium botulinum* type A. *Biochem. Biophys. Res. Commun.*, **22**, 750–756.

DasGupta, B.R. and Sugiyama, H. (1972), A common subunit structure in *Clostridium botulinum* type A, B and E toxins. *Biochem. Biophys. Res. Commun.*, **48**, 108–112.

Dickson, E.C. and Shevky, R. (1923a), Botulism, studies on the manner in which the toxin *Clostridium botulinum* acts upon the body. I. The effect upon the autonomic nervous system. *J. Exper. Med.*, **37**, 711–731.

Dickson, E.C. and Shevky, R. (1923b), Botulism, studies on the manner in which the toxin of *Clostridium botulinum* acts upon the body. II. The effect on the voluntary nervous system. *J. Exper. Med.*, **38**, 327–346.

Drachman, D.B. (1964), Atrophy of skeletal muscle in chick embryos treated with botulinum toxin. *Science*, **145**, 719–721.

Drachman, D.B. (1967), Is acetylcholine the trophic neuromuscular transmitter? *Arch. Neurol.*, **17**, 206–218.

Drachman, D.B. (1971), Botulinum toxin as a tool for research on the nervous system. In *Neuropoisons: Their Pathophysiological Actions.* Vol. 1. Poisons of Animal Origin (L.L. Simpson, Ed.), Plenum Press, New York and London.

Drachman, D.B. (1972), Neurotrophic regulation of muscle cholinesterase: Effects of botulinum toxin and denervation. *J. Physiol.* (Lond.), **226**, 619–627.

Dryden, W.F., Harvey, A.L. and Marshall, I.G. (1974a), Pharmacological studies on the bungarotoxins: Separation of the fractions and their neuromuscular activity. *Eur. J. Pharmacol.*, **26**, 256–261.

Dryden, W.F., Harvey, A.L. and Marshall, I.G. (1974b), Pharmacological studies on the bungarotoxins: An analysis of the site of action of nicotonic agonists. *Eur. J. Pharmacol.*, **26**, 262–267.

Duchen, L.W. (1970), Changes in motor innervation and cholinesterase localization induced by botulinum toxin in skeletal muscle of the mouse: Differences between fast and slow muscles. *J. Neurol. Neurosurg. Psychiat.*, **33**, 40–54.

Duchen, L.W. (1971a), An electron microscopic study of the changes induces by botulinum toxin in the motor endplates of slow and fast skeletal muscle fibres of the mouse. *J. Neurol. Sci.*, **14**, 47–60.

Duchen, L.W. (1971b), Changes in electronmicroscopic structure of slow and fast skeletal muscle fibres of the mouse after the local injection of botulinum toxin. *J. Neurol. Sci.*, **14**, 61–74.

Eccles, R.M. and Libet, B. (1961), Origin and blockade of the synaptic responses of curarized sympathetic ganglia. *J. Physiol.* (Lond.), **157**, 484–503.

Edmunds, C.W. and Long, P.H. (1923), Contribution to the pathologic physiology of botulism. *J.A.M.A.*, **81**, 542–547.

Eng, L.F., Uyeda, C.T., Chao, L.P. and Wolfgram, F. (1974), Antibody to bovine choline acetyltransferase and immunofluorescent localization of the enzyme in neurones. *Nature*, **250**, 243–245.

Fambrough, D.M. and Hartzell, H.C. (1972), Acetylcholine receptors: Number and distribution at neuromuscular junctions in rat diaphragm. *Science,* **176**, 189–191.

Fex, S., Sonesson, B., Thesleff, S. and Zelena, J. (1966), Nerve implants in botulinum poisoned mammalian muscle. *J. Physiol.* (Lond.), **184**, 872–882.

Gerwing, J., Dolman, C.E. and Bains, H.S. (1965), Isolation and characterization of a toxic moiety of low molecular weight from *Clostridium botulinum* type A. *J. Bacteriol.,* **89**, 1383–1386.

Giacobini, G., Filogamo, G., Weber, M., Boquet, P. and Changeux, J.P. (1973), Effects of snake α-neurotoxin on the development of innervated skeletal muscles in chick embryo. *Proc. Nat. Acad. Sci.,* **70**, 1708–1712.

Giacobini-Robecchi, M.G., Giacobini, F., Filogamo, G. and Changeux, J.P. (1975), Effects of the type A toxin from *Clostridium botulinum* on the development of skeletal muscles and of their innervation in chick embryo. *Br. Res.,* **83**, 107–121.

Guth, L. (1968), 'Trophic' influences of nerve on muscle. *Physiol. Rev.,* **48**, 645–687.

Guyton, A.C. and MacDonald, M.A. (1947), Physiology of botulinum toxin. *Arch. Neurol. Psychiat.,* **57**, 578–592.

Haga, T. and Noda, H. (1973), Choline uptake systems of rat brain synaptosomes. *Biochim. Biophys. Acta,* **291**, 564–575.

Hall, Z.W. (1973), Multiple forms of acetylcholinesterase and their distribution in endplate regions of rat diaphragm muscle. *J. Neurobiol.,* **4**, 343–361.

Harris, A.J. and Miledi, R. (1971), The effect of type D botulinum toxin on frog neuromuscular junctions. *J. Physiol.* (Lond.), **217**, 497–515.

Harry, J. (1962), Effect of cooling, local anaesthetic compounds and botulinum toxin on the responses of and the acetylcholine output from the electrically transmurally stimulated isolated guinea-pig ileum. *Brit. J. Pharmacol.,* **19**, 42–55.

Hartzell, H.C. and Fambrough, D.M. (1972), Acetylcholine receptors. *J. Gen. Physiol.,* **60**, 248–262.

Hauschild, A.H.W. and Hilsheimer, R. (1968), Heterogeneity of *Clostridium botulinum* type A toxin. *Can. J. Microbiol.,* **14**, 805–807.

Hebb, C. (1972), Biosynthesis of acetylcholine in nervous tissue. Physiol. Rev., **52**, 918–957.

Hebb, C.O., Krnjević, K. and Silver, A. (1964), Acetylcholine and choline acetyltransferase in the diaphragm of the rat. *J. Physiol.* (Lond.), **171**, 504–513.

Hilton, S.M. and Lewis, G.P. (1955), The cause of the vasodilation accompanying activity in the submandibular salivary gland. *J. Physiol.* (Lond.), **128**, 235–248.

Holman, M.E. and Spitzer, N.C. (1973), Action of botulinum toxin on transmission from sympathetic nerves to the vas deferens. *Brit. J. Pharmacol.,* **47**, 431–433.

Hughes, R. and Whaler, B.C. (1962), Influence of nerve-ending activity and of drugs on the rate of paralysis of rat diaphragm preparations by *Clostridium botulinum* type A toxin. *J. Physiol.* (Lond.), **160**, 221–233.

Jirmanova, I., Sobotkova, M., Thesleff, S. and Zelena, J. (1964), Atrophy in skeletal muscles poisoned with botulinum toxin. *Physiol. Bohemoslav.,* **13**, 467–472.

Kelly, R.B. and Brown, F.R. (1974), Biochemical and physiological properties of a purified snake venom neurotoxin which acts presynaptically. *J. Neurobiol.*, **5**, 135–150.

Kelly, R.B., Oberg, S.G., Strong, P.N. and Wagner, G.M. (in press), β-Bungarotoxin, a phospholipase which stimulates transmitter release. *Cold Spring Harbor Symp. Quant. Biol.*

Knox, J.N., Brown, W.P. and Spero, L. (1970), Molecular weight of type A botulinum toxin. *Infect. Immun.*, **1**, 205–206.

Kupfer, C. (1958), Selective block of synaptic transmission in ciliary ganglion by type A botulinus toxin in rabbits. *Proc. Soc. Biol. Med.*, **99**, 474–476.

Lamanna, C. (1948), Hemagglutination by botulinal toxin. *Proc. Soc. Exp. Biol. Med.*, **69**, 332–336.

Lamanna, C., McElroy, O.E. and Eklund, H.W. (1946), The purification and crystallization of *Clostridium botulinum* type A toxin. *Science*, **103**, 613–614.

Lee, C.Y. (1972), Chemistry and pharmacology of polypeptide toxins in snake venoms. *Ann. Rev. Pharmacol.*, **12**, 265–286.

Lee, C.Y., Chang, S.L., Kau, S.T. and Luh, S.H. (1972), Chromatographic separation of the venom of *Bungarus multicinctus* and characterization on its components. *J. Chromatogr.*, **72**, 71–82.

Lee, C.Y., Tseng, L.F. and Chiu, T.H. (1967), Influence of denervation on localization of neurotoxin from elapid venoms in rat diaphragm. *Nature*, **215**, 1177–1178.

Lowenthal, J.P. and Lamanna, C. (1953), Characterization of botulinal hemagglutination. *Am. J. Hyg.*, **57**, 46–59.

Marchbanks, R.M. (1975), The subcellular origin of the acetylcholine release at synapses. *Int. J. Biochem.*, **6**, 303–312.

Marshall, R. and Quinn, L.Y. (1967), *In vitro* acetylcholinesterase inhibition by type A botulinum toxin. *J. Bacteriol.*, **94**, 812–814.

Masland, R.L. and Gammon, G.D. (1949), The effect of botulinus toxin on the electromyogram. *J. Pharmacol. Exp. Ther.*, **97**, 499–506.

McConnell, M.G. and Simpson, L.L. (in press), The role of acetylcholine receptors and acetylcholinesterase activity in the development of denervation supersensitivity. *J. Pharmacol. Exp. Ther.*

Mikhailov, V.V. and Mikhailova, S.D. (1964), Mechanism of the respiratory paralysis in botulism, tetanus and diphtheria. *Bull. Exp. Biol. i. Med.*, **57**, 35–38.

Miledi, R. (1960), The acetylcholine sensitivity of frog muscle fibres after complete or partial denervation. *J. Physiol.* (Lond.), **151**, 1–23.

Miledi, R. and Potter, L.T. (1971), Acetylcholine receptors in muscle fibers. *Nature*, **233**, 599–603.

Pak, Z.P. (1959), The effect of botulinus toxin, type B, on the internal respiration of a guinea pig's brain. *Farmakol i Toksikol.*, **22**, 480–483.

Pak, Z.P. and Bulatova, T.I. (1962), Distribution of a labelled preparation of botulism toxin in the body of white mice. *Farmakol. i. Toksikol.*, **25**, 478–482.

Polley, E.H., Vick, J.A., Ciuchta, H.P., Fischetti, D.A., Macchitelli, F.H. and Montanelli, N. (1964), Botulinum toxin type A: Effects on central nervous system. *Science*, **147**, 1036–1037.

Porter, C.W., Chiu, T.H., Wieckowski, J. and Barnard, E.A. (1973), Types and locations of cholinergic receptor-like molecules in muscle fibers. *Nature,* **241**, 3–7.

Potter, L.T. (1970), Synthesis, storage and release of (^{14}C) acetylcholine in isolated rat diaphragm muscles. *J. Physiol.* (Lond.), **206**, 145–166.

Rand, M.J. and Whaler, B.C. (1965), Impairment of sympathetic transmission by botulinum toxin. *Nature,* **206**, 588–591.

Schubel, K. (1923), On botulinal toxin. *Arch. Exper. Path. u. Pharmacol.*, **96**, 193–259.

Shankland, D.L., Rose, J.A. and Donniger, C. (1971), The cholinergic nature of the cercal nerve-giant fiber synapse in the sixth abdominal ganglion of the American cockroach, *Periplaneta americana. J. Neurobiol.*, **2**, 247–262.

Simon, J.R. and Kuhar, M.J. (1975), Impulse-flow regulation of high affinity choline uptake in brain cholinergic nerve terminals. *Nature,* **255**, 162–163.

Simpson, L.L. (1971), Ionic requirements for the neuromuscular blocking action of botulinum toxin: Implications with regard to synaptic transmission. *Neuropharmacol.*, **10**, 673–684.

Simpson, L.L. (1973), Interaction between divalent cations and botulinum toxin type A in the paralysis of the rat phrenic nerve-hemidiaphragm preparation. *Neuropharmacol.*, **12**, 165–176.

Simpson, L.L. (1947a), Studies on the binding of botulinum toxin type A to the rat phrenic nerve-hemidiaphragm preparation. *Neuropharmacol.*, **13**, 683–691.

Simpson, Lance L. (1974b), The use of neuropoisons in the study of cholinergic transmission. *Ann. Rev. Pharmacol.*, **14**, 305–317.

Simpson, L.L. in press, The effects of acute and chronic botulinum toxin treatment on receptor number, receptor distribution and tissue sensitivity of the rat diaphragm. *J. Pharmacol. Exp. Ther.*

Simpson, L.L., Boroff, D.A. and Fleck, U. (1968), Examination of the possible pathophysiology of the central nervous system during botulinal poisoning. *Exper. Neurol.*, **22**, 85–95.

Simpson, L.L., deBalbian Verster, F. and Tapp, J.T. (1967), Electrophysiological manifestations of experimental botulinal intoxication: Depression of cortical EEG. *Exper. Neurol.*, **19**, 199–211.

Simpson, L.L. and Morimoto, H. (1969), Failure to inhibit *in vitro* or *in vivo* acetylcholinesterase with botulinum toxin type A. *J. Bacteriol.*, **97**, 571–575.

Simpson, L.L. and Rapport, M.M. (1971a), Ganglioside inactivation of botulinum toxin. *J. Neurochem.*, **18**, 1341–1343.

Simpson, L.L. and Rapport, M.M. (1971b), The binding of botulinum toxin to membrane lipids: Sphingolipids, steroids and fatty acids. *J. Neurochem.*, **18**, 1751–1759.

Spitzer, N. (1972), Miniature endplate potentials at mammalian neuromuscular junctions poisoned by botulinum toxin. *Nature*, **237**, 26–27.

Stevenson, J.W. (1951), Studies of the effect of botulinum toxin upon the bacterial acetylation of choline. *Canad. J. Pub. Health,* **42**, 68.

Stromblad, B.C. (1960), Cholinesterase activity in skeletal muscle after botulinum toxin. *Experientia,* **16**, 458–459.

Strong, P.N., Goerke, J., Oberg, S.G. and Kelly, R.B. (in press), β-Bungarotoxin, a presynaptic toxin with enzymatic activity. *Proc. Nat. Acad. Sci.*

Sumyk, G.B. and Yocum, C.F. (1968), Failure of type A botulinum toxin to inhibit acetylcholinesterase. *J. Bacteriol.*, **95**, 1970–1971.

Thesleff, S. (1960), Supersensitivity of skeletal muscle produced by botulinum toxin. *J. Physiol.* (Lond.), **151**, 598–607.

Torda, C. and Wolff, H.G. (1947), On the mechanism of paralysis resulting from toxin of *Clostridium botulinum. J. Pharmacol. Exp. Ther.*, **89**, 320–324.

Vinvenzi, F.F. (1967), Effect of botulinum toxin on autonomic nerves in a dually innervated tissue. *Nature,* **213**, 394–395.

Wernicke, J.F., Oberjat, T. and Howard, B.D. (1974), β-Neurotoxin reduces transmitter storage in brain synapses. *J. Neurochem.*, **22**, 781–788.

Westwood, D.A. and Whaler, B.C. (1968), Post-ganglionic paralysis of the guinea-pig hypogastric nerve-vas deferens preparation by *Clostridium botulinum* D. toxin. *Brit. J. Pharmacol.*, **33**, 21–31.

Winbury, M.M. (1959), Mechanism of the local vascular actions of 1, 1-dimethyl-4-phenylpiperazinium (DMOO), a potent ganglionic stimulant. *J. Physiol.* (Lond.), **147**, 1–13.

8 Batrachotoxin, a Selective Probe for Channels Modulating Sodium Conductances in Electrogenic Membranes

E. X. ALBUQUERQUE *and* J. W. DALY

Abbreviations

E = steady-state membrane potential in presence of concentration A of agonist (BTX)
E_0 = resting membrane potential in the absence of agonist
E_{Na} = the equilibrium potential for sodium ions
E_m = steady-state membrane potential at maximum depolarization by saturating concentrations of agonist
R = resting membrane resistance
g = mean conductance of a single sodium channel opened by agonist
N = number of sodium channels opened in the presence of agonist
N_m = maximum number of sodium channels blocked in the presence of antagonist
K_B = dissociation constant of antagonist
k = rate constant of association
t = fixed incubation time
N_B = number of sodium channels blocked by antagonist

Acknowledgements

The authors wish to thank Dr Jordan E. Warnick for his most constructive criticism of this manuscript. We also wish to thank Ms Mabel Zelle for her excellent assistance during revision of the manuscript.

The Specificity and Action of Animal, Bacterial and Plant Toxins
(*Receptors and Recognition,* Series B, Volume 1)
Edited by P. Cuatrecasas
Published in 1976 by Chapman and Hall, 11 Fetter Lane, London EC4P 4EE

8.1 INTRODUCTION

One of the major challenges of modern neurobiology is the elucidation of the molecular mechnisms involved in the control of ion fluxes across excitable membranes of nerves and muscles, both at rest and during electrogenic activity. A key to an understanding of the nature and control of the membrane macromolecules which modulate ionic conductances has been the discovery and pharmacological charaterization of various toxins which interact with specific components of the membrane entities which regulate ionic permeability. Although poorly understood, these membrane entities probably consist of macromolecules which bridge the membrane and directly control the ionic conductance. In some cases, there is an associated regulatory unit such as the acetylcholine receptor. The macromolecular unit controlling the ionic conductance has been referred to as a channel, ionophore or ion conductance modulator. In the present review, we shall deal with phenomena which are generally felt to be mechanistically consonant with the presence of channel whose functional activity is regulated by 'gates' rather than with phenomena involving some carrier molecule. The term channel will, therefore, be used in this review in referring to the membrane entity which modulates ionic conductances.

A variety of toxins have proven useful for investigating the control of membrane permeability. One of these toxins is a novel steroidal alkaloid, batrachotoxin. This compound, isolated from a small neotropical frog, has proven to be an invaluable tool for the study of channels regulating sodium permeability in electrogenic membranes. The pharmacological effects of batrachotoxin are the subject of the present review.

8.2 HISTORICAL BACKGROUND

Indians from the Pacific drainage of northwestern South America have for centuries used the skin secretions from a small brightly colored frog, the *kokoi*, as a source of poison to tip the blow darts which they use in hunting small game. These skin toxins had apparently evolved in the frog as a passive chemical defense against predators. The chemical and pharmacological properties of the skin secretion were first investigated by Posada-Arango in 1871. However, it required nearly a century before the use of modern techniques of isolation and characterization led to the pure chemically defined toxin (Marki and Witkop, 1963). The frogs contained only minute quantities of a toxin whose toxicity to mammals placed it among the most poisonous small molecules known to man. After a number of years of effort using chemical and spectroscopic analysis (Daly *et al.*, 1965), crystals of a relatively inactive congener of batrachotoxin (from the greek batrachos = frog) were converted

Table 8.1 Tissue distribution of batrachotoxinin A 20 α-2, 4-dimethyl-5-(α, α-ditri +; oethyl)-pyrrole-3-carboxylate ('5-Ethyl [^{3}H] BTX').
NIH white mice (25 g) were injected with 0.35 μg (14 μg kg^{-1}) of '5-ethyl [^{3}H] BTX' (0.85 x 10^6 cpm). Pooled tissues from three animals obtained at the time of death (30–35 min) were homogenized with 5 vol of 0.4 N perchloric acid. Total radioactivity was measured on an aliquot solubilized with hyamine hydroxide, while 'free' '5-ethyl [^{3}H] BTX' was measured in the perchloric acid supernatant after centrifugation.

Tissue	Total radioactivity cpm mg^{-1} tissue	Percentage present as free '5-ethyl [^{3}H] BTX'
Heart*	3400	43
Liver	8200	43
Kidney	7400	53
Intestine	2900	46
Spleen	7200	53
Brain	580	66
Muscle		
Forelimb	2400	47
Hindlimb**	16 000	37

* At the time of death, the heart contains 0.43% of the dose or about 15 n*M* '5-ethyl [^{3}H] BTX'.
** Near site of injection.

to a crystalline derivative, whose analysis by X-ray crystallography defined the structure of the steroidal portion of this class of toxins (Tokuyama *et al.*, 1968; Karle and Karle, 1969). Chemical and spectroscopic studies quickly led to the definition of the structures of the highly toxic congeners, batrachotoxin and homobatrachotoxin, both of which were deduced and then proven to be pyrrole carboxylate esters of the relatively inactive congener, batrachotoxinin-A. Determination of the tissue distribution of batrachotoxinin-A 20α-2, 4-dimethyl-5-(α, α-ditritioethyl)-pyrrole-3-carboxylate after injection of the labelled toxin into white mice indicated that the toxin reaches most of the vital organs (Table 8.1). Initial studies had aslo demonstrated a blockade of neuromuscular transmission and the appearance of muscle contracture (Marki and Witkop, 1963). Subsequent investigations were designed to delineate the basis of the toxicity of batrachotoxin and its pharmacological effects on nerve and muscle. The results of these investigations led to the definition of batrachotoxins as extremely potent, selective and virtually irreversible activators of 'sodium channels' in excitable membranes. The initial results of these studies have been previously reviewed (Albuquerque *et al.*, 1971a; Albuquerque, 1972) and therefore the present review will emphasize more recent findings.

8.3 CHEMISTRY

The structure of the batrachotoxins have proved unique in nature. Three of the naturally occurring batrachotoxins (batrachotoxin, homobatrachotoxin and batrachotoxinin-A) are shown in Fig. 8.1. The fourth naturally occurring compound of this series, pseudobatrachotoxin, is of unknown structure, but is readily converted to batrachotoxinin-A during isolation, presumably *via* hydrolysis of a labile ester moiety (Tokuyama *et al.*, 1969). Structures of batrachotoxins are unique in the following aspects: (i) the hemiketal function between carbons 3 and 9 which is important to biological activity (A similar hemiketal function between carbons 4 and

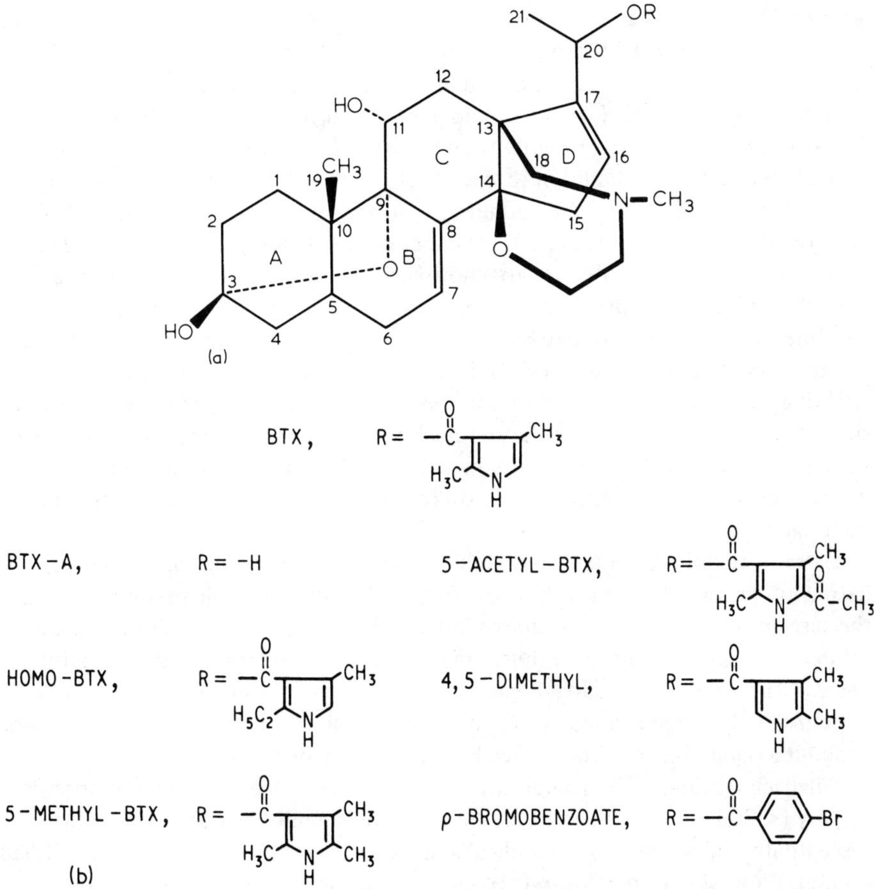

Fig. 8.1 (a) Steroidal ring system of batrachotoxin alkaloids. (b) Structure of the 20α-substituent. Batrachotoxin (BTX), homobatrachotoxin (Homo-BTX) and batrachotoxinin-A (BTX-A) occur naturally. The other compounds are analogs synthesized from BTX-A.

9 is present in *Veratrum* alkaloids, such as veratridine); (ii) the heterocyclic homomorpholine ring fused across the C/D rings of the steroidal system. The tertiary nitrogen present in this heterocyclic ring confers weakly basic properties to batrachotoxin (pK of about 7.5); (iii) the 20α-2, 4-dialkyl-pyrrole-3-carboxylate function present in batrachotoxin and homobatrachotoxin. The relatively inactive batrachotoxinin-A (BTX-A) lacks an ester substituent at the 20α-hydroxy group. It has, however, been used to prepare a variety of synthetic ester analogs of batrachotoxin (BTX) and homobatrachotoxin (homo BTX) (Tokuyama *et al.*, 1969). These include '5-methyl BTX' (BTX-A 20α-2, 4, 5-trimethylpyrrole-3-carboxylate), '5-acetyl BTX' (BTX-A 20α-2, 4-dimethyl-5-acetylpyrrole-3-carboxylate), a 4, 5-dimethyl analog of BTX (BTX-A 20α-4, 5-dimethylpyrrole-3-carboxylate) and the *p*-bromobenzoate ester of BTX-A (BTX-A 20α-*p*-bromobenzoate) (Fig. 8.1). Reduction of the hemiketal function of BTX has yielded the relatively inactive 'H_2-BTX' (3α, 9α-dihydro BTX). Quarternization of the tertiary nitrogen in the homomorpholine ring yielded 'quaternary BTX' (BTX N, N-dimethyliodide). The benzoate ester of BTX-A (BTX-A 20α-benzoate) has recently been prepared (Sundermier *et al.*, 1976, and unpublished results). In addition to these analogs, a total synthesis of batrachotoxinin-A and of 7, 8- and 15, 16-dihydro and tetrahydro analogs of batrachotoxinin-A has been obtained (Gössinger *et al.*, 1971; Graf *et al.*, 1971; Imhoff *et al.*, 1973). The crystal structure of 7, 8-dihydrobatrachotoxinin-A was shown to be identical to batrachotoxinin-A except for saturation at the 7, 8-double bond (Karle, 1972). The absolute stereochemistry of batrachotoxinin-A corresponds in all points of similarity to that of cholesterol (Gilardi, 1970). It is perhaps noteworthy that the cis A/B and C/D ring junctions in the batrachotoxins result in a globular-shaped molecular form similar to the globular shape of the steroidal portion of cardiac glycosides. It is also perhaps noteworthy that batrachotoxin and veratridine are similar in terms of the position of their tertiary nitrogen relative to general conformation of the 'steroidal' skeleton.

Batrachotoxinin-A is a relatively stable molecule, while the pyrrole-containing batrachotoxin and homobatrachotoxin are significantly less stable presumably due to the susceptibility of the pyrrole to oxidative attack. All of the batrachotoxins are soluble in organic solvents, and thus should penetrate biological membranes fairly readily. The quaternary analog of batrachotoxin is less lipid-soluble since it exists, of course, only in the cationic form. At physiological pH values batrachotoxin exists at about a one to one mixture of free base and protonated amine.

Wilzbach-labelling of batrachotoxin led to material with a relatively low specific activity (< 300 mCi mol^{-1}) and a rather poor radiostability probably due to the susceptibility of pyrroles to free radical and oxidative attack (Brown, G. unpublished results). Formation of the 20α-*p*-[^{3}H]-benzoate of batrachotoxinin-A has now provided a biologically active analog of batrachotoxin at a high specific activity (> 15 Ci mol^{-1}) for studies on binding and disposition of this class of compounds (Sundermier *et al.*, 1976).

8.4 PHARMACOLOGY

Batrachotoxin and its various analogs have now been studied in a variety of biological preparations. In most of these preparations, batrachotoxin elicits a time and concentration-dependent depolarization of excitable membranes. Most of the effects of batrachotoxin on the functional activity and morphology of nerves and of striated and cardiac muscles would appear to be both direct and indirect consequences of membrane depolarization. Based on a variety of evidence, the mechanism by which batrachotoxin depolarizes electrogenic tissue would appear to involve a selective activation of sodium channels. Thus, in mammalian and amphibian nerve-striated muscle preparations (Albuquerque *et al.*, 1971c, 1973b, 1976; Albuquerque and Warnick, 1972; Warnick *et al.*, 1971, 1975; Jansson *et al.*, 1974), dog heart Purkinje fibers (Hogan and Albuquerque, 1971) and dog heart papillary muscle (Shotzberger *et al.*, 1976), squid and lobster axons (Narahashi *et al.*, 1971a, b; Albuquerque *et al.*, 1971b, 1973a), myelinated nerve fiber of frog (Khodorov *et al.*, 1975), electroplax of the electric eel (Bartels de Bernal *et al.*, 1975; Barnard, Weber, Albuquerque and Changeux, paper in preparation), mammalian brain slices (Shimizu and Daly, 1972), and brain synaptosomes (Holz and Coyle, 1974) batrachotoxin-elicited depolarizations and/or the consequences of depolarization are blocked or reversed in a low sodium medium or by tetrodotoxin. The latter toxin is generally accepted to be a specific antagonist of activated sodium channels (Kao, 1966; Narahashi, 1974). In neuroblastoma cells which are capable of generating action potentials, batrachotoxin causes an increase in sodium conductance which is blocked by tetrodotoxin (Catterall, 1975a, b). The batrachotoxin-induced increase in sodium conductance in cultured muscle cells is likewise blocked by tetrodotoxin (Catterall, 1976). The only system in which sodium ions are primarily involved in electrical spike activity and in which batrachotoxin is ineffective in eliciting a depolarization is the sciatic nerve-sartorius muscle preparation from the frog that produces the toxin (Albuquerque *et al.*, 1973). Batrachotoxin at a concentration of 1.2 μM had no effect on either membrane potential, miniature endplate potential amplitude and frequency, spike generation or twitch of nerve-sartorius muscle preparation of the poison-dart frog *Phyllobates aurotaenia.* Tetrodotoxin or the absence of extracellular sodium ions blocked the intracellularly recorded muscle action potential. It was concluded that: (1) in sartorius muscle fibers of *P. aurotaenia* the early inward current during the action potential is carried by sodium ions; (2) the electrogenic membranes of the sartorius muscles of *P. aurotaenia* are insensitive to the sodium-dependent depolarizing action of batrachotoxin, but they remain sensitive to the blocking action of tetrodotoxin, and (3) the site of action of batrachotoxin may be absent in nerve and muscle membranes of *P. aurotaenia* (Albuquerque *et al.*, 1973). It is noteworthy that veratridine, another agent which increases sodium conductance, was much less effective in preparations of *P. aurotaenia* than in frog sartorius muscles of *Rana pipiens.* Veratridine (7.3 μM) produced a 20 mV membrane depolarization of surface fibers of stimulated sartorius muscles from *P. aurotaenia* after 1 hour of exposure to the drug, while in similar experiments with preparations from the frog *Rana pipiens* a 95 mV membrane depolarization occurred within 15 min. When the sartorius muscles of *Rana pipiens* were stimulated at a frequency of 0.2 Hz and exposed to batrachotoxin (40 nM), the indirectly elicited muscle twitch was blocked and all surface fibers were depolarized within 13 min; simultaneously there was a transient muscle contracture lasting 7–8 min. Increases in the frequency of stimulation have been shown to enhance the rate of depolarization elicited by the action of batrachotoxin on the sciatic nerve-muscle preparation of the frog and mammals (Warnick *et al.*, 1971; Albuquerque *et al.*, 1971, 1973b; Hogan and Albuquerque, 1971; Warnick *et al.*, 1975), squid giant axon (Narahashi *et al.*, 1971a, b), and Ranvier node of frog nerve (B. I. Khodorov, personal communication). In electroplax preparations from the electric eel, batrachotoxin causes depolarization only in stimulated tissue (Bartels de Bernal *et al.*, 1975).The enhancement in rates of batrachotoxin action by stimulation is perhaps due to increased penetration of the toxin to its site of action in the stimulated preparations (Albuquerque, Garrison, Warnick, Daly and Witkop, unpublished observations). Electron microscopic observations of batrachotoxin-treated sartorius muscles from *Rana pipiens* revealed damage to the presynaptic elements, whereas the nerve and muscles of *P. aurotaenia* were unaffected. The results of these comparative studies in *Rana pipiens* and *P. aurotaenia* suggested that although tetrodotoxin-sensitive sodium channels appeared involved in action potential generation in nerve and muscles from both species, the recognition sites for batrachotoxin and probably veratridine which result in activation of the sodium channel have a much lower affinity to the alkaloids in the *P. aurotaenia* species (Albuquerque *et al.*, 1973b).

In lobster or crayfish skeletal muscle where calcium ions are the main carrier of the peak inward current associated with action potentials, batrachotoxin has no effect (Albuquerque *et al.*, 1972). At concentrations as high as 0.6 μM, batrachotoxin had no effect on the membrane potential, action potential or membrane resistance of lobster and crayfish muscle fibers bathed in normal physiological solution or pretreated with 205 mM tetraethylammonium bromide and bathed in 236 mM tris (hydroxymethyl) aminoethane with 16 mM calcium chloride. The lack of effect of batrachotoxin on these muscles suggests that the protein or proteolipid component which regulates the ionic permeability to calcium is not affected by the toxin, and that recognition sites for batrachotoxin are associated only with sodium channels.

Batrachotoxin, like veratridine, has virtually no effect on sodium permeability, in variant neuroblastoma cells which lack action potentials (W. Catterall, personal communication). Similarly, in human erythrocytes neither batrachotoxin nor tetrodotoxin even at 1 μM had any effect on sodium permeability (J.D. Gardner, personal communication).

In early investigations the question was posed as to whether the mechanism of membrane depolarization induced by batrachotoxin bore any relationship to the membrane depolarization induced by cardiac glycosides such as ouabain, which inhibit Na^+–K^+-ATPase (Lee and Klaus, 1971). Batrachotoxin did not inhibit

Na^+–K^+-ATPase at concentrations that would have produced membrane depolarization of the diaphragm muscle within 5–10 min at 37°C (Daly *et al.*, 1972). At much higher concentrations (60 μM), batrachotoxin partially inhibited Na^+–K^+-ATPase from electroplax of the eel, *Electrophorus electricus.* In contrast to cardiac glycosides, the inhibition of Na^+–K^+-ATPase produced by batrachotoxin was not antagonized by potassium chloride. Batrachotoxin had no effect on ATP levels in stimulated nerve-muscle preparations at times when a sustained contracture was initiated by the drug. ATP levels of the sciatic nerve of the cat were little affected by batrachotoxin (Ochs and Worth, 1975). In support of the findings that batrachotoxin does not affect Na^+–K^+-ATPase, the toxin did not increase the short circuit current in isolated ventral skin of *Rana pipiens* (Albuquerque and Paganelli, unpublished results).

8.4.1 Effects of batrachotoxin and analogs on resting membrane potential

All available evidence thus indicates that batrachotoxin has a high and selective affinity for a site associated with the sodium channel for electrogenic membranes. However, the rate of membrane depolarization of various electrically excitable cells by batrachotoxin varies considerably. Undoubtedly, a number of factors contribute to the rate of membrane depolarization. Among these are (i) physical barriers impeding access of batrachotoxin to recognition sites; (ii) differing affinities and rate of association of batrachotoxin for the sites in different cell types; and (iii) number of sodium channels per membrane surface area and its relationship to activity of Na^+–K^+-ATPases which act in opposition to batrachotoxin-elicited depolarization. In addition, the action of batrachotoxin is influenced by factors such as rate of electrical stimulation of the preparations, temperature, calcium concentration and pH (Jansson *et al.*, 1974; Warnick *et al.*, 1975). Its effects are antagonized by high concentrations of calcium ions, presumably due to membrane-stabilizing activity of this ion. Batrachotoxin is more effective in nerve-muscle preparations at higher pH values suggesting that the uncharged free base form of this alkaloid is the more active agonist (Warnick *et al.*, 1975). Similar findings were also obtained on the electroplax preparations from the electric eel. In these preparations, batrachotoxin and its quaternary analog caused a reversible prolongation of action potentials at pH 8.5, but had no effect at pH 6.0 (E. Bartels de Bernal and T. Rosenberry, personal communication). The positively charged quaternary analog was 50 to 100-fold less active than the unprotonated batrachotoxin at pH 8.5. The results suggest that the free base form of batrachotoxin is the more active agonist, but that, in addition, an ionized membrane constituent with a pK of > 6.0 is critical to the action of batrachotoxins in electroplax preparations.

In general, batrachotoxin caused a marked membrane depolarization in most electrogenic tissues when present at only nanomolar concentrations. In squid axons, due perhaps to permeability barriers, batrachotoxin is relatively less effective when applied externally. During internal application of batrachotoxin (100 nM) so squid

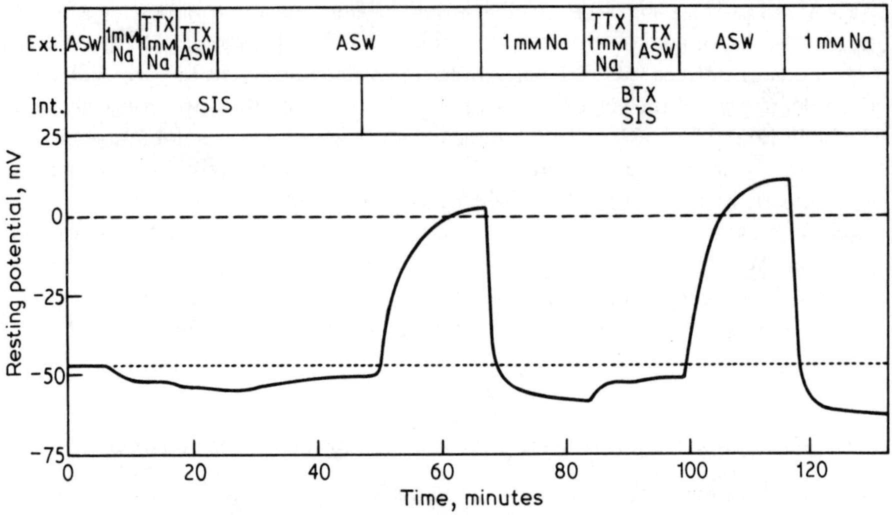

Fig. 8.2 Effects of internally applied batrachotoxin (550 nM, BTX), externally applied tetrodotoxin (1000 nM, TTX and externally applied 1 mM Na (1 mM) on the resting membrane potential of internally perfused squid giant axons. Standard internal solutions (SIS) contains 50 mM Na. ASW: artificial seawater. The dotted line represents the control resting membrane potential. From Narahashi *et al.*, 1971.

giant axons a marked membrane depolarization developed over a period of 1 h (Fig. 8.2). In sodium-free media or with tetrodotoxin present, membrane potential was restored in spite of the continued presence of batrachotoxin. At much lower concentrations, batrachotoxin (10 nM) rapidly depolarized rat striated muscles (Fig. 8.3; Albuquerque *et al.*, 1971c; Warnick *et al.*, 1971). Blockade of nerve action potentials in a nerve-diaphragm preparation due to batrachotoxin-elicited depolarization appeared to required a longer time period than depolarization of muscle (Jansson *et al.*, 1974). In cardiac Purkinje fibers and papillary muscle, batrachotoxin (2 nM) elicited a marked depolarization within less than 1 h (Shotzberger *et al.*, 1976). Thus, at least 50-fold differences in the apparent potencies of batrachotoxin pertain in different biological preparations. However, because of the complexity of the membrane systems which establish membrane potential, little can be proposed as to the affinity of batrachotoxin for the recognition sites in various cell types. Clearly, however, in nerve and striated muscle of the poison-dart frog the affinity of batrachotoxin for recognition sites is profoundly decreased or absent when compared to its affinity for sites in similar nerve-muscle preparations of the frog *Rana pipiens* (Albuquerque *et al.*, 1973b). On nerve-muscle preparations from another neotropical frog, *Dendrobates histrionicus,* batrachotoxin had similar effects to those observed with *Rana pipiens* (Albuquerque, Lapa, Onur

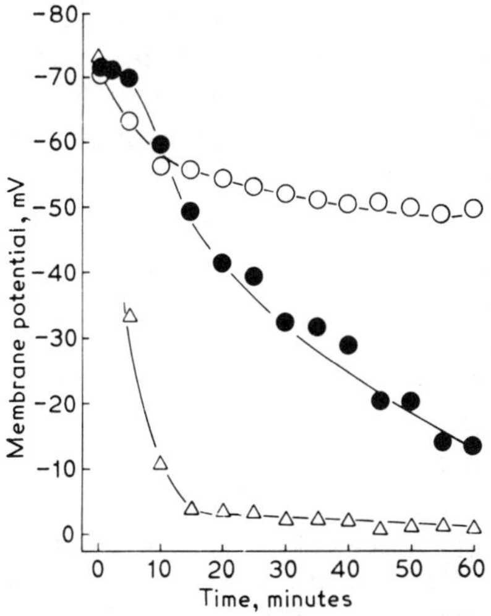

Fig. 8.3 Effects of batrachotoxin on the membrane potential of isolated rat diaphragm muscles. Each point is the mean obtained from 4–23 individual surface fibers from at least three experiments. (○), batrachotoxin 1 nM; (●), batrachotoxin 5 nM; (△), batrachotoxin 80 nM.

and Daly, unpublished observations). Thus, recognition sites for batrachotoxin are present in this related species of poison-dart frog which produces not batrachotoxin but instead histrionicotoxins (Daly *et al.*, 1971). Undoubtedly, recognition sites in electrogenic membranes from various sources will be found to differ significantly in their affinity for batrachotoxin, depending on the type of cell under study. It should be noted that in most nerve and muscle preparations the depolarizing effects of batrachotoxin have appeared to be relatively irreversible in nature. Thus, when batrachotoxin-depolarized preparations are washed, repolarized with tetrodotoxin or a low sodium medium and then placed in normal medium, a rapid depolarization occurs (Albuquerque *et al.*, 1971a). In part, the apparent irreversibility may reflect both a slow 'off reaction' for batrachotoxin because of permeability and other factors and the fact that in many preparations only a small percentage ($< 5\%$) of the sodium channels need to be activated by batrachotoxin to cause a rapid membrane depolarization. Studies with radioactive batrachotoxin will be necessary to define both affinity and reversibility of binding of this alkaloid to recognition sites (see below).

Several analogs of batrachotoxin were compared with regard to their effects on muscle membrane potential (Warnick *et al.*, 1975). All of these compounds except 'H_2-BTX' and batrachotoxinin-A had the ability to markedly depolarize the rat diaphragm muscle fibers. All the effective analogs caused similar maximal depolarizations, but either higher concentrations or longer times were required for less active compounds (see below, Fig. 8.7). The relative order was batrachotoxin > '5-methyl-

BTX' > homobatrachotoxin > 'quaternary BTX' > the '4,5-dimethyl' analog of batrachotoxin > '5-acetyl BTX' > the *p*-bromobenzoate ester of batrachotoxinin-A > 'H_2-BTX' > batrachotoxinin-A. Recently the 20α-benzoate of batrachotoxinin-A was found to be nearly as active as batrachotoxin itself in eliciting muscle depolarization (unpublished results). Thus, subtle differences in the nature of the 20α-ester substituent have a great effect on activation of sodium channels by batrachotoxin analogs.

8.4.2 Effect of batrachotoxin on membrane conductances

Only limited studies of the effects of batrachotoxin on membrane conductances have been undertaken. Such studies in nerve and muscle require the use of voltage clamp conditions. Nerve and muscle preparations treated with batrachotoxin and then repolarized by anodal current have been shown to be capable of generating an action potential upon depolarizing stimulation (Albuquerque *et al.*, 1971c; Albuquerque and Warnick, 1972; Hogan and Albuquerque, 1971; Khodorov *et al.*, 1975; Narahashi *et al.*, 1971a, b; Warnick *et al.*, 1971). Thus, in spite of the increase in sodium conductance elicited by batrachotoxin, sufficient sodium channels remain available for generation of a normal further increase in sodium conductance. Other toxins such as grayanotoxin exhibit similar properties (Narahashi, 1974; Narahashi and Seyama, 1974).

In squid giant axons repolarized under voltage clamp conditions, batrachotoxin had no effect on sodium conductances associated with action potentials, while significantly reducing potassium conductances (Narahashi *et al.*, 1971a, b). This result is, however, not incompatible with a correspondence of sodium channels sensitive to batrachotoxin action and sodium channels involved in action potential generation. In order for batrachotoxin to depolarize a squid axon by about 60 mV, the resting sodium permeability would have to increase by a factor of 100. If one assumes that the normal value of resting conductance in nerve membrane is about 3.0 mmho cm^{-2}, using the Hodgkin–Huxley equation the calculated resting sodium conductance is approximately 0.1 mmho cm^{-2}. Under voltage clamp conditions and in the presence of BTX, the membrane conductance increases from 10 mmho cm^{-2} to 100 mmho cm^{-2}. Thus, the net increase in sodium conductance caused by depolarizing stimulation is 90 mmho cm^{-2} in BTX-poisoned axons, as against 99.9 mmho cm^{-2} in normal axons. These observations can be explained with or without assuming separate sodium channels in the nerve membrane at rest and during activity.

Similarly, in experiments with Ranvier nodes of myelinated nerve fibers of the frog *Rana ribidunda,* batrachotoxin at 1.0×10^{-7} to 2.5×10^{-5} M caused a gradually increasing depolarization but the node still had reserve sodium channels capable of generating action potentials (Khodorov *et al.*, 1975). Under voltage clamp conditions with potassium conductances blocked by tetraethylammonium, batrachotoxin increased steady state sodium conductances and blocked inactivation of the transient sodium currents. Batrachotoxin-altered channels appeared to open at potentials more negative than those required for unaltered channels, to open more slowly and to have decreased sensitivity to blockade by local anesthetics (see below). The kinetic of the sodium conductance-increase of batrachotoxin-altered channels was no longer sigmoidal but instead became linear.

In cultured neuroblastoma cells, effects of batrachotoxin on sodium influx can be conveniently studied. The experimental paradigm involves inhibition of Na^+-K^+-ATPase with ouabain followed by kinetic studies on the effects of batrachotoxin and other parameters on initial rates of uptake of radioactive sodium ions (Catterall, 1975a, b). In this system, the apparent affinity constant for activation of sodium channels by batrachotoxin was $2-4 \times 10^{-7}$ M. Veratridine appeared to be a partial agonist at the same sites with an apparent affinity of $500 - 800 \times 10^{-7}$ M. Veratridine competitively inhibited batrachotoxin-elicited increases in sodium uptake. The effects of both toxins were readily reversible, although both the 'on' and 'off' reactions for batrachotoxin were markedly slower than those for veratridine. Aconitine appeared to be a weak partial agonist for sodium channels in neuroblastoma cells with and apparent affinity of 80×10^{-7} M for the same recognition site. A component of scorpion vemon (*Leiurus quinquestriatus*) was found to both potentiate and increase the apparent affinity of batrachotoxin, veratridine and aconitine for the site controlling sodium ion conductances (Catterall, 1975b; Catterall *et al.*, 1976). In the presence of scorpion venom veratridine and aconitine were now nearly complete agonists with their apparent affinity constants reduced by 50 and 10-fold, respectively. The apparent affinity constant for batrachotoxin was reduced 70-fold in the presence of scorpion venom. Two regulatory sites associated with the sodium channel were postulated; one interacting with batrachotoxin, veratridine and aconitine, the other with a peptide component of scorpion venom. The two sites appeared to interact cooperatively. Tetrodotoxin antagonized the action of the alkaloids and the scorpion venom by blockade of sodium channels at yet another site (Catterall, 1975a,b; Warnick *et al.*, 1975, 1976).

Batrachotoxin and veratridine have recently been shown to activate sodium channels in clonal muscle cells (Catterall, 1976). The results with the two alkaloids in cultured muscle cells were substantially identical to the data with cultured neuroblastoma cells. Tetrodotoxin was relatively ineffective in the muscle cells, thus reinforcing previous suggestions that the recognition site at the sodium channel is different for the two toxins (Albuquerque *et al.*, 1971 b; 1973b; Colguhoun *et al.*, 1972).

8.4.3 Effects of batrachotoxin and analogs on spontaneous and evoked transmitter release

The presynaptic effects of batrachotoxin have studied in detail (Albuquerque *et al.*, 1971; Jansson *et al.*, 1974). When phrenic nerve-diaphragm preparations of rats were exposed to physiological solution containing batrachotoxin (10 nM) the following changes occurred: (1) a progressive decrease in the amplitude of the extracellularly recorded nerve action potential due to depolarization of the motor nerve; (2) when

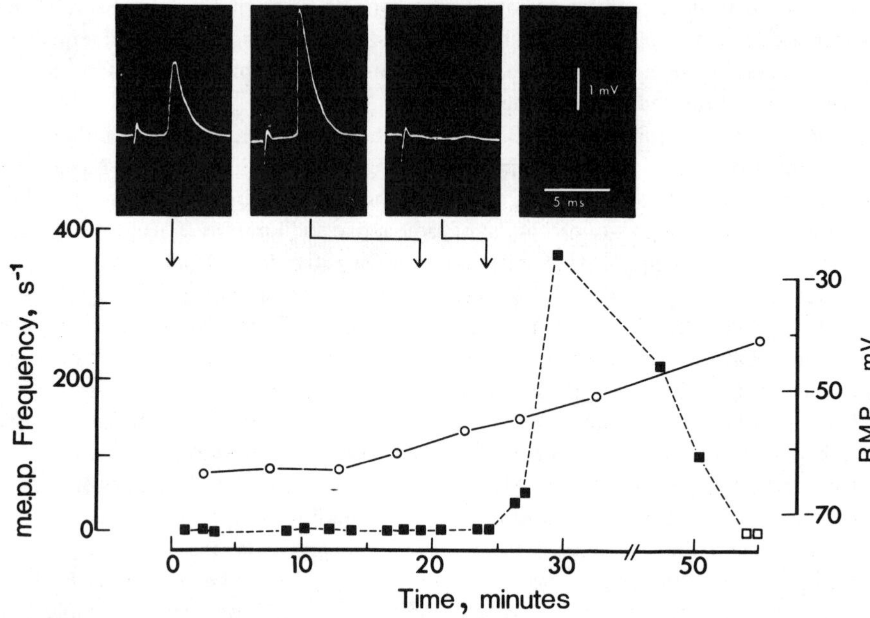

Fig. 8.4 Batrachotoxin (10 nM, BTX) induced block of the endplate potential (upper insert), muscle depolarization and transient increase and block of spontaneous miniature endplate potential (m.e.p.p.) frequency in rat diaphragm muscle. The resting membrane potentials of muscle are shown by open circles (○); the m.e.p.p. frequency by solid squares (■). Open squares (□) indicate that no spontaneous m.e.p.p.s. could be detected. Temperature, 34°C. From Jansson *et al.* (1974).

the nerve action potential reaches about half of its normal value, evoked transmitter release as reflected by endplate potentials was blocked, due probably to failure of the nerve action potential to invade the nerve terminal; (3) eight to ten minutes after block of the evoked transmitter release, spontaneous miniature endplate potential frequency increased to levels above 400 s^{-1} ; and (4) finally, there is complete block of spontaneous miniature endplate potentials, probably due to marked accumulation of sodium within the nerve terminal, dispersion of vesicles and depolarization of the nerve terminal. The time courses for blockade of evoked release of transmitter, spontaneous release and muscle depolarization are depicted in Fig. 8.4. When calcium was removed from the bathing solution and EGTA was applied to reduce the extracellular calcium concentration to less than 1×10^{-11} M, application of BTX to the nerve-muscle preparation at 28°C did not produce an increase in miniature endplate potential frequency (Fig. 8.5). However, at 37°C, the toxin produced a marked increase in spontaneous miniature endplate potentials even under calcium-free conditions. The pronounced temperature dependence observed for the BTX-induced increase in miniature endplate potential frequency under calcium-free conditions might be due

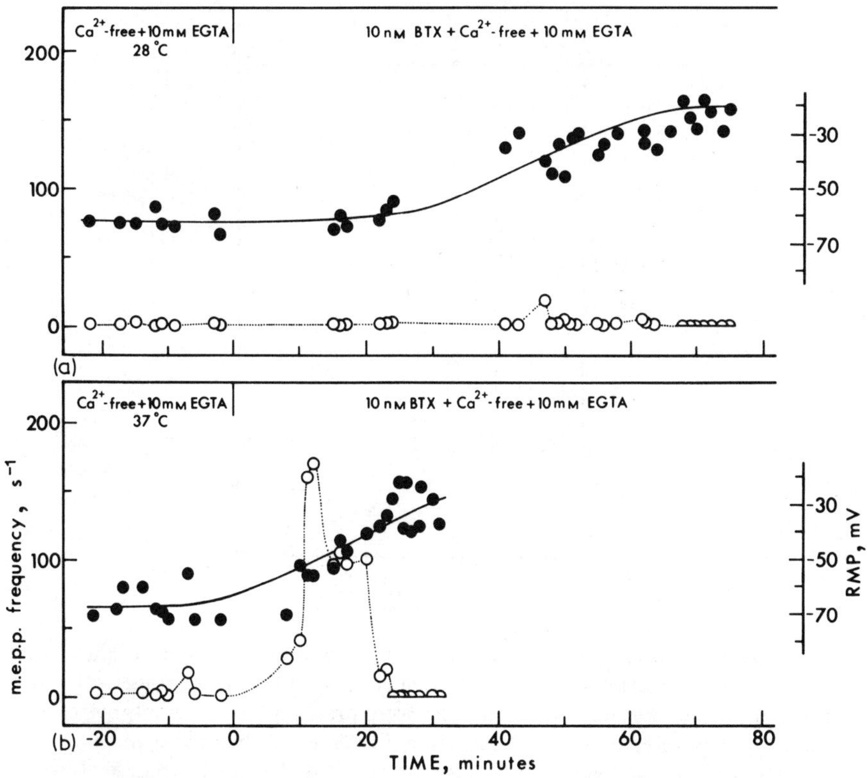

Fig. 8.5 Effect of batrachotoxin (10 nM, BTX) on the resting membrane potential (●) and spontaneous miniature endplate potential (m.e.p.p.) frequency (○) in rat diaphragm muscles at two different temperatures: 28°C (a) and 37°C (b). Muscles were bathed for one hour in physiological solutions containing no calcium and 10 mM EGTA, and then BTX was added at time zero. Half circles (⌓) indicate that no spontaneous m.e.p.p.s could be detected. Each point is the mean obtained from at least 40 single surface fibers of 4 diaphragm muscles. From Jansson *et al.* (1974).

to a phase transition of a lipid component involved at the calcium-binding site. Cholesterol and various phospholipids, known to be important membrane constituents, have been shown to have several transition points in the temperature range of 34–37°C (Luzzati and Husson, 1962; Chapman, 1967). As can be seen from the Arrhenius plot in Fig. 8.6, the increase in miniature endplate potential frequency induced by batrachotoxin in the absence of external calcium depends on a high energy of activation (50 Kcal deg^{-1}) in the temperature range of 35–37°C, with an abrupt change in the energy requirements at about 34°C. These studies using calcium-free conditions suggested that the sodium entering the nerve terminal during the action of batrachotoxin releases calcium necessary for the increase in spontaneous transmitter

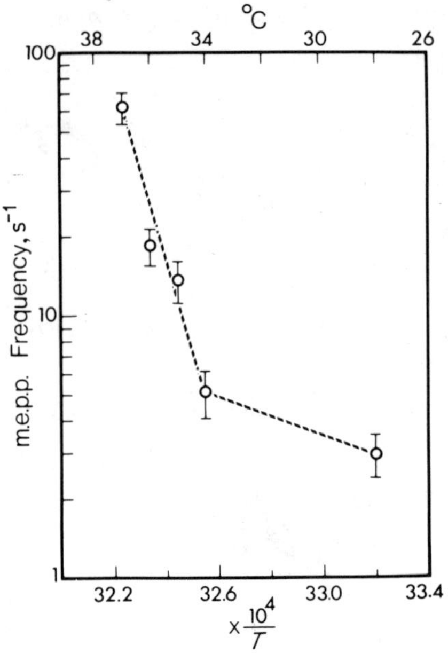

Fig. 8.6 Effect of temperature on the batrachotoxin-induced decrease in spontaneous miniature endplate potential (m.e.p.p.) frequency under calcium-free conditions. Rat diaphragm muscles were pre-equilibrated with physiological solution containing no calcium and 10 mM EGTA and then batrachotoxin (10 nM) was added. Upper abscissa, temperature in degrees Centigrade; lower abscissa, reciprocals of absolute temperature; ordinate, frequency of m.e.p.p.s expressed as number of m.e.p.p. s^{-1}. Each point is the mean ± S.E.M. obtained from 23 to 89 fibers in 3 to 6 different muscles, except the point at 36°C which is from just one experiment. From Jansson *et al.*, (1974).

release. Although in previous studies (Jansson *et al.*, 1974), the mitochondria were not considered as a source for the presynaptic calcium released by the action of batrachotoxin, Alnaes and Rahaminoff (1975) have recently suggested that mitochondrial stores of calcium may be involved in the transmitter release process. Further investigations will be required to determine the mechanism of interaction of sodium with calcium at either presynaptic or mitochondrial sites, and toxins such as tityustoxin and derivatives of batrachotoxin may be important for the elucidation of this process (Warnick *et al.*, 1975, 1976).

Several analogs of batrachotoxin were compared with regard to their effects on peak miniature endplate potential frequency (Warnick *et al.*, 1975). All analogs except 'H_2-BTX' and batrachotoxinin-A had the ability to transiently increase and subsequently block spontaneous miniature endplate potentials (Fig. 8.7). All compounds caused similar final effects, but either higher concentrations or longer times were required for the less active analogs. The activity in terms of presynaptic effects on spontaneous

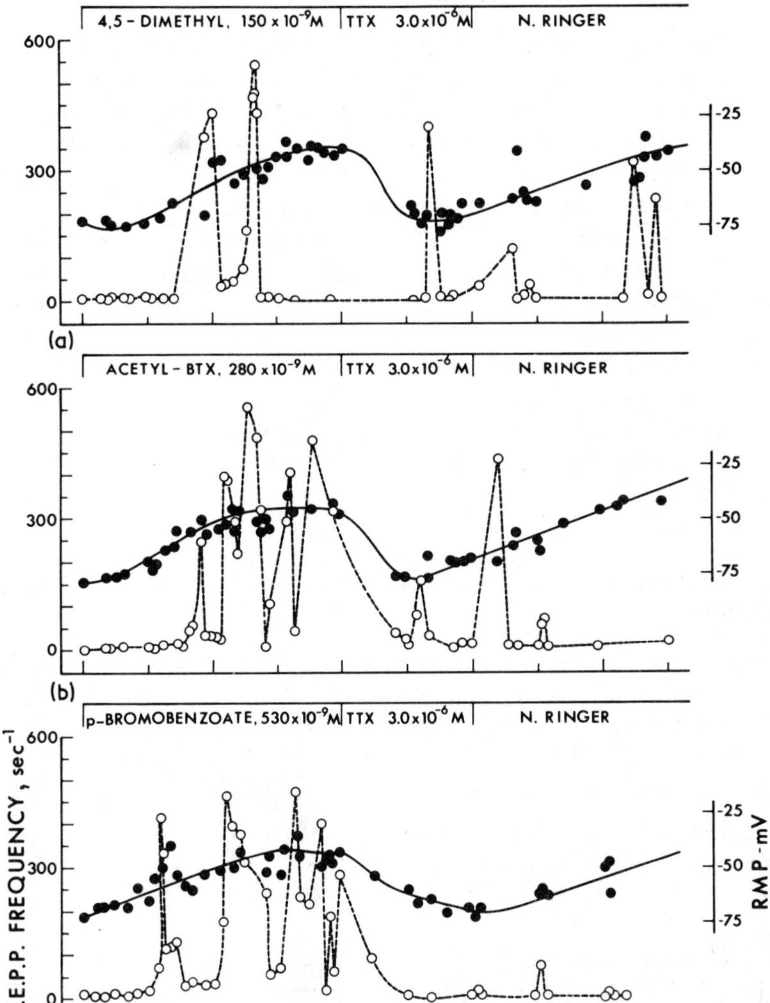

Fig. 8.7 Comparison of the time course of the effect of three batrachotoxin analogs on miniature endplate potential (m.e.p.p.) frequency (number of m.e.p.p.s s^{-1}; left ordinate) and resting membrane potential (RMP, right ordinate). Temperature, 37°C; m.e.p.p. frequency (○); RMP (●). Rat diaphragm muscles were exposed to the 4, 5-dimethyl analog of BTX (4, 5-dimethyl) in (a); '5-acetyl-BTX' in (b); and BTX-A 20α-*p*-bromobenzoate (*p*-bromobenzoate) in (c); after 60 min, the analog-containing solution was removed and physiological solution containing tetrodotoxin (TTX) was added to the muscle bath. Treatment with TTX was continued for 30 min, and then the muscles were washed with normal physiological solution (N. Ringer). Half circles (⌓) indicate that no spontaneous m.e.p.p.s were detected. From Warnick *et al.* (1975).

transmitter release were as follows: batrachotoxin > homobatrachotoxin > '5-methyl BTX' > the 4, 5-dimethyl analog of batrachotoxin > '5-acetyl BTX' > 'quaternary BTX' > the *p*-bromobenzoate ester of batrachotoxinin-A > 'H_2-BTX' > batrachotoxinin-A. Of the series of alkaloids, batrachotoxin, homobatrachotoxin and the *p*-bromobenzoate exhibited an apparent selectivity towards presynaptic rather than postsynaptic membranes. Batrachotoxin and homobatrachotoxin were, however, the most active compounds of the series at both pre- and postsynaptic membranes. Recently, the 20 α-benzoate ester of batrachotoxinin-A was found to be nearly as active as batrachotoxin in eliciting increases and subsequently blocking spontaneous miniature endplate potentials. Thus, the mere presence or absence of a *p*-bromo-substituent had profound effects on the activity of benzoate esters. In view of the virtual absence of activity of batrachotoxinin-A in nerve diaphragm (Warnick *et al.*, 1975), cardiac muscle (Shotzberger *et al.*, 1976, and unpublished observations) and brain slices (Huang *et al.*, 1972) a 20 α-ester function appears critical to activity of this class of compounds.

8.4.4 Effects of batrachotoxin on neuromuscular transmission and the contraction mechanism

Batrachotoxin produced a complete and irreversible block of the indirectly and of the directly elicited twitch of diaphragm muscle, simultaneously with membrane depolarization and cessation of spontaneous and stimulus-evoked transmitter release (Warnick *et al.*, 1971, Figs. 8.8, 8.9). The complete block of the directly elicited muscle twitch occurred only when the membrane potential was decreased to levels

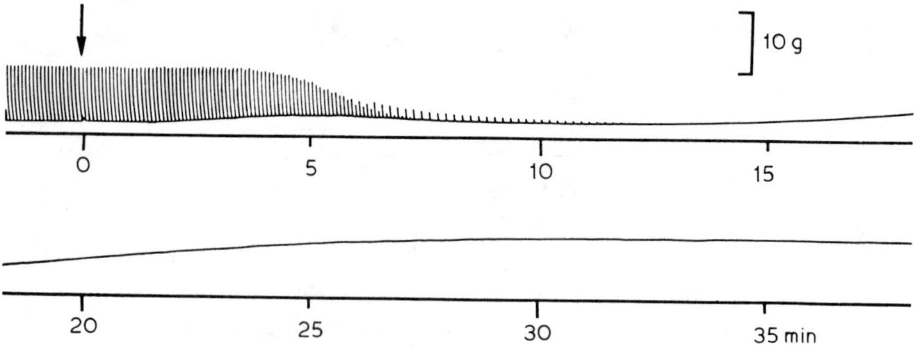

Fig. 8.8 Records showing the isometric twitch and contracture tension of rat diaphragm muscle in response to supramaximal direct and indirect stimulation in the presence of 20 nM batrachotoxin. Stimulus rate was 0.2 p.p.s. with direct and direct stimulation being applied alternatively. Response to indirect stimulation failed at 7 min, while the response to direct stimulation was detectable up to 12 min. From Warnick *et al.* (1971).

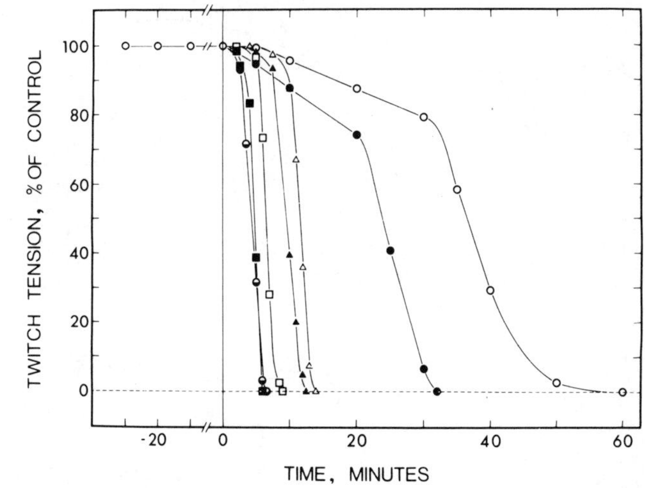

Fig. 8.9 Time course of batrachotoxin-induced blockade of neuromuscular transmission at various concentrations of the toxin. Each point is the average twitch tension of phrenic nerve-diaphragm preparations in three different experiments. (○), 2 nM; (●), 5 nM; (△), 10 nM; (▲), 20 nM; (□), 40 nM; (■), 75 nM; (◓), 150 nM. From Warnick *et al.*, (1971).

below −50 mV. In addition to its action on neuromuscular transmission, batrachotoxin caused a contraction of rat diaphragm muscle with two distinct phases in time (Warnick *et al.*, 1971). The first phase of contraction which occurred simultaneously with membrane depolarization was phasic in nature, i.e., tension was not maintained. After the decline in muscle tension, a later prolonged phase of muscle contracture was observed which was only partially correlated with the decrease in membrane potential. The first phase of contraction could be abolished by pre-treating the diaphragm with tetrodotoxin or a sodium-deficient physiological solution. Batrachotoxin caused a marked swelling and disruption of the terminal cisternae and sarcoplasmic reticulum, indicating that penetration of large amounts of sodium into the sarcoplasmic reticulum osmotically causes mechanical disruption of the organelles and release of calcium and other components. Thus, drugs such as caffeine which cause contraction by releasing calcium from the sarcoplasmic reticulum were ineffective when the preparation has been pretreated with batrachotoxin (Fig. 8.10). It is evident that the action of batrachotoxin on the contractile mechanism is due to a direct effect on sodium permeability which in turn affects the calcium stores located in the sarcoplasmic reticulum and terminal cisternae.

8.4.5 Interaction of batrachotoxin with cholinergic agonists at muscle endplates

Batrachotoxin had little or no effect on miniature endplate potentials or on transient potentials elicited by microiontophoretic application of acetylcholine at muscle

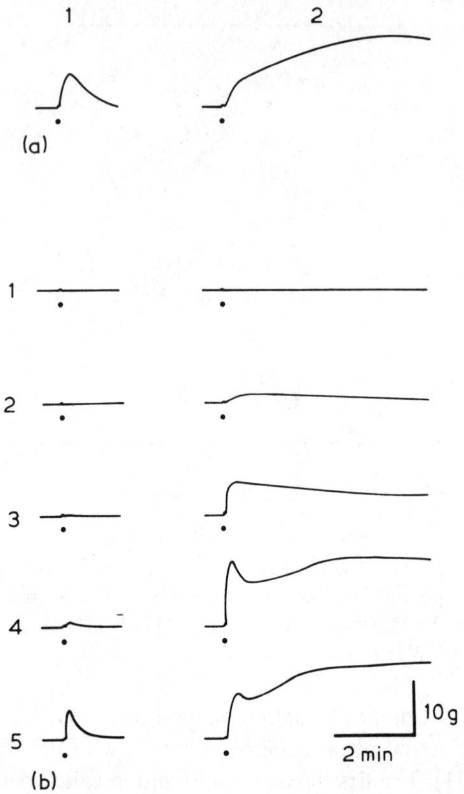

Fig. 8.10 Effects of different concentrations of batrachotoxin on muscle contracture in response to isotonic potassium chloride (1) and caffeine (2; 20 mM). (a) Response of rat diaphragm muscle before adding batrachotoxin; (b) After the muscle strip had been exposed to batrachotoxin for 2 h. In (b) 1–5 the concentrations of BTX used were 40 nM, 20 nM, 10 nM, 5 nM, and 2 nM, respectively. From Warnick *et al.* (1971).

endplates (Albuquerque *et al.*, 1971c; Warnick *et al.*, 1971; Hogan and Albuquerque, 1971). In fact, even when batrachotoxin was used in concentrations as high as 1.8 μM no protection against the irreversible binding of α-bungarotoxin to the recognition site for acetylcholine (Barnard *et al.*, 1975; Lee, 1972) was obtained (Albuquerque, unpublished observations, Garrison *et al.*, 1976). Thus, batrachotoxin does not appear to block cholinergic receptors. However, in muscle endplates, depolarized by batrachotoxin and repolarized by tetrodotoxin, 0.3 μM carbamylcholine no longer caused a pronounced depolarization of the endplate (Garrison *et al.*, 1976). Tetrodotoxin itself did not prevent carbamylcholine-elicited depolarizations in control preparations. Thus, batrachotoxin either blocks the coupling of activation of the acetylcholine-sensitive channel to an activation of endplate sodium channels, or converts these sodium channels from a tetrodotoxin-insensitive to a tetrodotoxin-sensitive state.

8.4.6 Effects of batrachotoxin and analogs on cardiac function

Batrachotoxin, one of the most potent cardiotoxins known, causes multifocal ventricular ectopic systoles, ventricular tachycardia and ventricular fibrillation (Kayaalp *et al.*, 1970). In isolated Purkinje fibers, batrachotoxin (2–10 nM)

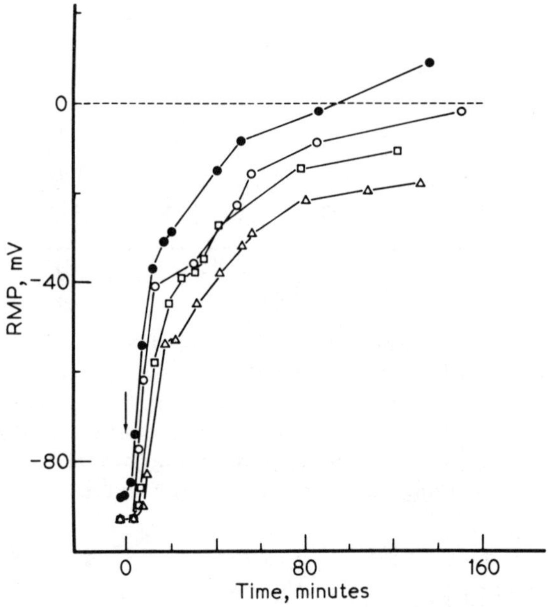

Fig. 8.11 Effect of different concentrations of batrachotoxin on the resting membrane potential (RMP) of Purkinje fibers of the dog heart. Batrachotoxin was added at zero time (↑) at the following concentrations: (●–●), 10 nM; (○–○), 6 nM; (□–□), 4 nM; (△–△), 2 nM. Each point represents the mean of at least 3 heart Purkinje fibers. From Hogan and Albuquerque (1971).

decreases the maximum diastolic membrane potential and increases the rate of depolarization that occurs during the diastolic interval, effects which result in the development of rapid spontaneous rhythms (Hogan and Albuquerque, 1971). As in other preparations depolarization by batrachotoxin in Purkinje fibers is concentration- and time-dependent (Fig. 8.11). Application of tetrodotoxin or low sodium physiological solution to the Purkinje fibers previously depolarized by batrachotoxin restores the membrane potential to control levels (Fig. 8.12), suggesting that the action of batrachotoxin is dependent upon a selective increase in the resting sodium conductance of the cell membrane. An augmentation of contractile force has also been observed when the Purkinje fiber preparation is exposed to batrachotoxin.

Exposure of cat papillary muscle to 2 nM batrachotoxin for 30 min increases the

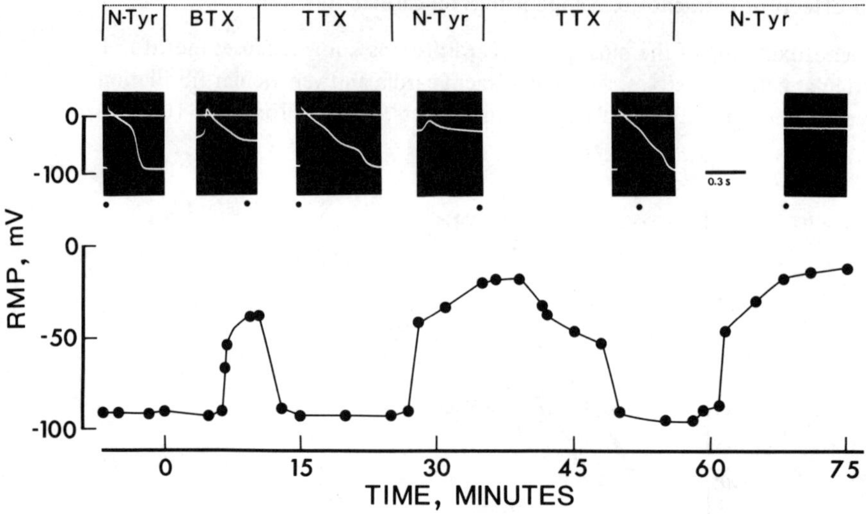

Fig. 8.12 Effect of batrachotoxin (2 nM, BTX) on resting membrane potential (RMP) of Purkinje heart fibers of the dog and reversal by tetrodotoxin (4.7 μM, TTX). Note the persistent effect of batrachotoxin when the Purkinje fiber is washed with normal Tyrode solution (N-Tyr). Representative action potentials are shown in the inserts. The dot under each insert indicates the time on the abscissa at which the recording was made. From Hogan and Albuquerque (1971).

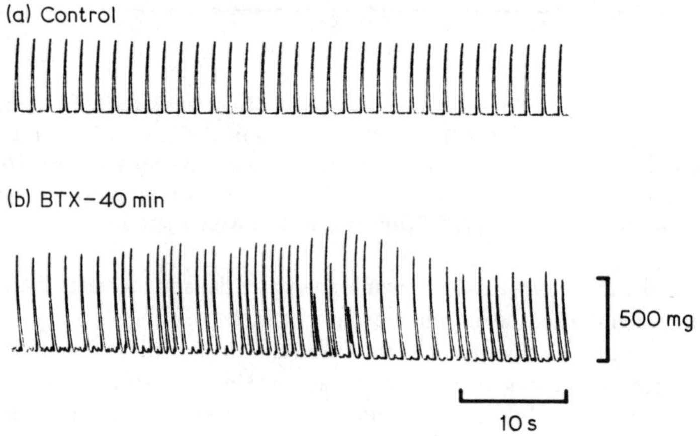

Fig. 8.13 Oscillographic recordings of the response of heart papillary muscles to stimulation at 40 p.p.m. in control (a) and during exposure to batrachotoxin (BTX, 2 nM) (b) for 40 min. From Shotzberger *et al.* (1976).

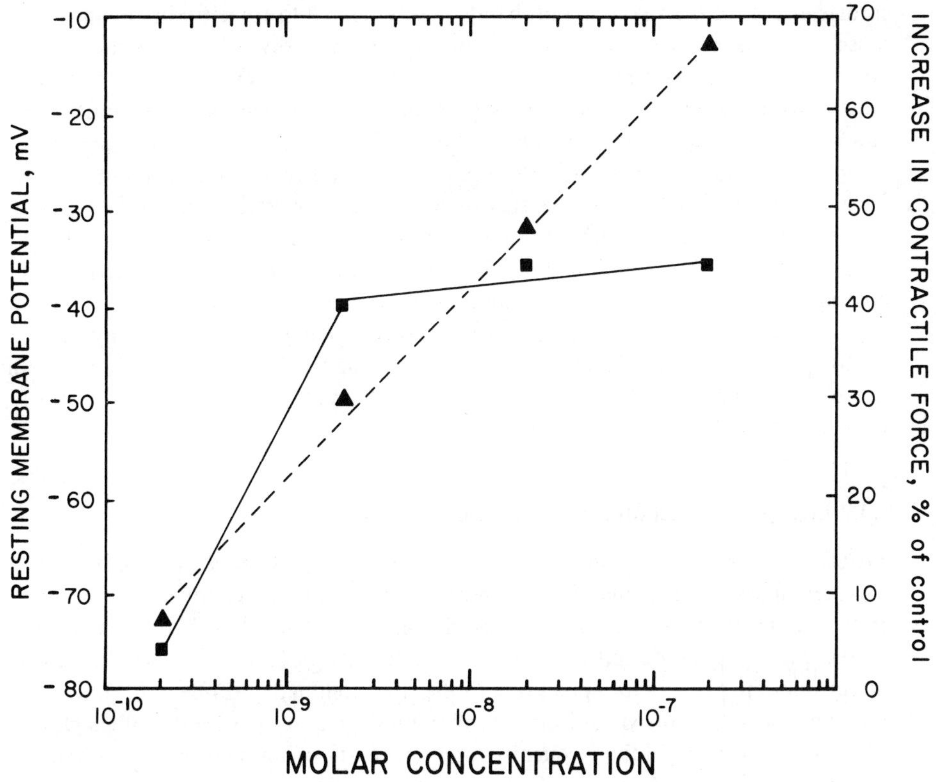

Fig. 8.14 Force of contraction (■—■) and resting membrane potential (▲—▲) of heart papillary muscles of the cat in the presence of different concentrations of batrachotoxin. Each point is the mean obtained from three to five preparations. Resting membrane potentials prior to addition of BTX (2 nM) were approximately −80 mV. Force of contraction was measured prior to any spontaneous activity induced by the toxin. From Shotzberger *et al.* (1976).

isometric force of contraction by approximately 50% of the initial force, decreases the potential difference across the cell membrane to approximately −50 mV and produces spontaneous contraction of the papillary muscle (Shotzberger *et al.*, 1976) (Figs. 8.13, 8.14). The positive inotropic effect of batrachotoxin is accompanied by an increase in the rate of force development. Tetrodotoxin antagonizes the effect of batrachotoxin, which again suggests that the actions of batrachotoxin are mediated by a selective increase in membrane permeability to sodium ions. The resultant increase in intracellular sodium ion concentration may increase the force of contraction through an augmentation of calcium influx *via* the sodium-calcium exchange system.

Several analogs of batrachotoxin have been compared with regard to toxicity and cardiac activity (Shotzberger *et al.*, 1976). Analogs with the highest toxicity *in vivo* also had the greatest cardiac effects in isolated papillary muscles, indicating that the cardiac effects of the toxins are probably closely related to toxicity. In papillary muscles from excised cat hearts, batrachotoxin and '5-methyl-BTX' were the most potent compounds with regard to production of membrane depolarization, positive inotropism and spontaneous activity. 'Quaternary BTX', the 4, 5-dimethyl analog of batrachotoxin, '5-acetyl-BTX' and batrachotoxinin-A were the least active of the compounds studied. The results indicated that there are two structural requirements for high toxicity of batrachotoxin analogs: (1) the 20α-pyrrole-3-carboxylate moiety with methyl substituents at the 2 and 4 positions, and (2) the 3α, 9α-hemiketal linkage. Recently, however, the 20α-benzoate of batrachotoxinin-A has been found possess equal toxicity to batrachotoxin in mice (unpublished results).

Batrachotoxin at a concentration of 140 nM completely abolished the rhythmic contractures of cultured muscle cells from embryonic chick heart within about 10 min (Roseen and Fuhrman, 1971). Prior to cessation of activity cells began to contract at extremely rapid and irregular rates.

8.4.7 Effects of batrachotoxin on ganglionic function

Batrachotoxin is a potent ganglionic blocking agent (Kayaalp *et al.*, 1970). Its effects appeared linked to blockade of preganglionic nerve conduction although a blockade of neurotransmitter release was probably also involved. Batrachotoxin appeared to depress the postsynaptic spikes due to fast conducting preganglionic fibers to a greater extent than the presynaptic spikes due to the slow conducting C fibers. In some experiments, batrachotoxin appeared to directly depress responsiveness of ganglion cells. Further studies on the mechanism and selectivity of action of batrachotoxin and batrachotoxin-analogs in ganglionic preparations are clearly needed.

8.4.8 Reactivity of the electroplax to batrachotoxin

Batrachotoxin has no effect on resting membrane potential in electroplax preparations from the electric eel, *Electrophorus electricus* (Bartels de Bernal *et al.*, 1975). However, the duration of sequentially evoked action potentials in this preparation increases markedly in the presence of batrachotoxin. High rates of electrical stimulation result with batrachotoxin-treated preparations in a complete depolarization which is sustained even during washing and subsequent treatment with *d*-tubocurarine. Tetrodotoxin at concentrations which have no effect on action potentials in electroplax causes a complete repolarization of the batrachotoxin-depolarized preparation. Local anesthetics such as tetracaine, diphenhydramine, lidocaine and procaine cause repolarization of batrachotoxin-depolarized preparations. The effect of batrachotoxin appeared virtually irreversible under all conditions.

In a batrachotoxin-treated electroplax preparation, previously depolarized by stimulation and then repolarized by tetrodotoxin, the partial depolarization normally elicited by 30 mM potassium ions or 20 μM carbamylcholine resulted in a complete and sustained depolarization similar to that which would have been elicited by electrical stimulation in batrachotoxin-treated electroplax (E. Bartels de Bernal and T. Rosenberry, personal communication). Batrachotoxin had no effect on the response to carbamylcholine unless, as was proposed, the toxin had gained access to binding sites regulating the action potential channel during a prior electrical stimulation. Repetitive treatments of electroplax with batrachotoxin and carbamylcholine did finally lead to a depolarization to −50 mV which was sustained in normal Ringers and reversed only by tetrodotoxin. Thus, repetitive application of carbamylcholine apparently resulted in formation of sufficient batrachotoxin-activated sodium channels to sustain a partial depolarization.

8.4.9 Effects of batrachotoxin on brain slices and brain synaptosomes

Depolarization of neurons in the central nervous system is undoubtedly elicited by batrachotoxin, but has not been studied directly. In brain slices, batrachotoxin, in common with other depolarizing conditions such as electrical stimulation, veratridine, ouabain and high concentrations of potassium ions, elicits an increase in levels of cyclic AMP (Shimizu and Daly, 1972; Huang *et al.*, 1972, 1973). The mechanism by which membrane depolarization enhances accumulations of cyclic AMP in this preparation is as yet uncertain, although various evidence indicates that it may be mediated by adenosine, a stimulant of brain cyclic AMP-generating systems, which accumulates extracellularly as a result of depolarization. This stimulatory effect of batrachotoxin on levels of cyclic AMP was blocked by tetrodotoxin. Analogs of batrachotoxin such as '5-methyl BTX' and batrachotoxinin-A *p*-bromobenzoate elicited accumulations of cyclic AMP. Batrachotoxinin-A even at very high concentrations had no apparent agonist or antagonist activity. The action of batrachotoxin in brain slices was relatively irreversible, while the action of the *p*-bromobenzoate analog appeared reversible.

With brain synaptosomal preparations, batrachotoxin and veratridine markedly inhibited uptake of radioactive dopamine (Holz and Coyle, 1974). Tetrodotoxin antagonized the inhibitory action of batrachotoxin and veratridine in synaptosomal preparations. The mechanism underlying inhibition of uptake of dopamine by batrachotoxin and veratridine was not defined, but clearly was linked to the effects of these alkaloids on sodium permeability.

8.4.10 Blockade of axoplasmic transport by batrachotoxin

Batrachotoxin has been shown to irreversibly block fast axoplasmic transport in cat sciatic nerve (Ochs and Worth, 1975). Its potency in this regard is some 5000 to 10 000-fold greater than the *Vinca* alkaloids, vinblastine and vincristine. The mechanism involved in blockade of axoplasmic flow by batrachotoxin has not been resolved. Tetrodotoxin prevented the effect of batrachotoxin but sodium-deficient medium did not. Thus, it remained uncertain as to whether the blockade of axoplasmic flow by batrachotoxin was due to its well-known depolarizing action. High concentrations of extracellular calcium ions also prevented the effect of batrachotoxin on axoplasmic flow.

The blockade of axoplasmic flow and hence ultimate suppression of release of putative neurotrophic factors by batrachotoxin has been utilized to study the mechanism whereby inactivity of motor neurons affects muscles. Repeated application of batrachotoxin (20 ng in 20 μl) into the subarachnoid space of the spinal cord of female rats at the level of the lumbar enlargement (L2-L4) resulted in marked physiological alteration of the hind limb muscles (Albuquerque *et al.*, 1975; Garcia *et al.*, 1975; Pence *et al.*, 1976). Batrachotoxin caused a flaccid paralysis a few minutes after injection, suggesting a depolarization of the spinal motoneuron. The animal remained paralyzed for about 4 days and on the third day after injection, animals were reinjected to maintain the paralysis. The axonal damage, e.g., a marked increase in neurofilaments, occurred within 10 h after injection of batrachotoxin, suggesting that the toxin will block axoplasmic flow within a few hours after injection into the subarachnoid space. Initial studies (Albuquerque *et al.*, 1975; Garcia *et al.*, 1975; Pence *et al.*, 1976) using light and electron microscopy revealed marked changes in neuronal perikarya, including central chromatolysis (10 days), peripheral chromatolysis and cytoplasmic shrinkage. Satellitosis and perineuronal vacuolation were also present in the spinal cord. The first electrophysiological alteration which occurred in the muscles of the paralyzed limbs was membrane depolarization. Twenty-four hours after batrachotoxin injection, the membranes of extensor muscles were depolarized by about 6 mV. In contrast to the extensor muscle, the membrane potentials of soleus muscles were not different from control values for at least 20 days. The delayed effect in the soleus muscles is not due to a lack of effect of batrachotoxin-injections on the soleus motoneurons, since histological and ultrastructural studies have shown that signs of degeneration such as swelling and disappearance of microtubules and a marked increase in neurofilaments were prominent as early as 10 h after injection of batrachotoxin in most of the motoneurons located at the L2-L4 region of the spinal cord. In both extensor and soleus muscles an increase in miniature endplate potential amplitude and half-decay time was noted. The miniature endplate potential frequency increased only slightly. It is possible that the block of axoplasmic flow by batrachotoxin influences a trophic factor controlling acetylcholinesterase, which could account for these alterations. Indirect action potentials could be evoked in most of the analyzed muscles. The rate of rise of the spikes was, however, slower than normal. Threshold, amplitude and rate of rise of the direct elicited action potentials were little affected. Tetrodotoxin-resistant action potentials developed in extensor muscles 5–9 days after batrachotoxin injection, while the soleus remained highly sensitive to the toxin. Extrajunctional acetylcholine sensitivity developed similarly in the extensor and soleus muscles 10 days after injection of batrachotoxin. Continuous direct stimulation of extensor and soleus muscles of rats injected with batrachotoxin reduced extrajunctional acetylcholine sensitivity and increased sensitivity to tetrodotoxin. The membrane potential and miniature endplate potential amplitudes, however, remained similar to those of the non-stimulated contralateral muscles. These findings suggested that direct stimulation cannot effectively reverse the alterations of electrophysiological properties produced by batrachotoxin-blockade of axoplasmic flow in the motor neuron at the level of the spinal cord.

8.4.11 Sites of interaction of batrachotoxin with electrogenic membranes

The nature of the macromolecules with which batrachotoxin interacts to elicit an increase in sodium ion permeability in electrogenic membranes have not yet been defined. However, with the successful preparation of radioactive benzoate of batrachotoxinin-A, binding and isolation studies have been initiated in our laboratories. Biochemical and electrophysiological observations did provide several important clues for the identification of the macromolecular compounds with which batrachotoxin reacts. For example, the depolarizing action of batrachotoxin in lobster giant axons can be blocked by pretreatment with sulfhydryl blocking agents such as *p*-chloromercuribenzene sulphonic acid or N-athylmaleimide or by sulfhydryl reducing agents such as reduced glutathione or dithiothreitol (Fig. 8.15, 8.16). The pretreatments did not prevent generation of action potentials and tetrodotoxin was still effective in blocking action potentials. The prevention by the chemical pretreatments of the depolarization normally elicited by batrachotoxin suggested that certain sulfhydryl groups and/or disulfide bonds are important to the action of batrachotoxin. Certainly, such results are indicative of the interaction of batrachotoxin with protein components associated with the sodium channel. Further support to the importance of proteins to the control of ion channels comes from the observations that proteolytic enzymes such as pronase will produce inactivation of sodium channels when applied internally in axons [see recent review by Armstrong (1974)].

Tetradotoxin antagonizes the effects of batrachotoxin and batrachotoxin analogs in many preparations. The antagonistic effect of tetrodotoxin is rapidly reversible while the effect of batrachotoxin and analogs is relatively irreversible (Fig. 8.17, 8.18). However, the antagonism between batrachotoxin and tetrodotoxin should not be construed to indicate competition for a common recognition site. Indeed, a variety of data indicates that batrachotoxin and tetrodotoxin interact with different sites associated with the same sodium channel:

(a) Batrachotoxin has no effect on the binding of tritiated tetrodotoxin to rabbit vagus nerve (Colquhoun *et al.*, 1972).

(b) The effect of batrachotoxin could be blocked by sulfhydryl reagents or reducing agents in axons without affecting tetrodotoxin-blockade of action potentials (Albuquerque *et al.*, 1971b).

(c) Batrachotoxin is an effective depolarizing agent in chronically denervated muscle, while tetrodotoxin has no effect either on potentials or batrachotoxin-elicited depolarization in denervated muscle (Figs. 8.19, 8.20).

(d) Tetrodotoxin blocks action potentials in nerve muscle preparations from the poison dart frog, while batrachotoxin has little or no effect in this preparation (Albuquerque *et al.*, 1973).

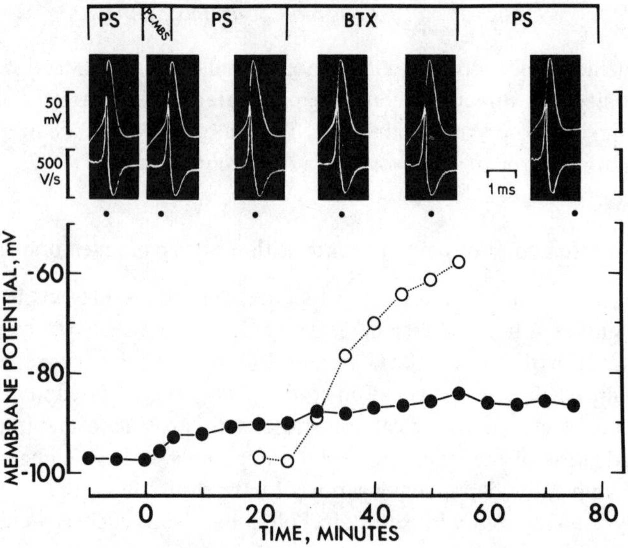

Fig. 8.15 Membrane and action potentials of lobster giant axons in control and during exposure to batrachotoxin (500 nM, BTX). One preparation (○–○) was pretreated only with physiological solution (PS); another (●–●) was pretreated with *p*-chloromercuribenzene sulphonic acid (1 mM, PCMBS). The dot under each insert is the time on the abscissa at which the record was made. From Albuquerque *et al.* (1971b).

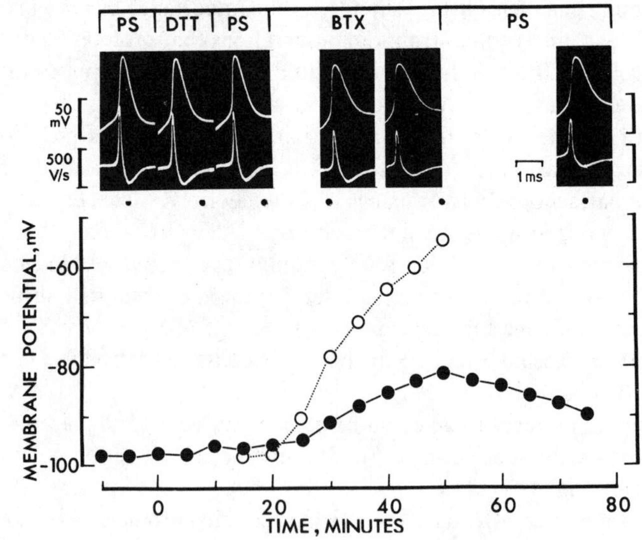

Fig. 8.16 Membrane and action potentials of lobster giant axons in control and during exposure to batrachotoxin (500 nM, BTX). One preparation (○–○) was pretreated only with physiological solution (PS); another (●–●) was pretreated with dithiothreitol (20 mM, DTT). The dot under each insert is the time on the abscissa at which the record was made. From Albuquerque *et al.* (1971b).

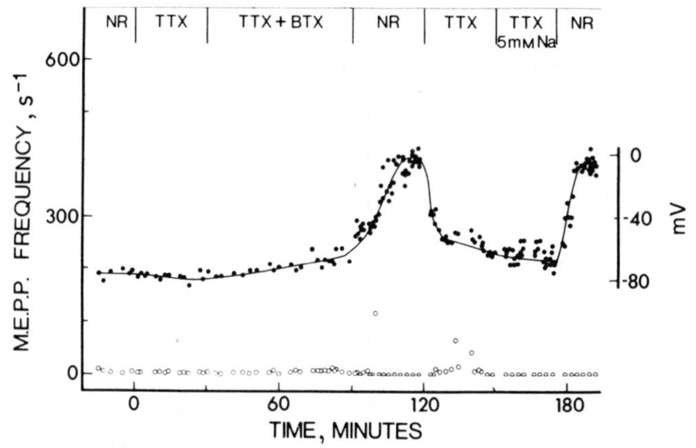

Fig. 8.17 Recording of resting membrane potential (RMP; ●) and spontaneous miniature endplate potential (m.e.p.p.) frequency (○) at endplate regions of rat diaphragm muscles in the presence of tetrodotoxin (3. 13 μM, TTX), TTX and batrachotoxin (150 nM, BTX) and TTX plus 5 mM Na. Note that pretreatment with TTX prior to exposure to the combination of BTX and TTX prevented the appearance of BTX effects on the RMP and spontaneous m.e.p.p. frequency, however, upon washing with normal Ringer's solution (NR), the effects of BTX were apparent. Half circles (◓) indicate that no spontaneous m.e.p.p.s could be detected. From Albuquerque *et al.* (1971c).

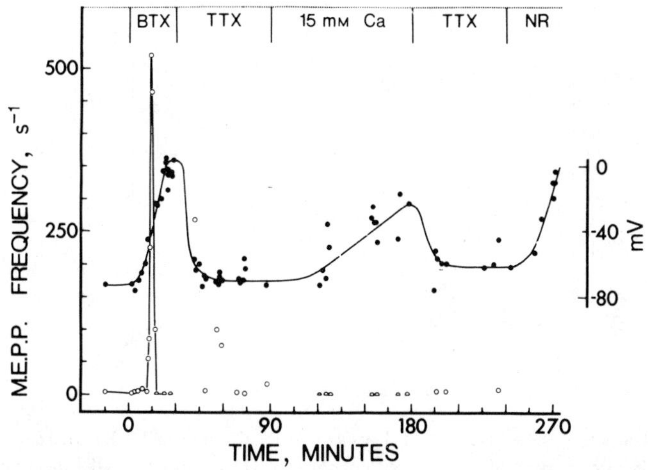

Fig. 8.18 Effect of batrachotoxin (150 nM) on resting membrane potential (RMP, ●) and spontaneous miniature endplate potential (m.e.p.p.) frequency (○) of rat diaphragm muscle. Repolarization of the muscle membrane by tetrodotoxin (3.13 μM, TTX), followed by a gradual depolarization in the presence of physiological solution containing 15 mM calcium chloride (Ca) and repolarization with TTX are shown. Note the appearance of the irreversible effects of BTX with the final wash with normal Ringer's solution (NR). Half circles (◓) indicate that no spontaneous transmitter release is detected. From Albuquerque *et al.* (1971c).

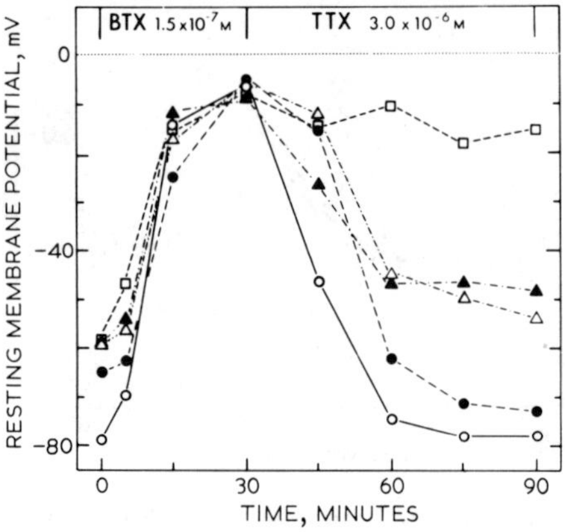

Fig. 8.19 Reversal by tetrodotoxin (TTX) of the batrachotoxin (BTX) induced depolarization of extensor digitorum longus muscles of the rat at different times after denervation. Innervated (○), 1-day (●), 2-day (△), 5-day (▲) and 10-day (□) denervated extensor digitorum longus muscles. Each point represents the mean obtained from 15–25 surface fibers in three muscles. Temperature, 28°C. From Albuquerque and Warnick (1972).

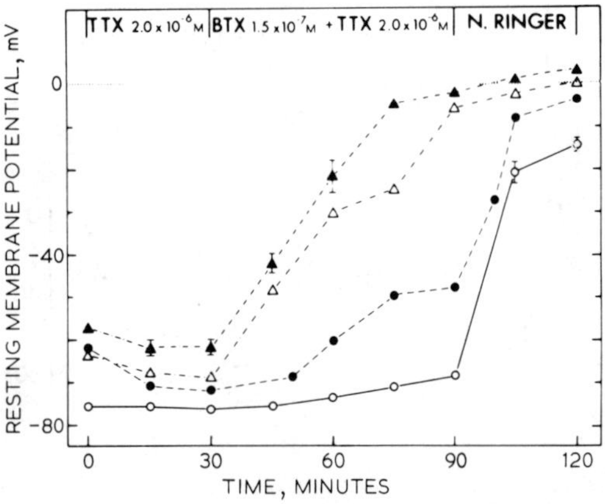

Fig. 8.20 Effect of batrachotoxin (BTX) and tetrodotoxin (TTX) on the resting membrane potential of control (innervated) and chronically denervated extensor digitorum longus muscles of the rat. Preparations were first exposed to TTX for 30 min, and then sufficient BTX was added to the TTX solution to make the concentration of BTX 150 nM. After one hour exposure of the muscles to both toxins, the preparations were washed for 30 min with normal Ringer's (N. Ringer) solution. Each symbol represents the mean value ± S.E.M. of 15–25 fibers obtained from at least three muscles in control innervated (○), 1-day (●), 5-day (△) and 10-day (▲) denervated extensor muscles. Where no S.E.M. appears the value was too small to be shown. Temperature, 28°C. From Albuquerque and Warnick (1972).

(e) The antagonism by tetrodotoxin of both batrachotoxin and veratridine-elicited increased in sodium conductances in neuroblastoma cells is noncompetitive (Catterall, 1975a).

Thus, the sites of interaction of batrachotoxin and tetrodotoxin with the sodium channel are clearly independent and distinct. In diaphragm muscle, the antagonism of batrachotoxin-elicited membrane depolarization by tetrodotoxin has been studied in detail (Albuquerque *et al.*, 1976). At low concentrations tetrodotoxin antagonized the dose-dependent membrane depolarization of fibers induced by batrachotoxin without altering its maximal response (Fig. 8.21). Only at 30-fold higher concentrations of tetrodotoxin was the maximal membrane depolarization elicited by batrachotoxin apparently reduced in magnitude. Because of the time dependency of batrachotoxin-elicited depolarizations and other factors, no quantitative analysis of competitive or noncompetitive interactions of batrachotoxin and tetrodotoxin in such a paradigm is possible. A theoretical analysis (Albuquerque *et al.*, 1976) of the interaction of the irreversible agonist (batrachotoxin) with sodium channels and the influence of a noncompetitive reversible antagonist (tetrodotoxin) is, however, possible given several basic assumptions. First, one can assume that the rate of binding of the irreversible agonist is directly proportional to its concentration A; i.e.,

$$N = N_m \, (1 - e^{-kAt})^*.$$

Secondly, one assumes that the equivalent circuit shown in Fig. 8. 22a adequately describes the steady-state properties of the membrane and that R and E_{Na} are materially unaltered as a consequence of the action of batrachotoxin. In fact, it is clear that any substantial increase in sodium permeability will result in a net inward movement of sodium ions and a consequent displacement of E_{Na}. It is also known that the membrane potential of batrachotoxin-treated muscles is fully restored on removal of sodium from the bathing solution (Albuquerque *et al.*, 1971; Albuquerque and Warnick, 1972; Hogan and Albuquerque, 1971). Thus, E_{Na} remains essentially constant and there is probably no secondary distribution of other ions. Lastly, tetrodotoxin is assumed to bind reversibly in blocking the sodium channel at a site other than that which binds batrachotoxin and no interaction between these sites occurs.

Based on these assumptions and the equivalent circuit of Fig. 8.22a if $E = E_m$ when $N = N_m$, then

$$\frac{E - E_0}{E_m - E} = (e^{kAt} - 1)\,(1 + N_m gR) \qquad (8.1)$$

and when $[E - E_0/E_m - E]$ is plotted against log batrachotoxin concentration yields a Hill plot (Fig. 8. 22b). If the number of sodium channels blocked by a given concentration of tetrodotoxin is N_B and if tetrodotoxin does not act at the same site

* (For definition of symbols see Abbreviations).

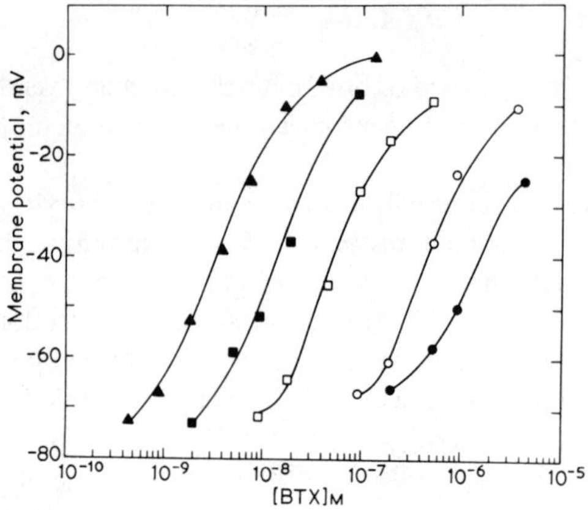

Fig. 8.21 Effect of tetrodotoxin on membrane depolarization induced by batrachotoxin (BTX) in surface fibers of rat diaphragm muscles. Each point is the membrane potential recorded 55–60 min after exposure to BTX at the concentration indicated on the abscissa plus the indicated concentration of TTX. Muscles were first treated with the TTX alone for 30 min. (▲) TTX-free; (■) 0.1 μM TTX; (□) 0.3 μM TTX; (○) 1.0 μM TTX; (●) 3.0 μM TTX. Standard errors of the means were too small to be shown. From Albuquerque *et al.*, (1976).

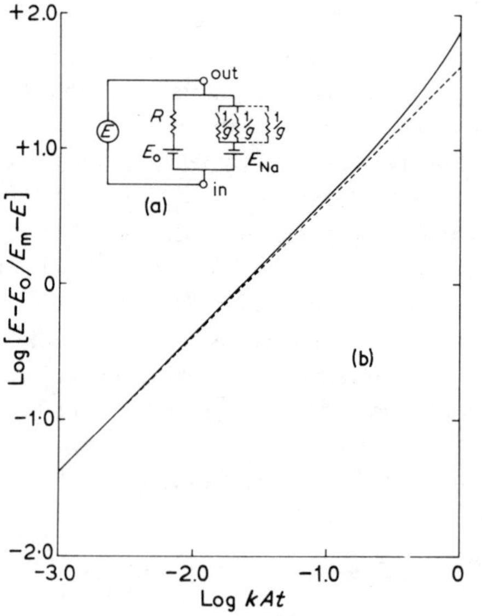

Fig. 8.22 Equivalent electrical circuit (a) of muscle membrane and theoretical Hill plot (b) of the data obtained from Equation 8.1. E, the steady-state membrane potential in the presence of concentration A of the agonist batrachotoxin; E_0, resting membrane potential in the absence of agonist; E_{Na}, equilibrium potential for Na^+; E_m, steady-state membrane potential at maximum depolarization by saturating concentrations of agonists; R, resting membrane resistance; g, mean conductance of a single sodium channel opened in the presence of agonist. The value 40 is assigned to $N_m gR$ (see Warnick *et al.*, 1976). Dashed line represents extrapolation of the linear portion of the theoretical curve (slope = 1.0).

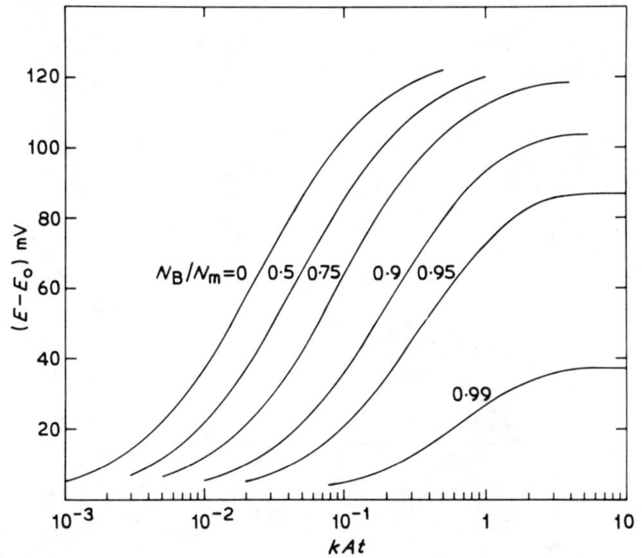

Fig. 8.23 Theoretical log dose-response curves obtained from Equation 8.2 for different values of N_B/N_m in the range of 0–0.99. N_m is the maximum number of sodium channels opened by saturating concentrations of agonist; N_B, number of sodium channels blocked in the presence of antagonist. E is the steady-state membrane potential in the presence of concentration A of the agonist batrachotoxin; E_0, resting membrane potential in the absence of agonist. A value of 40 is assigned to $N_m gR$ (see Warnick *et al.*, 1976). An arbitrary value of 130 mV is given to $(E_{Na} - E)$, based on approximate physiological parameters for mammalian skeletal muscle.

as batrachotoxin then

$$\frac{E_0 - E}{E_0 - E_{Na}} = \frac{N_m gR\,(1 - e^{-kAt})}{(1 - N_B/N_m)^{-1} + N_m gR\,(1 - e^{-kAt})} \tag{8.2}$$

and this equation when plotted for various values of H_B/N_m provides the curves in Fig. 8.23. These theoretical curves are quite similar to the experimental curves of Fig. 8.21. When $N_m gR$ is made equal to 40, corresponding to blockade of 90% of batrachotoxin-activated sodium channels the maximum response is depressed by only 20%.

Another result which follows from these equations concerns the ratio DR of the concentrations A_1 and A_2 of batrachotoxin that produce equal depolarization $(E_0 - E)$ in the presence and absence, respectively, of a given concentrations of TTX. It can readily be shown that

$$\frac{1 - e^{-kA_1 t}}{1 - e^{-kA_2 t}} = \left(1 - \frac{N_B}{N_m}\right) - 1\,. \tag{8.3}$$

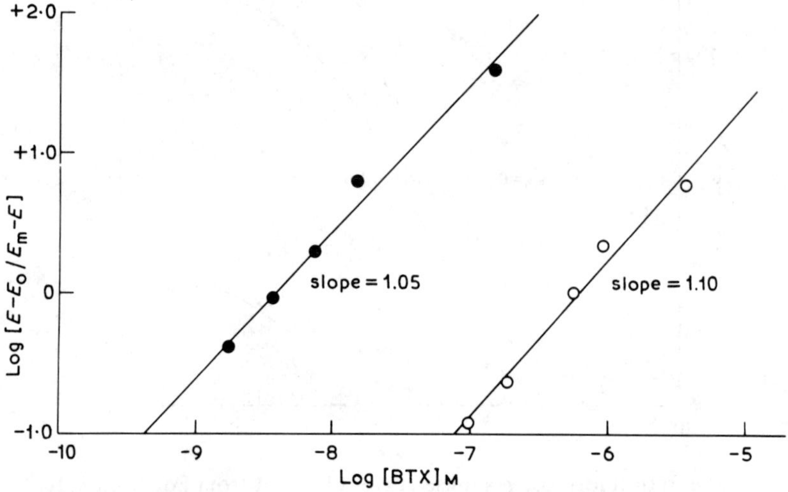

Fig. 8.24 Hill plots constructed from the data in Fig. 8.21 (●), batrachotoxin (BTX); (○), BTX plus tetrodotoxin (1.0 M). The correlation coefficient of both curves is 0.98. E is the mean membrane potential in the presence of toxins at the end of the 60-minute incubation period (see Fig. 8.21); E_0 is the membrane potential when no drug is present, and E_m is the membrane potential in the presence of saturating concentrations of BTX. From Albuquerque *et al.* (1976).

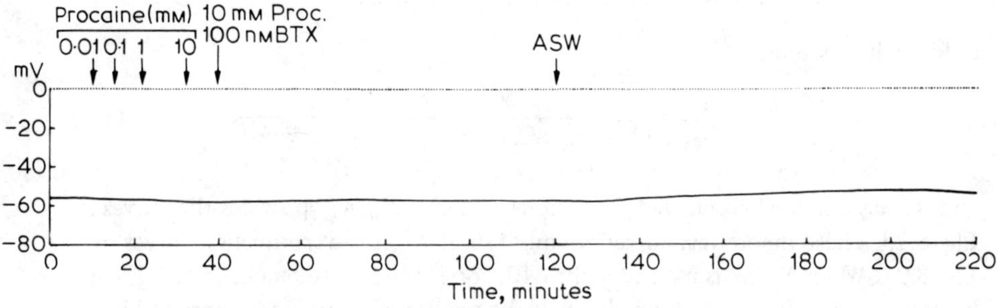

Fig. 8.25 Antagonism of batrachotoxin (BTX) action by procaine in the squid giant axon, Preparations were first exposed to increasing concentrations of procaine, then to procaine plus BTX and finally washed with artificial seawater (ASW). From Albuquerque *et al.* (1973a).

Applying mass action law to TTX binding gives

$$\frac{N_B}{N_m} = \frac{[TTX]}{K_B + [TTX]} \, . \tag{8.4}$$

From Equations 8.5 and 8.6,

$$\frac{1 - e^{-kA_1 t}}{1 - e^{-kA_2 t}} - 1 = \frac{[TTX]}{K_B} \, . \tag{8.5}$$

For small values of the concentration terms kAt, the left side of Equation 8.5 approaches ($DR - 1$), and the relationship is then indistinguishable from that derived by Gaddum *et al.*, (1955) on the basis of competitive antagonism. An estimate of the dissociation constant K_B for TTX can therefore be obtained by plotting log ($DR - 1$) against log [TTX] after the manner of Arunlakshana and Schild (1959). This plot should begin to deviate from linearity at high concentrations of TTX, but a linear relationship with unity slope obviously cannot be taken as evidence of competition in this case. This theoretical analysis of the experimental data of Fig. 8.21 was thus consonant with noncompetitive antagonism between an 'irreversible' (batrachotoxin) and reversible antagonist (tetrodotoxin) (Warnick *et al.*, 1975). Experimental Hill plots (Fig. 8.24) of some of the data of Fig. 8.21 revealed that cooperativity is probably not involved in the interaction of batrachotoxin with recognition sites on the sodium channel in the presence or absence of tetrodotoxin (Albuquerque *et al.*, 1976).

In addition to tetrodotoxin, a variety of local anesthetics have been found to antagonize or reverse the depolarizing action of batrachotoxin in different biological preparations. In the squid *Loligo paeli,* pretreatment with procaine followed by simultaneous exposure of the giant axon to batrachotoxin and procaine prevents the membrane depolarization which batrachotoxin would produce if applied alone (Albuquerque *et al.*, 1973a) (Fig. 8.25). If procaine is applied to axons while batrachotoxin depolarization is in progress, the membrane depolarization is halted, but subsequent washing never restores the membrane potential to control values (Fig. 8.26). Apparently procaine is able to prevent the access or binding of batrachotoxin to its recognition sites if the procaine is applied prior to and during exposure of the preparation to batrachotoxin. Procaine does not, however, appear to affect sodium channels already activated by batrachotoxin. Lidocaine, or the quaternary local anesthetic QX-572 were able to block sodium and potassium conductances but were ineffective in preventing the BTX-induced membrane depolarization (Albuquerque *et al.*, 1973a). These observations indicate that the effects of local anesthetics on normal ion channels and batrachotoxin-activated channels differ significantly and that the site of action of procaine in preventing batrachotoxin-induced membrane depolarization is probably different from the site at which procaine antagonizes increases in sodium conductances.

In diaphragm muscles of rats, procaine (1 mM) applied simultaneously with

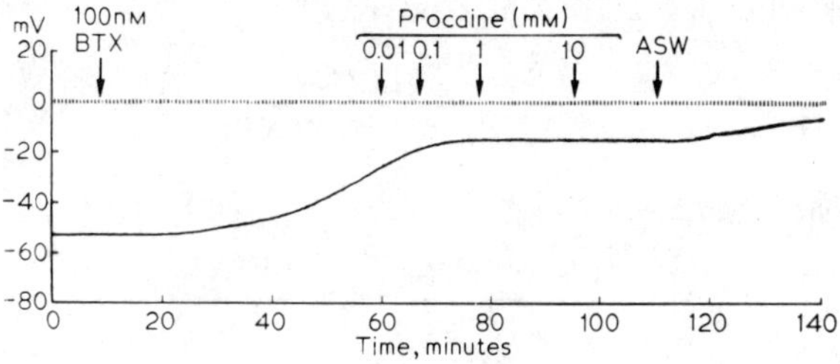

Fig. 8.26 Batrachotoxin (BTX) induced membrane depolarization of the squid giant axon. Procaine was added to the bathing solution at 60 minutes and the concentration of the local anesthetic increased stepwise to 10 mM by 100 min. Finally the axons were washed with artificial seawater (ASW). From Albuquerque *et al.* (1973a).

batrachotoxin (20 nM) prevents membrane depolarization (Fig. 8.27) and when applied during the course of batrachotoxin action the local anesthetic arrests and partially reverses the membrane depolarization in progress (Fig. 8.28). Although the local anesthetic itself blocked neuromuscular transmission, washing of the preparations after exposure to batrachotoxin and procaine resulted in the return of miniature endplate potentials. The effects of lidocaine in skeletal muscle of the rat were similar to those of procaine with respect to antagonism of the actions of batrachotoxin (Fig. 8.28). Thus, the actions of batrachotoxin on membrane potential and spontaneous transmitter release were no longer irreversible after an effective blockade by either lidocaine or procaine.

In electroplax preparations, the presence of tetracaine, procaine, lidocaine and diphenhydramine could repolarize preparations depolarized by batrachotoxin (Bartels de Bernal *et al.*, 1975). However, in contrast to results with muscle, the depolarizing action of batrachotoxin remains irreversible in the electroplax after treatment with local anesthetics.

In batrachotoxin-treated Ranvier node of frog sciatic nerve, procaine at 5×10^{-4} M completely blocked transient sodium conductances, while having little effect on steady state batrachotoxin-induced sodium conductances (Khodorov *et al.*, 1975). Thus, in Ranvier node as in squid axon (Albuquerque *et al.*, 1973a) procaine was relatively ineffective in blocking sodium channels activated by batrachotoxin. Only at very high concentrations did procaine, trimecaine and the quaternary local anesthetic QX-572, block batrachotoxin-induced sodium conductances in the Ranvier node (B. I. Khodorov, personal communication). The local anesthetic, benzocaine, had similar potency with regard to blockade of transient sodium conductances and batrachotoxin-induced conductances.

In toto, the results with squid axon, nerve-muscle and Ranvier node indicate that interactions of local anesthetics with normal and batrachotoxin-activated sodium channels differ markedly. In addition, it appears that the antagonistic interactions of local anesthetics with batrachotoxin vary with different preparations and in some

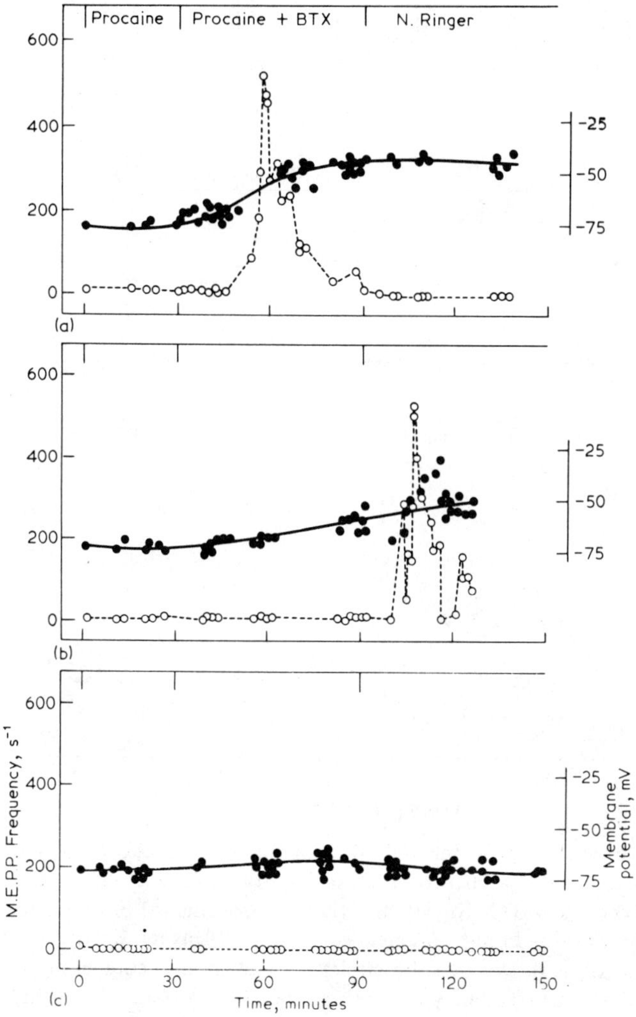

Fig. 8.27 Effect of procaine on the time course of batrachotoxin (BTX)-induced depolarization of the rat diaphragm muscle. The muscles were treated for 30 min with 0.1 mM procaine (a), 0.3 mM procaine (b) or 1.0 mM procaine (c), followed by procaine plus batrachotoxin (BTX, 10 nM) for 60 min, and then washed with physiological solution (N. Ringer). Membrane potential is shown by solid circles (●) and spontaneous miniature endplate potential (m.e.p.p.) frequency by open circles (○). Half-filled circles (◒) indicate no detectable spontaneous m.e.p.p.s. From Albuquerque *et al.* (1976).

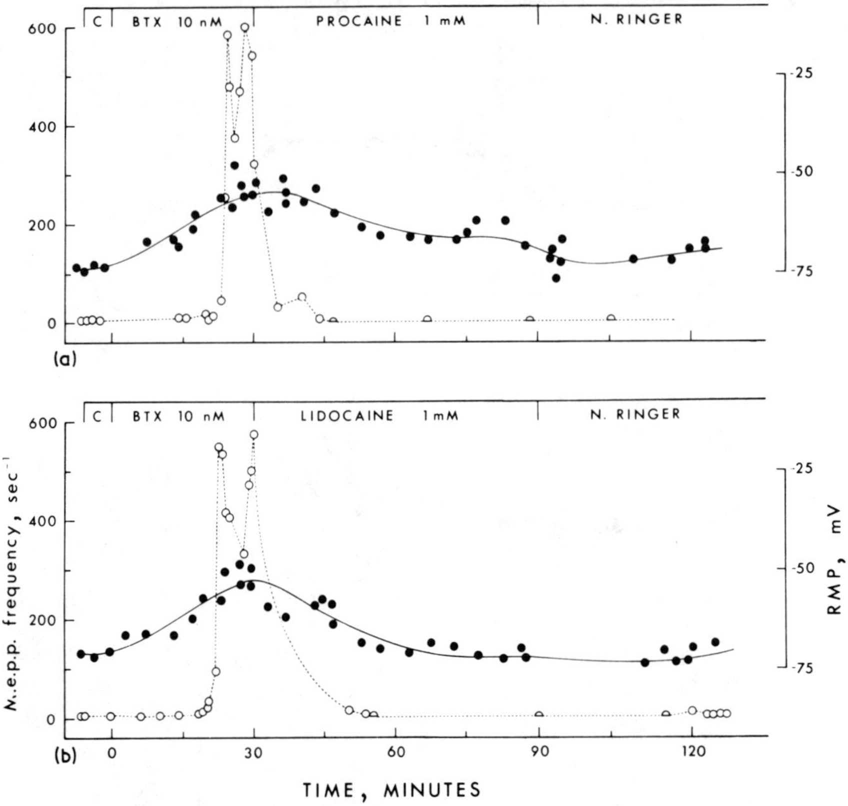

Fig. 8.28 Reversal by procaine and lidocaine of the batrachotoxin (BTX)-induced membrane depolarization in rat diaphragm muscles. The muscles were first exposed to BTX for 30 min, then to procaine (a) or lidocaine (b), followed by washing in physiological solution (N. Ringer). Resting membrane potential (RMP) is shown by solid circles (●) and spontaneous miniature endplate potential (m.e.p.p.) frequency by open circles (○). Half circles (◠) indicate no detectable spontaneous m.e.p.p.s. From Albuquerque *et al.* (1976).

preparations may reflect blockade of sodium channels activated by batrachotoxin and in other preparations a direct masking or alteration of batrachotoxin-binding sites. In diaphragm muscle, the antagonism of batrachotoxin-elicited membrane depolarization by procaine appeared on kinetic analysis noncompetitive in nature (Fig. 8.29), but such analysis cannot give insight into the nature of batrachotoxin-antagonist interactions in such a paradigm. Instead, the results indicate that procaine at the concentration used is not merely, like tetrodotoxin, blocking sodium channels activated by batrachotoxin, but is interacting with the system in a more complex manner. Further studies will be required to elucidate the mechanism involved in batrachotoxin – local anesthetic interactions in different biological preparations.

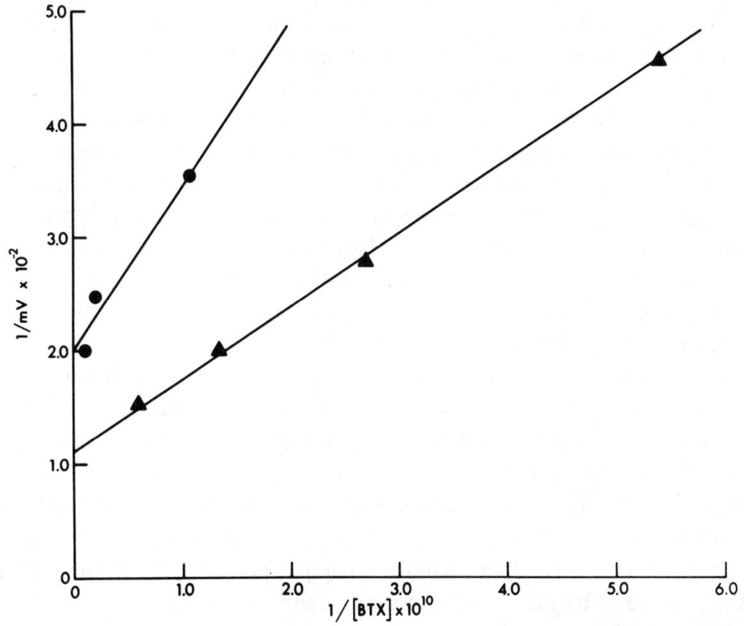

Fig. 8.29 Double-reciprocal plot of batrachotoxin (BTX)-induced membrane depolarization in the absence and presence of procaine. The data for BTX alone (▲) are the values obtained after 55–60 min and shown in Fig. 8.21. Closed circles (●) are for BTX plus procaine (0.25 mM). The values for BTX plus procaine were obtained from muscles exposed to procaine for 30 min and then to procaine plus BTX for 60 min. Recodings were made between 55 and 60 min. From Albuquerque *et al.* (1976).

In addition, possible differences in the selectivity and kinetics of normal sodium channels (Hille, 1972) and of batrachotoxin-activated channels should be investigated. Batrachotoxin-activated channels in Ranvier node appear to exhibit reduced selectivity towards ions (B.I. Khodorov, personal communication).

8.5 SUMMARY

After less than a decade of pharmacological studies, batrachotoxin has been firmly established as an important tool for the study of the mechanisms involved in activation of sodium channels in electrically excitable membranes. Many questions, however, remain unanswered and have had to await the preparation of a radioactive batrachotoxin derivative of high activity for investigation of binding at recognition sites associated with sodium channels. Such a derivative, batrachotoxinin-A

20α-[^{3}H] benzoate, has now been prepared and shown to bind to intact garfish olfactory nerve, nerve homogenates, and axonal membrane fractions (Sundermier *et al.*, 1976). Preliminary studies in our laboratory indicate that the density of specific batrachotoxin binding sites in the small fibers of the garfish nerves (Eldefrawi *et al.*, 1976, paper in preparation) is similar to that reported by Benzer and Raftery (1972) and Colquhoun *et al.*, (1972, 1974) for tetrodotoxin, i.e., 3.9–5.6 sites μm^2. Although these values may be subject to error due to unspecific binding of the toxins and difficulties in assessing extra-cellular space, they provide, nevertheless, very useful information regarding the density of sodium channels in a variety of preparations (Jaimovich *et al.*, 1976; Ritchie *et al.* 1976). Further studies of this type should provide answers to many questions: (i) the irreversibility of batrachotoxin-binding; (ii) the dependence of the binding of batrachotoxin on the transmembrane potential; (iii) the nature of antagonism of batrachotoxin action by tetrodotoxin, and by local anesthetics such as procaine, lidocaine, etc.; (iv) the interrelationship of veratridine and batrachotoxin-recognition sites; (v) the equivalency or lack thereof between sodium channels activated by batrachotoxin and sodium channels participating in action potential generation; (iv) the mechanisms by which batrachotoxin maintains the sodium channel in open conformation; and (vii) the feasibility of isolating the sodium channel. Such questions and the possible physiological significance of batrachotoxin-recognition sites to function of nerve and muscle remain challenges for the future.

REFERENCES

Albuquerque, E.X. (1972), *Fed. Proc.*, **31**, 1133-1138.

Albuquerque, E.X., Brookes, N., Onur, R. and Warnick, J.E. (1976), *Mol. Pharmacol.* **12**, 82.

Albuquerque, E.X., Daly, J.W. and Witkop, B. (1971a), *Science*, **172**, 995.

Albuquerque, E.X., Deshpande, S.S., Kauffman, F.C., Garcia, J. and Warnick, J.E. (1975), *Proceedings of the Sixth International Congress of Pharmacology*, Helsinki, Finland, Vol. 2, p. 185.

Albuquerque, E.X., Sasa, M., Avner, B.P. and Daly, J.W. (1971b), *Nature New Biol.*, **234**, 93.

Albuquerque, E.X., Sasa, M., and Sarvey, J.M. (1972), *Life Sci.*, **11**, 357.

Albuquerque, E.X., Seyama, I. and Narahashi, T. (1973a), *J. Pharmacol. exp. Therap.*, **184**, 308.

Albuquerque, E.X. and Warnick, J.E. (1972), *J. Pharmacol. exp. Therap.*, **180**, 638.

Albuquerque, E.X., Warnick, J.E. and Sansone, F.M. (1971c), *J. Pharmacol. exp. Therap.*, **176**, 511.

Albuquerque, E.X., Warnick, J.E., Sansone, F.M. and Daly, J. (1973b), *J. Pharmacol. exp. Therap.*, **184**, 315.

Alnaes, E. and Rahamimoff, R. (1975), *J. Physiol. (Lond.)*; **248**, 285.

Armstrong, C.M. (1974), *Quart. Rev. Biophys.*, **7**, 179.

Arunlakshana, O. and Schild, H.O. (1959), *Br. J. Pharmacol. Ther.*, **14**, 48.
Barnard, E.A., Dolly, J.O., Porter, C.W. and Albuquerque, E.X. (1975), *Exp. Neurol.*, **48**, 1.
Bartels de Bernal, E., Llana, M.A. and Diaz, E. (1975), *Acta Méd. Valle (Colombia)*, **6**, 74.
Benzer, T.I. and Raftery, M.A. (1972), *Proc. Nat. Acad. Sci. U.S.A.* **69**, 3634.
Catterall, W.A. (1975a), *J. Biol. Chem.*, **250**, 4053.
Catterall, W.A. (1975b), *Proc. Nat. Acad. Sci.*, **72**, 1782.
Catterall, W.A. (1976), *Biochem. Biophys. Res. Commun.*, (in press).
Catterall, W.A., Ray, R. and Morrow, C.S. (1976), *Proc. Nat. Acad. Sci.*, (in press).
Chapman, D. (1967), *In*: Thermobiology, ed. by A.H. Rose, pp. 123-146, Academic Press, New York.
Colquhoun, D., Henderson, R. and Ritchie, J.M. (1972), *J. Physiol. (Lond.)*, **227**, 95.
Colquhoun, D., Rang, H.P. and Ritchie, J.M. (1974), *J. Physiol. (Lond.)*, **240**, 199.
Daly, J., Albuquerque, E.X., Kauffman, F.C. and Oesch, F. (1972), *J. Neurochem.*, **19**, 2829.
Daly, J.W., Witkop, B., Bommer, D. and Biemann, K. (1965), *J. Am. Chem. Soc.*, **87**, 124.
Gaddum, J.H., Hameed, K.A., Hathway, D.E., and Stephens, F.F. (1955), *Q. J. Exp. Physiol. Cogn. Med. Sci.*, **40**, 49.
Garcia, J.H., Pence, R.S., Albuquerque, E.X., Deshpande, S. and Warnick, J.E. (1975), *Abstracts of the Sixth International Congress of Pharmacology*, Helsinki, Finland, p. 36.
Garrison, D.L., Albuquerque, E.X., Warnick, J.E., Daly, J. and Witkop, B. (1976), *Neurosci., Soc.*, (abstract, in press).
Gilardi, R.D. (1970), *Acta Cryst.*, **B26**, 440.
Gössinger, E., Graf, W., Imhof, R. and Wehrli, H. (1971), *Helv. Chim. Acta*, **54**, 2785.
Graf, W., Gössinger, E., Imhof, R. and Wehrli, H. (1971), *Helv. Chim. Acta*, **54**, 2789.
Hille, B. (1972), *J. Gen. Physiol.*, **59**, 637.
Hogan, P.M. and Albuquerque, E.X. (1971), *J. Pharmacol. exp. Therap.*, **176**, 529.
Holz, R.W. and Coyle, J.T. (1974), *Mol. Pharmacol.*, **10**, 746.
Huang, M., Gruenstein, E. and Daly, J.W. (1973), *Biochim. Biophys. Acta*, **329**, 147.
Huang, M., Shimizu, H. and Daly, J.W. (1972), *J. Med. Chem.*, **15**, 462.
Imhof, R., Gössinger, E., Graf, W., Berner-Fenz, L., Berner, H. and Wehrli, H. (1973), *Helv. Chim. Acta*, **56**, 139.
Jaimovich, E., Venosa, R.A., Shrager, P. and Horowicz, P. (1976), *J. Gen. Physiol.*, **67**, 399.
Jansson, S.-E., Albuquerque, E.X. and Daly, J. (1974), *J. Pharmacol. exp. Therap.*, **189**, 525.
Kao, C.Y. (1966), *Pharm. Rev.*, **18**, 997.
Karle, I.L. (1972), *Proc. Nat. Acad. Sci.*, **69**, 2932.
Karle, I.L. and Karle, J. (1969), *Acta Cryst.*, **B25**, 428.
Kayaalp, S.O., Albuquerque, E.X. and Warnick, J.E. (1970), *Eur. J. Pharmacol.*, **12**, 10.
Khodorov, B.I., Peganov, E.M., Revenko, S.V. and Shishkova, L.D. (1975), *Brain Res.*, **84**, 541.
Lee, C.Y. (1972), *Ann. Rev. Pharmacol.*, **12**, 265.

Lee, K.S. and Klaus, W. (1971), *Pharm. Rev.*, **23**, 193.
Luzzati, V. and Husson, F. (1962), *J. Cell Biol.*, **12**, 207.
Marki, F. and Witkop, B. (1963), *Experientia*, **19**, 329.
Narahashi, T. (1974), *Physiol. Rev.*, **54**, 813.
Narahashi, T., Albuquerque, E.X. and Deguchi, T. (1971a), *J. Gen. Physiol.*, **58**, 54.
Narahashi, T., Deguchi, T. and Albuquerque, E.X. (1971b), *Nature New Biol.*, **228**, 221.
Narahashi, T. and Seyema, I. (1974), *J. Physiol. (Lond.)*, **242**, 471.
Ochs, S. and Worth, R. (1975), *Science*, **187**, 1087.
Pence, R.S., Garcia, J.H. and Albuquerque, E.X. (1976), *American Association of Neuropathologists*, June, 1976 (in press).
Posada-Arango, A. (1871), *Arch. Méd. Navale*, **16**, 203.
Ritchie, J.M., Rogart, R.B. and Strichartz, 6. (1976), *J. Physiol. (Lond.)*, **258**, 99P.
Rosen, J.S. and Fuhrman, F.A. (1971), *Toxicon*, **9**, 411.
Shimizu, H. and Daly, J.W. (1972), *Eur. J. Pharmacol.*, **17**, 240.
Shotzberger, G.S., Albuquerque, E.X. and Daly, J.W. (1976), *J. Pharmacol. exp. Therap.*, **196**, 433.
Sundermier, H., Dolly, J.O., Albuquerque, E.X., Brown, G., Burgermeister, W., Daly, J. and Witkop, B. (1976), *Neuro Sci. Soc.*, (abstract, in press).
Tokuyama, T., Daly, J. and Witkop, B. (1969), *J. Am. Chem. Soc.*, **91**, 3931.
Tokuyama, T., Daly, J., Witkop, B., Karle, I.L. and Karle, J. (1968), *J. Am. Chem. Soc.*, **90**, 1917.
Warnick, J.E., Albuquerque, E.X. and Diniz, C.R. (1976), *J. Pharmacol. exp. Therap.*, **198**, 155-167.
Warnick, J.E., Albuquerque, E.X., Onur, R., Jansson, S.-E., Daly, J., Tokuyama, T. and Witkop, B. (1975), *J. Pharmacol. exp. Therap.*, **193**, 232.
Warnick, J.E., Albuquerque, E.X. and Sansone, F.M. (1971), *J. Pharmacol. exp. Therap.*, **176**, 497.

Index (Chapters 1–7)

Index (Chapter 8)